A Guide to Chi-Squared Testing

WILEY SERIES IN PROBABILITY AND STATISTICS

Established by WALTER A. SHEWHART and SAMUEL S. WILKS

A complete list of the titles in this series appears at the end of this volume

This text is printed on acid-free paper.

Published simultaneously in Canada.

Library of Congress Cataloging-in-Publication Data

Greenwood, P. E., 1937–
A guide to chi-squared testing / Priscilla E. Greenwood, Mikhail S. Nikulin.
p. cm. — (Wiley series in probability and statistics. Applied probability and statistics)
"A Wiley-Interscience publication."
Includes bibliographical references and indexes.
ISBN 0-471-55779-X (cloth : alk. paper)
1. Chi-square test. I. Nikulin, M. S. (Mikhail Stepanovich) II. Title. III. Series.
QA277.3.G74 1996
519.5'6—dc20 95-52754

Printed in the United States of America

10 9 8 7 6 5 4 3 2 1

A Guide to Chi-Squared Testing

PRISCILLA E. GREENWOOD
Department of Mathematics
University of British Columbia
Vancouver, Canada

MIKHAIL S. NIKULIN
Steklov Mathematical Institute
St. Petersburg, Russia
and
Université Bordeaux 2
Bordeaux, France

A Wiley-Interscience Publication
JOHN WILEY & SONS, INC.
New York • Chichester • Brisbane • Toronto • Singapore

Contents

Preface and Introduction

What Sort of Guide is This?

This is a guide to χ^2 testing in the following sense: Suppose you have decided to use a χ^2 test. Then how is this test done? What choices are there to be made and how can you sensibly make these choices? How do you deal with various special circumstances? What kinds of hypotheses can be tested where the necessary computations have already been done and the complexities have already been explored? You might want to choose one of these.

There is not much literature in book form on χ^2 testing. The book of Lancaster (1969) is a detailed and comprehensive account, to that date, which rightly claims to be the first such. It remains unrivaled. There are survey articles on χ^2 and books on goodness-of-fit statistics which include χ^2 but lack the authority of Lancaster.

This book is less ambitious than Lancaster's but complements it in some ways. We assume the reader has less statistical knowledge and less time to invest. We focus mainly on more recent information, not dealt with by Lancaster or elsewhere in book form. The monographs of Read and Cressie (1988) and Rayner and Best (1989) complement this one by considering broader but related classes of statistics. Someone who is undecided on what test to use will want to consult these books, and in addition the monograph of Drost (1988) with its extensive study of asymptotic power.

We assume that the possessors of data have had a basic course in statistics and expect that they may at some point carry this book to their statistical consultants. The consultants, we hope, will want to have their own copies.

Goodness-of-Fit Tests

An experimenter has data and an idea about the process that produced it. The data should have, according to the idea and corresponding model, a probability distribution in a particular family. If this is not true, the idea is wrong. So the experimenter wants to test the hypothesis that the data come from a probability distribution belonging to a certain family. A goodness-of-fit test is needed.

There are many ways to test goodness-of-fit, several of them are described in Section 7 of this guide, though our main focus is on the χ^2 family of tests. Some books that study a variety of tests are those of Martynov (1977), Feinburg (1980), Fleiss (1981), Agresti (1990), D'Agostino and Stephens (1986), Linhart and Zucchini (1986), and Behnen and Neuhaus (1989). There is no general theory that tells us how to choose a test. There are many considerations, and as usual one must make one's own decisions. When the amount of data is small, no test is very good, and when there is a lot of high-quality data almost any test, properly designed and applied, will give good results. So one reasonable criterion is whether we know how to construct the test. This often involves estimation and then finding the distribution of the test statistic. In the case of χ^2-type tests we construct a test statistic such that its distribution is asymptotically χ^2 and such that one can also compute power against various alternatives – this is the subject of this guide. For many other tests, however, the only method currently available to get an idea of the distribution of the test statistic is by simulation, and computation of power is even further out of reach.

Here are some additional points relevant to choosing χ^2 testing.

When data is sparse relative to the dimensionality of the parameter space, as in sparse contingency tables, χ^2-type tests are not going to be useful.

While many tests do not have effective multivariate versions, χ^2 tests are adaptable to testing multivariate distributions.

Limits of precision in measurement and recording give us grouped, not really continuous, data. For other reasons, data may come to us in grouped form, making χ^2-testing appropriate.

The χ^2-type tests are general purpose tests and often used for initial testing. Additional reasons for choosing χ^2-testing will appear as we go along.

Objectives

χ^2 testing remains an art. No attempt at a complete table of set-ups and optimal tests can be made. Each practical situation has its own wrinkles. Still, by collecting and organizing existing knowledge, we hope to provide a basis for successful χ^2 testing. If the user of this guide should find that his particular problem is not covered, it is likely that by scanning a bit he will find something that comes close and can be adapted to his needs. He will also obtain an orientation to the literature.

χ^2 testing is quite easy to understand in principle, and yet, as everyone who has applied a χ^2 test to real data knows, each situation demands that several decisions be made. That these decisions are not so trivial, is suggested by a glance through this book. Worse, though, is that the textbook literature is often inaccurate or misleading as well as sketchy. If a parameter must be estimated before constructing the test, then special attention is necessary. One need not master the theory to construct and perform tests. However, we have supplied most of the mathematics necessary to justify the recommendations and tried to keep it simple and self-contained. The typical real problem does not fit exactly with existing theory. Having the mathematics at hand, one may see how to change things to fit one's own problem.

This is not a survey of problems in which the χ^2 distribution plays a role. Rather our emphasis is on proper application of χ^2 tests in standard hypothesis-testing situations.

The reader will notice that our bibliography includes many references to journals and books published in the former Soviet Union. One of the aims of this guide is to provide easy access to this information.

Every publication has its boundaries. Topics which should be inside find themselves excluded. For us some of these topics are the effect of dependence on χ^2 testing (Moore, 1982), functional versions of χ^2 tests (as studied by Bosq, 1978, 1983; and Blacher, 1988, 1990), and χ^2 tests designed for censored data and other survival analysis problems (e.g., Chen, 1975; Turnbull and Weiss, 1978; Mihalko and Moore, 1980; Akritas, 1988; Habib and Thomas, 1986; Hjort, 1984, 1990; Hollander and Peña, 1992; Li and Doss, 1993; Kim, 1993; Bagdonavičius and Nikulin, 1994; and others).

Outline of the Book

Chapter 1 introduces Pearson's statistic and explains its relation to the χ^2 distribution. A brief refresher on relevant facts about, for example, the multivariate normal distribution can be found in the Appendix. We state some limit theorems. In this chapter we have concentrated most of the speculative material. The difficult but essential topic of what to do when n is small is reviewed in Section 2. In Section 3, we broach the matter of alternatives and power, and face the fact that simple hypotheses, strictly speaking, are nonsense. Examples occupy most of Sections 4 and 5. In Section 6, we talk about choosing the number of classes, a question that leads back to power and alternatives. In Section 7, we delve into the general subject of measures of fit. Several of them are described in detail. We pay special attention to other measures of fit, besides Pearson's, which are asymptotically χ^2, and finish Chapter 1 with several remarks, in Section 8, about how the theory of estimation impinges on our topic. This readies us for Chapter 2.

Chapter 2 addresses the question: what happens if we must first estimate a parameter, θ, and then apply a χ^2 test to the same data? Using the same data twice is a well-known method of getting into trouble. In this setting, the first relatively mild trouble, discovered by Fisher, is that for three familiar methods of estimating θ, using grouped data, the end result is a statistic that is asymptotically χ^2 but with lost *degrees of freedom*. This phenomenon, explored in Section 9, is associated with any method of estimation that is efficient relative to the grouped data. Some typical applications are to testing for homogeneity, for example, whether the various outcomes of an experiment are equiprobable, and testing for independence of two criteria from a contingency table. All goes well if one is aware what χ^2 law is in effect. Later in Section 15 we return to the question: what about other *$\sqrt{n}$-consistent estimates*? It turns out that, under regularity conditions, the same χ^2 limit is obtained for a modified statistic.

However, grouping the data wastes information for the estimation of θ. If we greedily try to do better by using the ungrouped data, the resulting statistic is not asymptotically χ^2 and even depends on θ, in general, as Chernoff and Lehmann (1954) discovered and we report in Section 10. The need to base cell boundaries on an estimate of θ may force us to use ungrouped data for the estimate. We pursue this point with random cell boundaries and display similar unsatisfactory results in Section 11. Sections 10 and 11 show exactly what can go wrong and will enhance the reader's happiness over the useful information in Sections 12, 13, and 14.

A statistic that surmounts all the problems raised in Sections 10 and 11 is presented in Section 12. The *magic touch* amounts to a random normalization, which can be understood as a projection [see Eq. (12.6)]. This statistic can be constructed from frequencies for random cells based on any efficient estimate of θ and turns out to be asymptotically χ^2 with no degrees of freedom lost.

More good news appears in Section 13 for the large class of examples where the parameter to be estimated before a test can be constructed is either a location (the mean) or scale (the standard deviation) parameter, or both. We put the topic of testing for normality in its own section, Section 14, because it arises so often.

We continue with the problem of first estimating θ and then applying a χ^2 test. A general and powerful method of estimation is the Rao-Blackwell approach based on sufficient statistics, which can be used with certain exponential families of distributions. In Chapter 3, we consider an exponential family of distributions that contains most of the familiar examples and many more. The discrete exponential family is laid out in Section 16, and in Section 17 we devote some space to the construction of "best unbiased" estimators for this rich family. In Section 18 we construct χ^2 statistics for this class of examples. A multivariate version of all this is summarized in Section 19, and in Section 20 we consider a continuous exponential family analogous to the discrete one presented in Section 16. The final section of Chapter 3, Section 21, is about another type of exponential family, having *power variance function*.

Chapter 4 shows the χ^2 test at work in some additional settings.

Section 22 is about testing whether several data sets come from the same distribution within a parameter family. In Section 23, we describe a test to select those data sets that are similar to data from a control group, when we have data sets from several treatments. Finally, in Section 24, we look at a model for ratings by various judges of the contestants in a sports competition. A χ^2 test detects nonobjectivity on the part of some judge. An additional test determines which contestants are indistinguishable on the basis of the model.

A grant from the Natural Science and Engineering Council of Canada supported visits by Nikulin to Vancouver, and a U.S. – U.S.S.R. Academies of Science exchange fellowship enabled Greenwood to visit St. Petersburg (Leningrad). We thank these agencies for making our collaboration possible. Janet Clark converted the manuscript to TEX.

A Guide to Chi-Squared Testing

CHAPTER 1

The Chi-Squared Test of Pearson

1. PEARSON'S STATISTIC

The Multinomial Distribution

First, we describe Pearson's test. That Pearson's statistic has an asymptotically χ^2 distribution follows from the central limit theorem and some matrix algebra. We assume that the central limit theorem is familiar and write down the matrix algebra. Supplementary information is in the Appendix. An additional limit theorem shows how the law of Pearson's statistic approximates a normal law when the number of classes is large. The examples in this section are classical: Buffon tests a coin and Von Mises looks at the decimal expansion of π.

We consider a sequence of independent experiments, or trials, indexed by $k = 1, 2, \ldots$. Suppose that the outcome of each trial is one of r incompatible events $A_1, \ldots, A_r$, with positive probabilities

$$p_i = \boldsymbol{P}(A_i), \qquad i = 1, \ldots, r,$$

where $p_1 + \cdots + p_r = 1$. For each trial, we define the random variable

$$\mu_{ki} = \begin{cases} 1, & \text{if } A_i \text{ occurs on the trial with number } k, \\ 0, & \text{otherwise.} \end{cases}$$

Then $\{\mu_{ki}\}$, $i = 1, 2, \ldots, r$; $k = 1, 2, \ldots$, is a sequence of independent random variables, and

$$\begin{aligned} \boldsymbol{P}\{\mu_{ki} = 1\} &= p_i, \qquad \boldsymbol{P}\{\mu_{ki} = 0\} = 1 - p_i, \\ \boldsymbol{E}\mu_{ki} &= p_i \qquad \mathbf{Var}\,\mu_{ki} = p_i(1 - p_i) \qquad i = 1, \ldots, r \end{aligned} \tag{1.1}$$

for any $k = 1, 2, \ldots$. Let us denote

$$\boldsymbol{\mu}_k = (\mu_{k1}, \mu_{k2}, \ldots, \mu_{kr})^T, \qquad k = 1, 2, \ldots \quad .$$

From the definition of μ_{ki}, it follows that each vector $\boldsymbol{\mu}_k$ has only one nonzero component, which is equal to 1 and located in position i, if the event A_i occurs at the kth trial. For any positive integer n, we write

$$\boldsymbol{\nu} = \sum_{k=1}^{n} \boldsymbol{\mu}_k = (\nu_1, \ldots, \nu_r)^T, \tag{1.2}$$

where

$$\nu_i = \sum_{k=1}^{n} \mu_{ki}.$$

It is evident that the random variables $\nu_1, \ldots, \nu_r$ satisfy the condition

$$|\boldsymbol{\nu}| = \nu_1 + \nu_2 + \cdots + \nu_r = n.$$

The random variable ν_i is called *the frequency of the event or class* A_i in a series of n trials. It is evident that the realization space of ν_i is the set $\boldsymbol{X}_0^n = \{0, 1, \ldots, n\}$, where n is the total number of trials. The vector $\boldsymbol{\nu}$ is called the *vector of frequencies* arising in a sequence of n trials. If n is the number of trials, then

$$\ell(\boldsymbol{p}) = \prod_{i=1}^{r} \prod_{k=1}^{n} p_i^{\mu_{ki}} = \prod_{i=1}^{r} p_i^{\nu_i} = \exp\left\{\sum_{i=1}^{r} \nu_i \ln p_i\right\} \tag{1.3}$$

is called the likelihood function for the vector constructed from the sample $\{\mu_{ki}\}$. The vector of frequencies, $\boldsymbol{\nu} = (\nu_1, \ldots, \nu_r)^T$, is a *sufficient* statistic for $\boldsymbol{p}$; that is, it contains all the information in the data about $\boldsymbol{p}$.

The distribution of $\boldsymbol{\nu}$ is *multinomial* with parameters n and $\boldsymbol{p}$; that is,

$$\begin{aligned} \boldsymbol{P}\{\boldsymbol{\nu} = \boldsymbol{x}\} &= \boldsymbol{P}\{\nu_1 = x_1, \nu_2 = x_2, \ldots, \nu_r = x_r\} \\ &= \frac{n!}{x_1! \ldots x_r!} p_1^{x_1} p_2^{x_2} \ldots p_r^{x_r}, \end{aligned}$$

for any integer vector $\boldsymbol{x} = (x_1, \ldots, x_r)^T$ such that $0 \le x_i \le n$,

$x_1+\cdots+x_r=n$. From Eqs. (1.1) and (1.2), it follows that

$$\begin{aligned}\boldsymbol{P}\{\nu_i \geq x_i\} &= \frac{n!}{(x_i-1)!(n-x_i)!}\int_0^{p_i} t^{x_i-1}(1-t)^{n-x_i}\,dt \\ &= I_{p_i}(x_i, n-x_i-1),\end{aligned}$$

and for $i \neq j$

$$\boldsymbol{P}\{\nu_i \geq x_i, \nu_j \geq x_j\} = \frac{n!}{(x_i-1)!(x_j-1)!(n-x_i-x_j)!}\int_0^{p_i} dt$$

$$\int_0^{p_j} t^{x_i-1}z^{x_j-1}(1-t-z)^{n-x_i-x_j}\,dz, \qquad (1.4)$$

where

$$I_u(a,b) = \frac{\Gamma(a+b)}{\Gamma(a)\Gamma(b)}\int_0^u t^{a-1}(1-t)^{b-1}\,dt, \qquad a>0,\, b>0, \qquad 0 \leq u \leq 1,$$

From Eqs. (1.1) to (1.4) one can easily obtain

$$\boldsymbol{E}\boldsymbol{\nu} = n\boldsymbol{p}, \quad \textbf{Var}\ \boldsymbol{\nu} = \boldsymbol{E}(\boldsymbol{\nu}-n\boldsymbol{p})(\boldsymbol{\nu}-n\boldsymbol{p})^T = n\boldsymbol{\Sigma} = n[\boldsymbol{D}-\boldsymbol{p}\boldsymbol{p}^T], \quad (1.5)$$

where $\boldsymbol{D}$ is a diagonal matrix with $p_1,\ldots,p_r$ on the main diagonal, $\boldsymbol{\Sigma} = \boldsymbol{D}-\boldsymbol{p}\boldsymbol{p}^T$ and rank $\boldsymbol{\Sigma} = r-1$.

The vector of expectations $\boldsymbol{E}\boldsymbol{\nu}=n\boldsymbol{p}=(np_1,\ldots,np_r)^T$ is often called *the vector of expected frequencies*, and the products np_i, $i=1,\ldots,r$, are the expected values of the frequencies $\nu_1,\ldots,\nu_r$. The difference vector

$$\boldsymbol{\nu}-n\boldsymbol{p} = (\nu_1-np_1,\ldots,\nu_r-np_r)^T$$

is called the *vector of deviations*. The law of large numbers tells us that as n increases the statistics ν_i/n converge to their expected values, p_i.

The Statistic of Pearson

Let us denote

$$\boldsymbol{X}_n = \left(\frac{\nu_1-np_1}{\sqrt{np_1}},\ldots,\frac{\nu_r-np_r}{\sqrt{np_r}}\right)^T. \qquad (1.6)$$

One might guess that for large n the square of the length of the vector $\boldsymbol{X}_n$

$$X^2 = X_n^2 = \boldsymbol{X}_n^T \boldsymbol{X}_n = \sum_{i=1}^{r} \frac{(\nu_i - np_i)^2}{np_i} \tag{1.7}$$

would not be large. This idea was used by K. Pearson in 1900 to construct the famous chi-squared test of a hypothesis, based on X_n^2, called the *statistic of Pearson*. Suppose that the vector of probabilities $\boldsymbol{p} = (p_1, \ldots, p_r)^T$ is unknown and there is a hypothesis

$$H_0 : \boldsymbol{p} = \boldsymbol{p}^{(0)} = (p_1^{(0)}, \ldots, p_r^{(0)})^T$$

according to which the probabilities p_i of the events A_i assume certain values $p_i^{(0)}$, $i = 1, 2, \ldots, r$; then to test the conformity of the hypothesis H_0 with the value of the vector of frequencies $\boldsymbol{\nu}$ observed in the course of n trials, K. Pearson proposed using the statistic

$$X^2 = \sum_{i=1}^{r} \frac{(\nu_i - np_i^{(0)})^2}{np_i^{(0)}} = \frac{1}{n} \sum_{i=1}^{r} \frac{\nu_i^2}{p_i^{(0)}} - n. \tag{1.8}$$

Formula (1.8) holds because

$$X^2 = \sum_{i=1}^{r} \frac{\nu_i^2 - 2np_i^{(0)}\nu_i + n^2[p_i^{(0)}]^2}{np_i^{(0)}} = \sum_{i=1}^{r} \frac{\nu_i^2}{np_i^{(0)}} - 2\sum_{i=1}^{r} \nu_i + n\sum_{i=1}^{r} p_i^{(0)}$$

and

$$|\boldsymbol{\nu}| = \nu_1 + \cdots + \nu_r = n, \qquad |\boldsymbol{p}^{(0)}| = p_1^{(0)} + \cdots + p_r^{(0)} = 1.$$

We confine our attention to Pearson's statistic (1.8), but there are natural competitors (see Section 7, and, e.g., the books of Read and Cressie, 1988; D'Agostino and Stephens, 1986; Drost, 1988; Bolshev, 1987).

If $r = 2$, then under H_0 the statistic X^2 has the form

$$X^2 = \sum_{i=1}^{2} \frac{(\nu_i - np_i^{(0)})^2}{np_i^{(0)}} = \frac{(\nu_1 - np_1^{(0)})^2}{np_1^{(0)}} + \frac{[(n-\nu_1) - n(1-p_1^{(0)})]^2}{n(1-p_1^{(0)})}, \tag{1.9}$$

because $\nu_1+\nu_2=n$ and $p_1^{(0)}+p_2^{(0)}=1$. From Eq. (1.9), we obtain

$$X^2=\frac{(\nu_1-np_1^{(0)})^2}{np_1^{(0)}}+\frac{(\nu_2-np_2^{(0)})^2}{n(1-p_1^{(0)})}=\frac{(\nu_1-np_1^{(0)})^2}{np_1^{(0)}(1-p_1^{(0)})}, \tag{1.10}$$

using $1/x+1/(1-x)=1/(x(1-x))$, for any $x\in(0,1)$.

For a fixed hypothesis H_0 and a given number n of trials, the statistic X^2 has a discrete probability distribution. As Pearson (1932) and Haldane (1937) showed,

$$\boldsymbol{E}X^2=r-1,\quad \mathbf{Var}\ X^2=2(r-1)+\frac{1}{n}\Big(\sum_{i=1}^{r}\frac{1}{p_i^{(0)}}-r^2-2r+2\Big), \tag{1.11}$$

and consequently, if, for large values of n, we have

$$\sum_{i=1}^{r}\frac{1}{p_i^{(0)}}=o(n)\quad\text{and}\quad r=o(n),$$

where $o(f(n))$ means that $o(f(n))/f(n)\to 0$ as $n\to\infty$, then

$$\mathbf{Var}\ X^2=2(r-1)+o(1),\qquad n\to\infty.$$

In particular, if

$$p_1^{(0)}=p_2^{(0)}=\cdots=p_r^{(0)}=\frac{1}{r},$$

Pearson's statistic Eq. (1.8) has the form

$$X^2=\frac{r}{n}\sum_{i=1}^{r}\nu_i^2-n=\frac{r}{n}(n^2-2\sum_{i>j}\nu_i\nu_j)-n=(r-1)n-\frac{2r}{n}\sum_{i>j}\nu_i\nu_j. \tag{1.12}$$

From Eq. (1.12), one can observe that the values of X^2 lie in the interval $[0,n(r-1)]$ and differ from one another by multiples of $2r/n$. It is clear that the statistic X^2 assumes its maximal value, $n(r-1)$, if some frequency, for example ν_1, is equal to n and the other $\nu_j=0$, $j=2,\ldots,r$, whereas the statistic X^2 assumes its minimal value if all the frequencies ν_i are equal. This is, of course, not always possible, but if n is a multiple of r and all $\nu_i=n/r$, then one sees from Eq. (1.12) that the statistic

X^2 assumes the value 0. Finally, we note that in the equiprobable case, that is, if $p_1 = \ldots = p_r = 1/r$, it follows from Eq. (1.11) that

$$\boldsymbol{E}X^2 = r-1, \qquad \textbf{Var } X^2 = 2(r-1)(1-1/n). \tag{1.13}$$

Pearson's Test

The procedure for testing the hypothesis H_0 consists of comparing the experimental values of Pearson's statistic X^2 with the *critical value* c_α:

If $X^2 \geq c_\alpha$, then the hypothesis H_0 is rejected, whereas
if $X^2 < c_\alpha$, then the hypothesis H_0 is not rejected,

and we say in the latter case that the results of the observations $\boldsymbol{\nu} = (\nu_1, \ldots, \nu_r)^T$ do not contradict the hypothesis H_0 advanced about the probabilities $p_1, p_2, \ldots, p_r$ of the events $A_1, A_2, \ldots, A_r$. The critical value c_α is determined so that the probability of rejecting the tested hypothesis H_0, when it is valid, that is, *the probability of error of the first kind*, does not exceed a given *level of significance* α $(0 < \alpha < 0.5)$.

By virtue of the discreteness of the statistic X^2, for a given level α there does not always exist a *critical value* c_α, such that the equality

$$\boldsymbol{P}\{X^2 \geq c_\alpha \mid H_0\} = \alpha \tag{1.14}$$

holds. If there is no c_α, such that Eq. (1.14) holds, one can always find unique "upper" and "lower" critical values u_α and l_α, which satisfy the conditions:

$$\begin{gathered} l_\alpha < u_\alpha, \qquad \boldsymbol{P}\{X^2 = l_\alpha \mid H_0\} > 0, \\ \boldsymbol{P}\{X^2 = u_\alpha \mid H_0\} > 0, \qquad \boldsymbol{P}\{l_\alpha < X^2 < u_\alpha \mid H_0\} = 0, \\ \boldsymbol{P}\{X^2 \geq l_\alpha \mid H_0\} > \alpha > \boldsymbol{P}\{X^2 \geq u_\alpha \mid H_0\} > 0. \end{gathered} \tag{1.15}$$

For definiteness, one often assumes that c_α is the smallest possible value of the statistic, X^2, satisfying the condition

$$\boldsymbol{P}\{X^2 \geq c_\alpha \mid H_0\} \leq \alpha, \tag{1.16}$$

that is, that $c_\alpha = u_\alpha$. Clearly, Eq. (1.14) corresponds to the situation when $c_\alpha = u_\alpha = l_\alpha$.

Katti (1973) tabulated the critical values of the statistic X^2 in the equiprobable case for $r = 2(1)10$ (the notation means "from 2 to 10 in steps of 1"), $n = 3(1)25$, corresponding to levels of significance $\alpha = 0.1$, $\alpha = 0.05$, and $\alpha = 0.01$. Smith et al. (1979) made tables of upper and lower critical values u_α and l_α, including the equiprobable case, corresponding to levels of significance $\alpha = 0.1, 0.05, 0.01, 0.005$ for

$$\begin{aligned} r &= 2(1)4, & n &= 26(1)50; \\ r &= 5, & n &= 26(1)40; \\ r &= 6(1)10, & n &= 26(1)30. \end{aligned}$$

Because the probability distribution of the statistic X^2 becomes more complex as n increases, instead of the critical value c_α one may use an approximate value $\tilde{c}_\alpha$, the solution of an inequality, obtained from the original inequality (1.12) by replacement of the probability $\boldsymbol{P}\{X_n^2 \geq c_\alpha \mid H_0\}$ by a suitable approximate expression. Pearson proceeded in precisely this way, using the fact that, for large n,

$$\boldsymbol{P}\{X_n^2 \geq x \mid H_0\} \approx \boldsymbol{P}\{\chi_{r-1}^2 \geq x\}, \tag{1.17}$$

where χ_{r-1}^2 is a random variable, which has the chi-squared distribution with $r-1$ degrees of freedom. Following Pearson, as $\tilde{c}_\alpha$, one uses the root $\chi_{r-1,1-\alpha}^2$ of the equation

$$\alpha = \boldsymbol{P}\{\chi_{r-1}^2 \geq x\} = \frac{1}{2^{(r-1)/2}\Gamma(\frac{r-1}{2})} \int_x^\infty z^{\frac{r-1}{2}} e^{-z/2}\, dz, \quad x > 0, \tag{1.18}$$

that is, one sets $\tilde{c}_\alpha$ equal to $\chi_{r-1,1-\alpha}^2$. The number $\chi_{r-1,1-\alpha}^2$ is called *the quantile of level* $1-\alpha$ or the α-*upper quantile* of the chi-squared distribution.

The procedure just described is called *Pearson's test* or the χ^2 *test*. An account of the history and influence of the test was written by Plackett (1983) and more recently by Stigler (1992); see also Hamdan, 1963; Kempthorne, 1966; Roscoe and Byars, 1971; Tate and Hyer, 1973; Radlow and Alf, 1975; Horn, 1977; Koehler and Larntz, 1980; Holtzman and Good, 1986). Hoaglin (1977) gives information on using the tables of percentage points for the chi-squared distribution. More extensive tables for central and noncentral chi-squared distributions, with algorithms, are given by Mardia and Zemroch (1978).

Pearson's Test for Continuous F

Let us show here how Pearson's test might be applied when we have i.i.d. (independent and identically distributed) random variables $X_1, \ldots, X_n$ and we wish to test the hypothesis

$$H_0 : \boldsymbol{P}\{X_i \leq x\} = F(x),$$

where the distribution function $F(x)$ has density $f(x)$ with respect to some measure, dx. If F is discontinuous at x, let $f(x) = \boldsymbol{P}(X_i = x)$. We fix a measurable partition $J_1, J_2, \ldots, J_r$ of the real line such that

$$p_i = \int_{J_i} dF(x) = \int_{J_i} f(x)\,dx = \boldsymbol{P}\{X_1 \in J_i\} > 0, \quad i = 1, 2, \ldots, r. \tag{1.19}$$

The integral is a sum if F has jumps. It is evident that $p_1 + \cdots + p_r = 1$. Further, let $\boldsymbol{\nu} = (\nu_1, \ldots, \nu_r)^T$ be the frequency vector obtained by counting the number of X_j, $j = 1, \ldots, n$, which fall in each set J_i. The observations $X_1, \ldots, X_n$ are referred to as the *ungrouped data*, and the frequencies, $\nu_1, \nu_2, \ldots, \nu_r$ are called the *grouped data*. The sets J_i are the *grouping sets* or *grouping classes*. Often $J_i = [a_{i-1}; a_i)$, $i = 1, \ldots, r$, where $a_1, a_2, \ldots, a_{r-1}$ are points on the real line such that

$$-\infty = a_0 < a_1 < a_2 < \cdots < a_{r-1} < a_r = +\infty.$$

In this case $J_1, \ldots, J_r$ are called *grouping intervals*, and

$$\nu_i = \int_{a_{i-1}}^{a_i} dF_n(x),$$

where $F_n(x)$ is the empirical function, obtained from the data $X_1, X_2, \ldots, X_n$:

$$F_n(x) = \sum_{i=1}^{n} I(X_i \leq x). \tag{1.20}$$

As before, if H_0 is true then the vector $\boldsymbol{\nu} = (\nu_1, \ldots, \nu_r)^T$ has the multinomial distribution with parameters n and $\boldsymbol{p}$, as in Eq. (1.4), and hence to test H_0 one can use Pearson's statistic X^2.

Example: Testing Normality. Let $\boldsymbol{\nu} = (\nu_{-1}, \nu_0, \nu_1)^T$ be a vector of frequencies obtained by grouping the elements X_i of a sample $X_1, X_2, \ldots, X_{50}$ of size $n = 50$ over three intervals

$$(-\infty, -0.5),\ [-0.5, 0.5),\ [0.5, +\infty).$$

We want to test the simple hypothesis $H_0 : X_i \sim N(0, 1/4)$, according to which the X_i have the normal distribution with parameters $\boldsymbol{E}X_i = 0$ and $\mathbf{Var}\ X_i = \sigma^2 = 0.25$, if

$$\nu_{-1} = 8,\ \nu_0 = 29,\ \nu_1 = 13.$$

To test H_0 we apply the chi-squared test, based on the statistic

$$X_n^2 = \sum_{i=-1}^{1} \frac{(\nu_i - np_i)^2}{np_i},$$

which in this case is

$$X_n^2 = \frac{1}{n}\left[\frac{\nu_{-1}^2}{p_{-1}} + \frac{\nu_0^2}{p_0} + \frac{\nu_1^2}{p_1} - n^2\right],$$

where

$$p_{-1} = \frac{1}{0.5\sqrt{2\pi}} \int_{-\infty}^{-0.5} \exp\{-\frac{x^2}{0.5}\}\, dx$$

$$= \frac{1}{\sqrt{2\pi}} \int_{-\infty}^{-1} e^{-y^2/2}\, dy = \Phi(-1) = 0.1587,$$

$$p_1 = p_{-1} = 0.1587 \text{ (by symmetry)},$$

and hence

$$p_0 = 1 - p_1 - p_{-1} = 0.6826.$$

Since $n = 50$, we obtain

$$X_n^2 = \frac{1}{50}\left[\frac{64}{0.1587} + \frac{841}{0.6826} + \frac{169}{0.1587} - 2500\right] = 4.005.$$

Suppose that $\alpha = 0.1$. Since

$$X_n^2 = 4.005 < \chi^2_{2,0.9} = 4.605,$$

we accept H_0.

Note that since

$$\boldsymbol{P}\{X_n^2 \geq 4.005 \mid H_0\} \approx \boldsymbol{P}\{\chi_2^2 \geq 4\} = 0.135,$$

we can say that the hypothesis, H_0 is consistent for any level of significance $\alpha < 0.135$.

This example is only a first illustration. The problem of testing for normality is considered in Section 14.

Example: Buffon Example. Buffon tossed a coin $n = 4,040$ times and obtained 2,048 heads. If H_0 says $p = 1/2$, then according to Eq. (1.9), $\boldsymbol{E}X_n^2 = 1$, **Var** $X_n^2 = 1.998$, and the experimental value of $X_n^2 = 0.776$ is consistent with the hypothesis. The corresponding chi-squared probability $\boldsymbol{P}(\chi_1^2 \geq 0.776)$ is approximately equal to 0.38.

Continuity Correction

If $r = 2$, then Eq. (1.10) is

$$X_n^2 = \frac{(\nu_1 - np_1)^2}{np_1(1-p_1)}. \tag{1.21}$$

In this case, the random variable ν_1 has a binomial distribution with parameters n and p_1, that is, for any $k = 0, 1, \ldots, n$

$$\boldsymbol{P}\{\nu_1 = k\} = \binom{n}{k} p_1^k (1-p_1)^{n-k}, \tag{1.22}$$

and $\boldsymbol{E}\nu_1 = np_1$, **Var** $\nu_1 = np_1(1-p_1)$, where $1 > p_i > 0$. If $n \to \infty$, then by the de Moivre-Laplace Theorem we have

$$\lim_{n\to\infty} \boldsymbol{P}\left\{\frac{\nu_1 - np_1}{\sqrt{np_1(1-p_1)}} \le x\right\} = \Phi(x),$$

that is, ν_1 is asymptotically normal with parameters np_1 and $np_1(1-p_1)$. We use the notation ν_1 is $AN_1(np_1, np_1(1-p_1))$. Formula (1.23) means that for large n the random variable $\mu_n = (\nu_1 - np_1)/\sqrt{np_1(1-p_1)}$ is distributed approximately as a standard normal random variable, and the statistic X^2 is distributed approximately as χ_1^2. This justifies the approximation (1.17) in the case $r = 2$.

In addition to Eq. (1.20), the theorem of de Moivre-Laplace tells us that, for any p_1, $0 < p_1 < 1$,

$$\boldsymbol{P}\{\nu_1 \le x\} = \Phi\left(\frac{x - np_1 + c}{\sqrt{np_1(1-p_1)}}\right) + O\left(\frac{1}{\sqrt{n}}\right), \tag{1.23}$$

where c may be any number and is often taken to be zero. Sometimes c is taken to be 0.5 and called a "continuity correction." One can justify this by noticing that the random variable $n - \nu_1$ also has the binomial distribution, but with the parameters n and $1 - p_1$, and, because

$$\boldsymbol{P}\{\nu_1 \le x\} + \boldsymbol{P}\{\nu_1 \ge x + 1\} = 1,$$

we have

$$P\{\nu_1 \le x\} + P\{n - \nu_1 \le n - x - 1\} = 1.$$

If we now use the normal approximation (1.23) for each term of the last equation, we obtain

$$\Phi\left(\frac{x - np_1 + c}{\sqrt{np_1(1-p_1)}}\right) + \Phi\left(-\frac{x - np_1 + (1-c)}{\sqrt{np_1(1-p_1)}}\right) \simeq 1,$$

and this sum is exactly 1 when $c = 0.5$, because

$$\Phi(z) + \Phi(-z) \equiv 1, \qquad z \in \boldsymbol{R}^1.$$

We shall use Eq. (1.21) with the continuity correction $c = 0.5$. For large n, Eq. (1.23) gives us

$$P\{\nu_1 \leq x\} \approx \Phi\left(\frac{x - np_1 + 0.5}{\sqrt{np_1(1-p_1)}}\right). \tag{1.24}$$

The Central Limit Theorem and the Chi-Squared Limit

In the general case, if $r \geq 2$, we obtain a justification for Eq. (1.15) from the Lindeberg-Levy multivariate central limit theorem and a theorem of Serfling (1980, p. 128-130), which is an extension of results of C.R. Rao (1973) and Ogasawara and Takahashi (1951) concerning the limit distribution of a quadratic form in multivariate normal vectors.

From the Lindeberg-Levy central limit theorem, it follows that when $n \to \infty$, the vector $(1/\sqrt{n})(\boldsymbol{\nu} - n\boldsymbol{p})$ has limiting normal distribution $N(\mathbf{0}_r, \boldsymbol{\Sigma})$, that is

$$\frac{1}{\sqrt{n}}(\boldsymbol{\nu} - n\boldsymbol{p}) \text{ is } AN(\mathbf{0}_r, \boldsymbol{\Sigma}), \tag{1.25}$$

where $\boldsymbol{\Sigma} = \boldsymbol{D} - \boldsymbol{p}\boldsymbol{p}^T$ [see Eq.(1.5)]. Since rank $\boldsymbol{\Sigma} = r - 1$, this normal law is degenerate.

Now we cite a theorem from the book by Serfling (1980), which allows us to construct a statistic of chi-squared type (see also Mäkeläinen, 1966, Johnson and Kotz, 1970, pp. 178–179, Morris, 1975, and the Appendix).

Theorem 1.1. *Let $\boldsymbol{X} = (X_1, \ldots, X_r)^T$ be a normal random vector,*

$$\boldsymbol{E}\boldsymbol{X} = \boldsymbol{\mu}, \qquad \boldsymbol{E}(\boldsymbol{X} - \boldsymbol{\mu})(\boldsymbol{X} - \boldsymbol{\mu})^T = \boldsymbol{\Sigma},$$

and let $\boldsymbol{C}$ be a symmetric matrix of order r. Assume that, for $\boldsymbol{\eta} = (\eta_1, \ldots, \eta_r)^T$,

$$\boldsymbol{\Sigma}\boldsymbol{\eta} = \mathbf{0}_r \quad \text{implies} \quad \boldsymbol{\eta}^T\boldsymbol{\mu} = \eta_1\mu_1 + \cdots + \eta_r\mu_r = 0.$$

Then $\boldsymbol{X}^T\boldsymbol{C}\boldsymbol{X}$ has a (possibly noncentral) chi-squared distribution, if and only if

$$\boldsymbol{\Sigma}\boldsymbol{C}\boldsymbol{\Sigma}\boldsymbol{C}\boldsymbol{\Sigma} = \boldsymbol{\Sigma}\boldsymbol{C}\boldsymbol{\Sigma}, \tag{1.26}$$

in which case the number of degrees of freedom is the trace of the matrix $\boldsymbol{C\Sigma}$ *and the noncentrality parameter is* $\boldsymbol{\mu}^T\boldsymbol{C}\boldsymbol{\mu}$.

If rank $\boldsymbol{\Sigma} = r$ (i.e. the vector $\boldsymbol{X}$ has a nondegenerate normal distribution) then one can choose $\boldsymbol{C}=\boldsymbol{\Sigma}^{-1}$, and in this case rank $(\boldsymbol{C\Sigma})=$ rank $\boldsymbol{I}_r = r$, where

$$\boldsymbol{I}_r = \begin{bmatrix} 1 & 0 & \dots & 0 \\ 0 & 1 & \dots & 0 \\ \vdots & \vdots & \ddots & \vdots \\ 0 & 0 & \dots & 1 \end{bmatrix}_{r\times r}$$

If rank $\boldsymbol{\Sigma}=m\leq r-1$, it means that the normal distribution is degenerate. One then chooses $\boldsymbol{C} = \boldsymbol{\Sigma}^-$, where $\boldsymbol{\Sigma}^-$ is a generalized inverse matrix of $\boldsymbol{\Sigma}$ of rank m, which satisfies Eq. (1.26). In this case rank $\Sigma^-\Sigma = m$.

Now we apply Theorem 1.1 to justify Eq. (1.15). To construct $\boldsymbol{\Sigma}^-$, let us consider the matrix $\widetilde{\boldsymbol{\Sigma}}$,

$$\widetilde{\boldsymbol{\Sigma}} = \begin{bmatrix} p_1(1-p_1) & -p_1p_2 & \dots & -p_1p_{r-1} \\ -p_2p_1 & p_2(1-p_2) & \dots & -p_2p_{r-1} \\ \vdots & \vdots & & \vdots \\ -p_{r-1}p_1 & -p_{r-1}p_2 & \dots & p_{r-1}(1-p_{r-1}) \end{bmatrix}. \tag{1.27}$$

Rank $\widetilde{\boldsymbol{\Sigma}} = r-1$, and its inverse is

$$\widetilde{\boldsymbol{\Sigma}}^{-1} = \begin{bmatrix} \frac{1}{p_1}+\frac{1}{p_r} & \frac{1}{p_r} & \dots & \frac{1}{p_r} \\ \frac{1}{p_r} & \frac{1}{p_2}+\frac{1}{p_r} & \dots & \frac{1}{p_r} \\ \vdots & \vdots & \ddots & \vdots \\ \frac{1}{p_r} & \frac{1}{p_r} & \dots & \frac{1}{p_{r-1}}+\frac{1}{p_r} \end{bmatrix}. \tag{1.28}$$

By direct calculation one can verify that

$$\widetilde{\boldsymbol{\Sigma}}\widetilde{\boldsymbol{\Sigma}}^{-1} = \boldsymbol{I}_{r-1},$$

Now we construct

$$\boldsymbol{\Sigma}^{-} = \begin{bmatrix} \frac{1}{p_1}+\frac{1}{p_r} & \frac{1}{p_r} & \cdots & \frac{1}{p_r} & 0 \\ \frac{1}{p_r} & \frac{1}{p_2}+\frac{1}{p_r} & \cdots & \frac{1}{p_r} & 0 \\ \vdots & \vdots & & \vdots & \vdots \\ \frac{1}{p_r} & \frac{1}{p_r} & \cdots & \frac{1}{p_{r-1}}+\frac{1}{p_r} & 0 \\ 0 & 0 & \cdots & 0 & 0 \end{bmatrix}. \tag{1.29}$$

It is evident that

$$\boldsymbol{\Sigma}^{-}\boldsymbol{\Sigma} = \boldsymbol{\Sigma}\boldsymbol{\Sigma}^{-} = \begin{bmatrix} 1 & 0 & 0 & \cdots & 0 & 0 \\ 0 & 1 & 0 & \cdots & 0 & 0 \\ 0 & 0 & 1 & \cdots & 0 & 0 \\ \vdots & \vdots & \vdots & \ddots & \vdots & \cdots \\ 0 & 0 & 0 & \cdots & 1 & 0 \\ 0 & 0 & 0 & \cdots & 0 & 0 \end{bmatrix}_{r\times r},$$

the matrix $\boldsymbol{C} = \boldsymbol{\Sigma}^{-}$ satisfies Eq. (1.26), and the trace of $\boldsymbol{\Sigma}^{-}\boldsymbol{\Sigma}$ is $r-1$. One can verify that

$$X_n^2 = \sum_{i=1}^{r} \frac{(\nu_i - np_i)^2}{np_i} = \frac{1}{n}(\boldsymbol{\nu}-n\boldsymbol{p})^T\boldsymbol{\Sigma}^{-}(\boldsymbol{\nu}-n\boldsymbol{p}), \tag{1.30}$$

and hence from the central limit theorem and Theorem 1.1 we obtain

$$\lim_{n\to\infty} \boldsymbol{P}\{X_n^2 \geq x \mid H_0\} = \boldsymbol{P}\{\chi_{r-1}^2 \geq x\}, \tag{1.31}$$

which justifies formula (1.17). One can easily verify that Eq. (1.30) holds with

$$\boldsymbol{\Sigma}^{-} = \boldsymbol{D}^{-1} + \frac{1}{p_r}\mathbf{1}_r \cdot \mathbf{1}_r^T.$$

Tumanyan's Theorems

The statistic X^2 does not depend on which column and row of $\boldsymbol{\Sigma}$ are omitted to obtain $\widetilde{\boldsymbol{\Sigma}}$. In the approach of Pearson, one starts with a *given* r. In practice, the statistician chooses r. Since Theorem 1.1 holds for any $r = 2, 3, \ldots$, we have a collection of different tests, one for each value of r. In fact, r can increase with n, subject to a restriction that none of the probabilities p_i associated with the grouping sets J_i, goes to zero too fast. Even with r fixed, the sets J_i may change and their measures p_i may be reduced as n increases. Furthermore, the statement (1.17) or (1.31) depends on x. For practical use one needs to know that the approximation holds uniformly in x for each r.

The following theorems of Tumanyan (1956) show us the role of r in the asymptotics and specify how the law of the X^2 statistic is approximated by a normal law when $r \to \infty$.

Theorem 1.2 (From Tumanyan, 1956). *If* $\min_i np_i^{(0)} \to \infty$, *then independently of the change of* n, r *and* $p_i^{(0)}$:

$$\sup_{|x|<\infty} |\boldsymbol{P}\{X_n^2 \geq x | H_0\} - \boldsymbol{P}(\chi_{r-1}^2 \geq x)| \to 0. \tag{1.32}$$

As the number of trials or the number n of random variables, $X_1, \ldots, X_n$, increases, the number of events or classes r can also increase without bound, and the probabilities $p_i^{(0)}$ can tend to 0, if nevertheless, $\min_i np_i^{(0)} \to \infty$. The X_n^2 statistic is then approximated as follows:

Theorem 1.3 (from Tumanyan, 1956). *If* $\min np_i^{(0)} \to \infty$ *when* $n \to \infty$, *then as* $r \to \infty$

$$\sup_{|x|<\infty} \left| \boldsymbol{P}\{X_n^2 \geq x \mid H_0\} - 1 + \Phi\left(\frac{x-r+1}{\sqrt{2(r-1)}}\right) \right| \to 0, \tag{1.33}$$

where $\Phi(x)$ *is the distribution function of the standard normal law.*

The relation (1.33) follows directly from (1.32) when $r \to \infty$, because we have

$$\sup_{|x|<\infty} \left| P(x; r-1) - 1 + \Phi\left(\frac{x-r+1}{\sqrt{2(r-1)}}\right) \right| \to 0, \qquad (r \to \infty). \tag{1.34}$$

If $\min_i np_i^{(0)}$ is large enough, Theorem 1.2 justifies using, in place of c_α from Eq. (1.16), the root $\tilde{c}_\alpha$ of

$$P(x; r-1) = \alpha. \tag{1.35}$$

Furthermore, from Theorem 1.3, it follows that if $\min_i np_i^{(0)}$ and r are sufficiently large, one can take

$$(\tilde{c}_\alpha)^2 = \sqrt{2(r-1)}\Psi(1-\alpha)+r-1, \tag{1.36}$$

where

$$\Psi(x) = \Phi^{-1}(x).$$

Sharper results in this direction have been obtained (e.g., Theorem 5.1 by Morris, 1975; Theorem II.22 by Drost, 1988). The behavior of Pearson's statistic as r increases was studied in the classical paper of Mann and Wald (1942). Cohen and Sackrowitz (1975), following up the work of Morris, studied the asymptotic optimality of Pearson's test with equiprobable cell probabilities.

Testing Uniformity: von Mises Example

In von Mises (1964) example, consider the first 2,035 digits following the decimal point in the decimal expansion of π. We group these in pairs of successive digits, without overlapping, to form $1,017 = n$ pairs. We hypothesize that these n pairs are data from independent observations, and that the pairs have a uniform distribution over the collection $00, \ldots, 99$. Then $p_i = 0.01$, $i = 1, \ldots, 100 = r$. Computation gives us that Pearson's statistic $X^2 = 100.31$. From Eq. (1.11), we see that $\boldsymbol{E}X^2 = 99$, $\mathbf{Var}\ X^2 \simeq 14$, so the "data" are quite consistent with the hypothesis. If we follow the same procedure using the first 2,020 digits of the decimal expansion of e, $\boldsymbol{E}X_n^2$ is the same as before and $\mathbf{Var}\ X^2$ is essentially the same. This time the "data" gives us $X^2 = 67.14$. Since $99-\sqrt{198} \simeq 85$, this value of X^2 is remarkably small.

More About the Distribution of X^2

One can compute $\boldsymbol{E}\chi^2_{r-1} = r-1$, $\mathbf{Var}\ \chi^2_{r-1} = 2(r-1)$. From Eq. (1.11) we see that if $H_0 : p_i = p_i^{(0)}$ is true then

$$\boldsymbol{E}\{X^2 \mid H_0\} = \boldsymbol{E}\chi^2_{r-1} = r-1 \text{ and } \mathbf{Var}\ X^2 \to \mathbf{Var}\ \chi^2_{r-1},$$

if $\sum_{i=1}^r 1/p_i^{(0)} = o(n)$ when $n \to \infty$. Moreover, as follows from

Theorem 1.2, if $H_0 : p_i = p_i^{(0)}$ is valid the statistic X^2 always has the same chi-squared distribution with $r-1$ degrees of freedom in the limit as $n \to \infty$. This limit distribution depends only on r and not on the construction of the classes $J_1, J_2, \ldots, J_r$ or on the distribution function $F(x)$ of X_i. We emphasize that the *prelimit distributions of the statistic X^2 for different r and different n are distinct.* Furthermore, from Eqs. (1.11) and (1.13), it follows that for given r and n the statistic X^2 has minimal variance, equal to $2(r-1)(1-1/n)$, when $p_1^{(0)} = \cdots = p_r^{(0)} = 1/r$,

$$p_i^{(0)} = \boldsymbol{P}\{X_j \in J_i \mid H_0\} = \int_{J_i} dF(x),$$

that is, in the equiprobable case, which is a reason for constructing our statistic over equiprobable grouping classes. Finally, it follows from Eq. (1.11) that for moderate n, when $\frac{1}{n}\sum_{i=1}^{r} 1/p_i^{(0)}$ is sufficiently large, the variance of the statistic X^2 can turn out to be essentially larger than the variance of the approximating distribution, which in turn can influence statistical inference about H_0. (H_0 can be rejected more often than would follow from a given level of significance α.)

The Cochran and Yarnold Limit

We conclude this section with a result of Yarnold (1970), concerning the approximation of the distribution of X^2 without the condition $\min_i np_i^{(0)} \to \infty$, which appeared in Theorem 1.2. Let us assume that by adjusting the classes $J_1, J_2, \ldots, J_r$ as $n \to \infty$ we have

$$\begin{aligned} np_i^{(0)} &\to m_i \quad &&\text{for } i = 1, 2, \ldots, k, \\ np_i^{(0)} &\to \infty \quad &&\text{for } i = k+1, \ldots, r. \end{aligned}$$

Then, as shown by Yarnold (1970), if the hypothesis $H_0{:}\boldsymbol{P}\{X_1{<}x\}{=}F(x)$ is true, and hence $p_i^{(0)} = \int_{J_i} dF(x)$, the statistic X^2 has, in the limit as $n \to \infty$, the so-called $C(\boldsymbol{m})$ distribution with parameters $r-1$ and $\boldsymbol{m} = (m_1, \ldots, m_k)^T$.

The distribution $C(\boldsymbol{m})$, first considered by Cochran (1952), is the probability distribution of the random variable

$$\sum_{i=1}^{k} \frac{(U_i - m_i)^2}{m_i} + \chi^2_{r-k-1}, \tag{1.37}$$

where U_i has a Poisson distribution with parameter m_i, $i = 1, \ldots, k$, and the random variables $U_1, U_2, \ldots, U_k$, χ^2_{r-k-1} are independent. Lawal (1980) has given percentage points of the $C(\boldsymbol{m})$ distribution for some values of the parameters. More information is in Lawal and Upton (1980). About Poisson approximation see also Holtzman and Good (1986).

If n is sufficiently large and $\min_i np_i^{(0)} \geq 10$, say, then under H_0 the statistic

$$\sum_{i=1}^{k} \frac{(U_i - m_i)^2}{m_i}$$

is distributed approximately as χ^2_k and hence in this case X^2 is distributed approximately as χ^2_{r-1} in accordance with Eq. (1.16).

2. SOME RECOMMENDATIONS WHEN n IS SMALL

Now we turn to the question of the choice of the critical value $\tilde{c}_\alpha$ in the case of application of a chi-squared type test for the verification of the simple hypothesis,

$$H_0 : p_i = p_i^{(0)}, \quad i = 1, \ldots, r,$$

when the number of observations $X_1, \ldots, X_n$ is not large or when $\min np_i^{(0)}$ is small. In this section, we collect some of the lore regarding this point based on experience and numerical experiments, together with several remarks and examples. For more detail see Kempthorne (1966), Tate and Hyer (1973), and Rayner and Best (1989).

The Approximation Problem

One might think of trying to refine the limit theorem for Pearson's statistic. However, the distribution of Pearson's statistic is nonlattice. Moreover the central moments of X^2 differ from the corresponding moments of the approximating chi-squared distribution with $r-1$ degrees of freedom by quantities of order $O(1/n)$, whereas Tumanyan (1956) showed that the jumps of the distribution function of the statistic X^2 have order $O(1/\sqrt{n})$. It follows that refinement of the limit theorems by further moment expansion does not lead to a more accurate choice of critical value $\tilde{c}_\alpha$ for small n (see Greenwood and Nikulin, 1988). The situation is even worse for composite hypotheses.

Cochran's Recommendations

We begin with Cochran's influential paper (1952). Cochran investigated the behavior of approximated statistical tests based on the rule of rejection of H_0 when $X^2 > \tilde{c}_\alpha$, where the level of significance is contained in the interval $0.01 < \alpha < 0.005$. According to Cochran, an approximate critical value, $\tilde{c}_\alpha$, is considered acceptable if the true level of significance of the corresponding approximate criterion,

$$\tilde{\alpha} = \boldsymbol{P}\{X^2 \geq \tilde{c}_\alpha \mid H_0\},$$

differs from the true level of significance α^* of the exact criterion, assuming $\alpha^* < \alpha$,

$$\alpha^* = \boldsymbol{P}\{X^2 \geq c_\alpha \mid H_0\}$$

by no more than 20%. For example, if $\alpha^* = 0.05$, Cochran admits a relative error of 20%, and for $\alpha^* = 0.01$, the relative error may be even 50%. Cochran's basic conclusions are as follows.

1. If the chi-squared criterion based on the statistic X^2 is used for testing a hypothesis according to which the distribution of the random variable X_i has one maximum, then usually as a result of the choice of grouping intervals, one or two of the expected frequencies np_i turn out to be small. If all $np_i \geq 5$, except for one, which is less than 5, then the approximate test will be acceptable for $\alpha = 0.05$ and $\min np_i \geq 0.5$. If $\alpha = 0.01$, then the approximate criterion will also be acceptable but under the condition that $r \geq 7$. If two expected frequencies are small, then it is necessary that they be not less than 1, and all the other np_i should be no less than 5.

Thus, if

$$n = 27,\ r = 7, \qquad p_1^{(0)} = p_7^{(0)} = \frac{1}{27}, \qquad p_2^{(0)} = \cdots = p_6^{(0)} = \frac{5}{27},$$

then approximate criteria constructed with the help of a chi-square approximation will be acceptable according to Cochran.

2. If all np_i are small and $r < 20$, then it is recommended to calculate besides X^2, the X^2 statistic computed using frequency vectors adjacent to the observed vector. One should look at all possible combinations of the type $\nu_1+1, \nu_2-1, \nu_3, \ldots, \nu_r; \nu_1, \nu_2+1, \nu_3-1, \ldots, \nu_r; \ldots$ etc. We denote by $\tilde{X}^2$ the maximum value thus obtained among those less than X^2. If it turns out that $P(\tilde{X}^2+X^2)/2, r-1 \leq \alpha$, then the hypothesis H_0 should be rejected. This and other methods of *correction by continuity* are described in detail by Haber (1980).

3. If all the expected frequencies are small and $r>50$, then to calculate the approximate critical value, $\tilde{c}_\alpha$, one can use the fact that in this case the statistic X^2 is the sum of a large number of weakly dependent variables, and for X^2 the central limit theorem is valid, and consequently as $\tilde{c}_\alpha$ one can choose

$$\tilde{c}_\alpha = \sqrt{\mathbf{Var} X^2}\, \Psi(1-\alpha) + \boldsymbol{E} X^2.$$

4. To choose the number r of grouping intervals one should be guided by the rule given by Williams (1950). According to this rule, it is desirable (in order not to lower the power of the criterion needlessly) to have $r = 16$ for $n = 200$, $r = 20$ for $n = 400$, $r = 30$ for $n = 1{,}000$, and so on, where it is desirable to choose the classes so that all p_i are approximately equal.

Other Recommendations

If n is small, one wants r to be as small as possible. If there is an alternative hypothesis one can construct a test with $r=2$ by the method described in Section 6. Then the relevant central limit theorem is the de Moivre-Laplace theorem.

Interesting investigations of empirical character about the relations among n, r, and α were made by Roscoe and Byars (1971), who say that the mean expected value, equal to n/r, is more important than the number of small expected frequencies. Haber (1980) considers that one can replace the traditional condition $\min np_i^{(0)} \geq 5$ by the less restrictive one: $n/r \geq 5$. In many publications (e.g. Vessereau, 1958), it is noted on the basis of experimental evidence that the size of the sample, n, possibly plays a more decisive role in the quality of approximation of the limiting distribution by the statistic X^2, than expected frequency values. Yarnold's investigation (1970) led him to formulate the following rule in terms of $\min np_i^{(0)}$, for when one can approximate the law of X^2 by a chi-squared distribution: "If the number of grouping intervals $r \geq 3$, and k is the number of intervals for which $np_i^{(0)} < 5$, then $\min_i np_i^{(0)}$ can be as small as $5k/r$." Yarnold (1970) gave a lognormal approximation for the distribution of the statistic X^2, which is recommended when $\min_i np_i^{(0)} > k/\sqrt{(r-1)^3}$. We also refer the reader to Hutchinson (1979), which contains a large bibliography on questions of application of chi-square type criteria when the expected frequencies are small, and to Harrison (1985), Diaconis and Efron (1985), Feinberg (1980), Read and Cressie (1988), Rayner and Best (1989), D'Agostino and Stephens (1986), and Agresti (1984). A more extensive review of recommendations when n is small is in the paper of Greenwood and Nikulin (1988).

3. CHI-SQUARED TEST AGAINST A SIMPLE ALTERNATIVE

The Noncentral Chi-Squared Distribution

Now, let us consider a testing problem with a simple alternative hypothesis. Suppose one is testing the hypothesis $H_0 : p_i = p_i^{(0)}$, $i = 1, 2, \ldots, r$, and the competing simple hypothesis is $H_n : p_i = p_i^{(n)}$, $i = 1, 2, \ldots, r$. For testing H_0 against H_n, one may use a chi-squared test of level α, based on the statistic X^2. It is assumed that $p_1^{(n)} + \cdots + p_r^{(n)} = 1$, $p_i^{(0)} - p_i^{(n)} = \delta_i/\sqrt{n}$, where δ_i are independent of n. Fix α and let the critical value, $\tilde{c}_\alpha$, be the quantile $\chi^2_{r-1,1-\alpha}$ of level $1-\alpha$ of a chi-squared distribution with $r-1$ degrees of freedom, that is,

$$\boldsymbol{P}\{\chi^2_{r-1} > \chi^2_{r-1,1-\alpha}\} = \alpha.$$

The hypothesis H_0 is rejected, according to our criterion, if $X^2 > \chi^2_{r-1,1-\alpha}$. One can write

$$\begin{aligned} X^2 &= \sum_{i=1}^{r} \frac{(\nu_i - np_i^{(0)})^2}{np_i^{(0)}} \\ &= \sum_{i=1}^{r} \left[\frac{\nu_i - np_i^{(n)}}{np_i^{(n)}} \sqrt{1 + \frac{\delta_i}{p_i^{(n)}\sqrt{n}}} + \frac{p_i^{(n)} - p_i^{(0)}}{\sqrt{p_i^{(0)}}} \sqrt{n} \right]^2, \end{aligned} \tag{3.1}$$

from which it follows that, if H_n is true,

$$\lim_{n\to\infty} \boldsymbol{P}\{X^2 > x \mid H_n\} = \boldsymbol{P}\{\chi^2_{r-1}(\lambda) > x\}, \tag{3.2}$$

where $\chi^2_{r-1}(\lambda)$ is a random variable having a noncentral chi-squared distribution with $r-1$ degrees of freedom and noncentrality parameter

$$\lambda = \sum_{i=1}^{r} \frac{\delta_i^2}{p_i^{(0)}}, \qquad \delta_1 + \cdots + \delta_r = 0. \tag{3.3}$$

Next we study a particular class of alternatives in more detail. Recall from Section 1 that the vector of frequencies, $\boldsymbol{\nu}$ has the multinomial distribution with parameters n and $\boldsymbol{p}$. The likelihood function is

$$\ell(\boldsymbol{p}) = \frac{n!}{\nu_1!\nu_2!\dots\nu_r!} p_1^{\nu_1} p_2^{\nu_2} \dots p_r^{\nu_r}, \tag{3.4}$$

where $\boldsymbol{p} = (p_1, \dots, p_r)^T$, $\boldsymbol{p}^T\mathbf{1}_r = 1$, all $p_i > 0$, $\mathbf{1}_r = (1, \dots, 1)^T$, the unit vector of order r. We know that

$$\boldsymbol{E}\boldsymbol{\nu} = n\boldsymbol{p}, \qquad \mathbf{Var}\ \boldsymbol{\nu} = n[\boldsymbol{D} - \boldsymbol{p}\boldsymbol{p}^T] = n\boldsymbol{\Sigma}, \qquad \text{rank } \boldsymbol{\Sigma} = r-1,$$

where $\boldsymbol{D}$ is the diagonal matrix with elements $p_1, \dots, p_r$ on the main diagonal. Furthermore, we know that

$$\frac{1}{\sqrt{n}}(\boldsymbol{\nu} - n\boldsymbol{p}) \text{ is } AN(\mathbf{0}_r, \boldsymbol{\Sigma}) \text{ as } n \to \infty \tag{3.5}$$

and the statistic

$$X^2 = \frac{1}{n}(\boldsymbol{\nu}-n\boldsymbol{p})^T \boldsymbol{\Sigma}^-(\boldsymbol{\nu}-n\boldsymbol{p}) = \sum_{i=1}^{r} \frac{(\nu_i - np_i)^2}{np_i} \tag{3.6}$$

is distributed as χ^2_{r-1} in the limit as $n \to \infty$.

Pearson's Statistic Under an Alternative Hypothesis

As an alternative hypothesis, we introduce a sequence of vectors $\{\boldsymbol{p}^{(n)}\}$, $\boldsymbol{p}^{(n)} = (p_1^{(n)}, \ldots, p_r^{(n)})^T$, such that $\mathbf{1}_r^T \boldsymbol{p}^{(n)} = 1$, all $p_i^{(n)} > 0$, and

$$\lambda_n = n \sum_{i=1}^{r} \frac{(p_i - p_i^{(n)})^2}{p_i} = \text{const} = \lambda \text{ as } n \to \infty. \tag{3.7}$$

Then

$$\frac{1}{\sqrt{n}}(\boldsymbol{\nu}-n\boldsymbol{p}^{(n)}) \text{ is } AN(\sqrt{n}(\boldsymbol{p}-\boldsymbol{p}^{(n)}), \boldsymbol{\Sigma}), \tag{3.8}$$

as $n \to \infty$, and hence the statistic

$$X^2 = \frac{1}{n}(\boldsymbol{\nu}-n\boldsymbol{p}^{(n)})^T \boldsymbol{\Sigma}^{-1}(\boldsymbol{\nu}-n\boldsymbol{p}^{(n)}) = \sum_{i=1}^{r} \frac{(\nu_i - np_i^{(n)})^2}{np_i} \tag{3.9}$$

is distributed in the limit as $n \to \infty$ as $\chi^2_{r-1}(\lambda)$ [see Eq. (3.2)]. We can take

$$\boldsymbol{\Sigma}^- = \boldsymbol{D}^{-1} + \frac{1}{p_r}\mathbf{1}_r\mathbf{1}_r^T, \qquad \mathbf{1}_r^T\boldsymbol{p} = \mathbf{1}_r^T\boldsymbol{p}^{(n)} = 1. \tag{3.10}$$

If $\lambda = 0$, that is, $\boldsymbol{p} = \boldsymbol{p}^{(n)}$, then we have $\chi^2_{r-1}(0) = \chi^2_{r-1}$ the central chi-squared distribution with $r-1$ degreees of freedom.

We state a proposition concerning the behavior of the frequency vector $\boldsymbol{\nu}$, when $\boldsymbol{p} = \boldsymbol{p}^{(n)}$ as in Eq. (3.7).

Proposition 3.1. *Let $\{\boldsymbol{p}^{(n)}\}$ be the sequence defined above for which Eq. (3.7) holds*

$$\lambda_n = n \sum_{i=1}^{r} \frac{(p_i - p_i^{(n)})^2}{p_i^{(n)}} = const = \lambda \ as \ n \to \infty. \tag{3.11}$$

Then the statistic

$$X^2 = \sum_{i=1}^{r} \frac{(\nu_i - np_i^{(n)})^2}{np_i^{(n)}} \tag{3.12}$$

is distributed as $\chi^2_{r-1}(\lambda)$ *in the limit as* $n \to \infty$.

Voinov, Neumann, and Nikulin (1986) evaluated quantiles and noncentrality parameters using the Lagrange method. Fix (1944) tabulated the noncentral χ^2 distribution.

Consistency

Before discussing the power of a test of the form "accept H_0 if $\phi_n < c$," we want to know that the test is at least *consistent*, that is, that for an alternative H_1, $\mathbf{P}\{\phi_n \geq c \mid H_1\} \to 1$ as $n \to \infty$. Sufficient conditions for consistency of a test in terms of $\mathbf{E}_1\phi_n$ and $\mathbf{Var}_1\phi_n$, the mean and variance under H_1, can be obtained using Chebychev's inequality, for example, if

$$\mathbf{E}_1\phi_n = \alpha n + \beta \text{ and } \mathbf{Var}_1\phi_n = \gamma n + \delta, \qquad \alpha \neq 0,$$

then for large enough n, $c - \alpha n - \beta$ has the same sign as α and

$$\begin{aligned} \mathbf{P}\{\phi_n < c \mid H_1\} &= \mathbf{P}\{(\phi_n - \alpha n - \beta)^2 < (c - \alpha n - \beta)^2\} \\ &\leq \frac{\gamma n + \delta}{(c - \alpha n - \beta)^2} \to 0 \end{aligned}$$

as $n \to \infty$. Hence $\mathbf{P}\{\phi_n \geq c \mid H_1\} \to 1$. We can see from Eq. (3.1) that if the alternative hypothesis $H_1 = H_n$ is fixed then $\mathbf{E}_1 X^2$ and $\mathbf{Var}_1 X^2$ both grow asymptotically linearly with n, so we have consistency of a chi-squared test.

Power of the Test

To test $H_0 : \mathbf{p} = \mathbf{p}^{(0)} = (p_1^{(0)}, \ldots, p_r^{(0)})^T$, $\mathbf{1}_r^T \mathbf{p}^{(0)} = 1$, all $p_i^{(0)} > 0$, against the alternative $H_n : \mathbf{p} = \mathbf{p}^{(n)}$, $\mathbf{p}^{(n)}$ as defined above, we choose as critical value $c_\alpha = \chi^2_{r-1,1-\alpha}$, the α-upper quantile of the chi-squared distribution with $r-1$ degrees of freedom

$$\mathbf{P}\{\chi^2_{r-1} > c_\alpha\} = \alpha. \tag{3.13}$$

According to Pearson, the hypothesis H_0 should be rejected if

$$X^2 = \sum_{i=1}^{r} \frac{(\nu_i - np_i^{(0)})^2}{np_i^{(0)}} \geq c_\alpha,$$

otherwise H_0 is not rejected. Of course, as $n \to \infty$

$$\boldsymbol{P}\{X^2 \geq c_\alpha \mid H_0\} \to \alpha \qquad \text{and} \qquad \boldsymbol{P}\{X^2 < c_\alpha \mid H_0\} \to 1-\alpha. \quad (3.14)$$

On the other hand, if H_n, according to which $\boldsymbol{p} = \boldsymbol{p}^{(n)}$, is true, then, from Proposition 3.1, we obtain

$$\beta(\lambda) = \boldsymbol{P}\{X^2 < c_\alpha \mid H_n\} = \boldsymbol{P}\{\chi^2_{r-1}(\lambda) < c_\alpha\} + o(1), \quad n \to \infty, \quad (3.15)$$

where λ is defined by Eq. (3.11). If r and λ are such that $r+\lambda$ is large enough, then from the normal approximation of the distribution $\chi^2_{r-1}(\lambda)$ it follows (see the Appendix) that

$$\beta(\lambda) \approx \Phi\left[\frac{c_\alpha - (r+\lambda-1)}{\sqrt{2(r-1+2\lambda)}}\right]. \quad (3.16)$$

The function $1-\beta(\lambda)$, $\lambda \geq 0$, is called the *power* of the test based on X^2 to test H_0 against H_n.

Approximation of Patnaik

Suppose that $p_1 = p_i = \cdots = p_r = 1/r$. Then

$$\lambda = nr \sum_{i=1}^{r} \left(\frac{1}{r} - p_i^{(n)}\right)^2 = nrd_n^2, \quad d_n^2 = \sum_{i=1}^{r} (\frac{1}{r} - p_i^{(n)})^2,$$

and hence, from the asymptotic formula of Patnaik (1949), we have as $\lambda \to 0$ [i.e., the hypotheses H_n are such that $d_n^2 = o(1/n)$ as $n \to \infty$],

$$\begin{aligned}\boldsymbol{P}\{\chi^2_{r-1}(\lambda) \geq c_\alpha\} &\sim \boldsymbol{P}\{\chi^2_{r-1} \geq c_\alpha\left(1 - \frac{\lambda}{r-1}\right)\} \\ &= \alpha + \frac{\lambda c_\alpha}{r-1} g_{r-1}(c_\alpha)(1+o(1)),\end{aligned} \quad (3.17)$$

where g_{r-1} is the density function of the chi-squared distribution with $r-1$ degrees of freedom (see Pearson, 1959, and the Appendix). If $\lambda c_\alpha \to 0$ and $c_\alpha \to \infty$ as $n \to \infty$, then from Eq. (3.17) and the asymptotic relation

$$2g_{r-1}(c_\alpha) \sim \boldsymbol{P}\{\chi^2_{r-1} \geq c_\alpha\} = \alpha,$$

we obtain the asymptotic formula for the power

$$1-\beta(\lambda) = \alpha + \frac{\alpha\lambda}{2(r-1)}\, c_\alpha(1+o(1)). \tag{3.18}$$

Normal Approximation

$(p_1 = p_2 = \ldots = p_r = 1/r)$

If r is large and fixed (i.e., $r \geq 30$), then, from Tumanyan's Theorem (Theorem 1.3) and Proposition 3.1, we have

$$\begin{aligned} \boldsymbol{P}\{X^2 \geq c_\alpha \mid H_n\} &\approx \boldsymbol{P}\{\chi^2_{r-1}(\lambda) \geq c_\alpha\} \\ &\approx 1-\Phi\left\{\frac{c_\alpha - r + 1 - \lambda}{\sqrt{2(r-1+2\lambda)}}\right\}, \end{aligned} \tag{3.19}$$

where $\lambda = nrd_n^2$. Now suppose that with r fixed, $n \to \infty$ and $\lambda = o(1)$. Then from Eq. (3.19), it follows that

$$\begin{aligned} 1-\beta(\lambda) = \boldsymbol{P}\{X^2 \geq c_\alpha \mid H_n\} &\simeq 1-\Phi\left\{\frac{(c_\alpha - r+1)[1-\frac{\lambda}{c_\alpha - r+1}]}{\sqrt{2(r-1)}\sqrt{1+\frac{2\lambda}{r-1}}}\right\} \\ &= 1-\Phi\left\{\frac{c_\alpha - r+1}{\sqrt{2(r-1)}} - \frac{\lambda c_\alpha}{(r-1)\sqrt{2(r-1)}}\right\} \\ &= 1-\Phi\left(\frac{c_\alpha - r+1}{\sqrt{2(r-1)}}\right) + \frac{\lambda c_\alpha}{(r-1)\sqrt{2(r-1)}}\varphi\left(\frac{c_\alpha - r+1}{\sqrt{2(r-1)}}\right) \\ &\approx \alpha + \varphi(u_\alpha)\frac{\lambda}{\sqrt{2(r-1)}}, \end{aligned} \tag{3.20}$$

where u_α is the α-quantile of the standard normal distribution and ϕ is its density.

A Closer Look at H_0; Choice of Critical Value

In connection with alternatives, let us point out that hypotheses are never really simple. One of the first to notice this was Berkson (1938). Practice shows that if the minimal expected frequency $\min np_i^{(0)}$ is sufficiently large for large n, a statistical criterion based on X^2 will reject H_0. The point is that the hypothesis H_0 is given in terms of equalities, $p_i = p_i^{(0)}$, whereas in "reality" only approximate relations, $p_i \approx p_i^{(0)}$, $i = 1, 2, \dots, r$ hold. This means that one should test not the simple hypothesis $H_0 : p_i = p_i^{(0)}$, which defines one point $(p_1^{(0)}, \dots, p_r^{(0)})^T$ belonging to the simplex

$$p_1 > 0, \dots, p_r > 0, \qquad p_1 + \cdots + p_r = 1,$$

but a complex hypothesis defining a domain in the simplex, which can be expressed by the inequality

$$n \sum_{i=1}^{r} \frac{(p_i - p_i^{(0)})^2}{p_i^{(0)}} < \lambda_0,$$

where λ_0 is a constant ($\lambda_0 > 0$) which depends on how precise the approximate relations $p_i \approx p_i^{(0)}$ are. With such an approach, even in the case of validity of H_0, the statistic X^2 has limit as $n \to \infty$, with $\min_i np_i^{(0)}$ sufficiently large, a noncentral chi-square distribution with $r-1$ degrees of freedom and parameter of noncentrality

$$\lambda = n \sum_{i=1}^{r} \frac{(p_i - p_i^{(0)})^2}{p_i^{(0)}}, \tag{3.21}$$

where $\lambda \leq \lambda_0$. The problem of testing the simple hypothesis H_0 itself against the simple alternatives H_n becomes the problem of testing the complex hypothesis $H_0 : 0 \leq \lambda \leq \lambda_0$ against complex alternatives $H_n : \lambda > \lambda_0$. One can see that in this case instead of the critical value of $\chi^2_{r-1,1-\alpha}$ one should take the quantile $c_\alpha = \chi^2_{r-1,1-\alpha}(\lambda_0)$ of level $(1-\alpha)$ of the noncentral chi-squared distribution with $r-1$ degrees of freedom and parameter of noncentrality λ_0.

In many cases of practical interest one can choose the parameter λ_0 equal to

$$\lambda_0 = \frac{1}{nr^2} \sum_{i=1}^{r} \frac{1}{p_i^{(0)}}. \tag{3.22}$$

In particular, if $p_i^{(0)} = \cdots = p_r^{(0)} = 1/r$, then $\lambda_0 = 1/n$, where n is the number of trials. Having made this observation about the simple H_0, we will return to talking about it as if it made sense.

Some Remarks

One should keep in mind when comparing various virtues of chi-squared type criteria, that as a rule these virtues have asymptotic character. Thus, for example, even when a chi-squared criterion is most powerful in the problem of testing H_0against H_1, the extremality of the power of this criterion is typically an asymptotic property and consequently may be relevant only for large n.

Chi-squared type criteria generally let one test only certain facets of the hypothesis H_0 . Nevertheless, if the power of the chi-squared test used in the problem of testing H_0 against an alternative H_1 is sufficiently high, and these facets essentially contradict the interesting aspects of the alternative H_1, then application of a chi-square test should be recognized as completely appropriate.

4. THE CONTINUITY CORRECTION FOR PEARSON'S TEST WITH ONE DEGREE OF FREEDOM AND THE 2×2 TABLE

The Corrected Test

If $r = 2$, as in Eq. (1.21), then, as $n \to \infty$,

$$X_n^2 = \frac{(\nu_1 - np_1^{(0)})^2}{np_1^{(0)}(1-p_1^{(0)})} \approx \chi_1^2. \tag{4.1}$$

From the de Moivre-Laplace theorem we have, with the continuity correction 0.5 [see Eq. (1.23) and discussion following], as $n \to \infty$

$$\boldsymbol{P}\{\nu_1 \le m \mid H_0\} = \Phi\left(\frac{m+0.5-np_1^{(0)}}{\sqrt{np_1^{(0)}(1-p_1^{(0)})}}\right) + O\left(\frac{1}{\sqrt{n}}\right) \tag{4.2}$$

and

$$\begin{aligned}\boldsymbol{P}\{\nu_1 \ge M \mid H_0\} &= 1-\boldsymbol{P}\{\nu_1 \le M-1 \mid H_0\}\\ &= \Phi\left(-\frac{M-0.5-np_1^{(0)}}{\sqrt{np_1^{(0)}(1-p_1^{(0)})}}\right) + O\left(\frac{1}{\sqrt{n}}\right).\end{aligned} \tag{4.3}$$

From Eqs. (4.2) and (4.3) it follows that if we want the test based on the statistic X^2 to have significance level $\approx \alpha$, $0 < \alpha < 0.5$, then we should reject H_0 if

$$\Phi\left(\frac{\nu_1+0.5-np_1^{(0)}}{\sqrt{np_1^{(0)}(1-p_1^{(0)})}}\right) \le \frac{\alpha}{2} \quad \text{or} \quad \Phi\left(-\frac{\nu_1-0.5-np_1^{(0)}}{\sqrt{np_1^{(0)}(1-p_1^{(0)})}}\right) \le \frac{\alpha}{2}. \tag{4.4}$$

One rejects H_0 if one of two events happens:

$$\frac{\nu_1-np_1^{(0)}}{\sqrt{np_1^{(0)}(1-p_1^{(0)})}} \le \Psi(\frac{\alpha}{2}) - \frac{1}{2\sqrt{np_1^{(0)}(1-p_1^{(0)})}} \tag{4.5}$$

or

$$\frac{\nu_1-np_1^{(0)}}{\sqrt{np_1^{(0)}(1-p_1^{(0)})}} \ge -\Psi(\frac{\alpha}{2}) + \frac{1}{2\sqrt{np_1^{(0)}(1-p_1^{(0)})}}, \tag{4.6}$$

where $\Psi(\cdot)$ is the inverse function of $\Phi(\cdot)$. From Eq. (4.6), we obtain that H_0 must be rejected if

$$X_n^2 \ge \left[\Psi(1-\frac{\alpha}{2}) + \frac{1}{2\sqrt{np_1^{(0)}(1-p_1^{(0)})}}\right]^2. \tag{4.7}$$

We used here that

$$\Psi(z) + \Psi(1-z) \equiv 0, \quad z \in (0,1).$$

From formula (4.7), one sees that in case $r = 2$ the choice

$$c_\alpha = \left[\Psi(1-\frac{\alpha}{2}) + \frac{1}{2\sqrt{np_1^{(0)}(1-p_1^{(0)})}} \right]^2 \tag{4.8}$$

produces a χ^2-test of level $\approx \alpha$.

Cramer's Example and Continuity Correction

We consider next (see also Bolshev and Smirnov, 1968) an example shown by Cramer (1946, pp. 444–445). Suppose N elements are divided randomly into two groups of n and $N-n$ elements and suppose that in the entire set of N objects there are M elements posessing some property Y, and the remaining elements do not have this property. The resulting division can be written in the form of a 2×2 table (μ is the number of elements with property Y, belonging to the first group):

	With Property Y	Without Property Y	Total
Sample or 1st group	μ	$n-\mu$	n
Residual or 2nd group	$M-\mu$	$N-n-M+\mu$	$N-n$
Total	M	$N-M$	N

If the division of the elements into two groups is really random and does not depend on possession of property Y, then

$$\boldsymbol{E}\mu = \frac{nM}{N} \quad \text{and} \quad \textbf{Var}\ \mu = \frac{nM(N-n)(N-M)}{N^2(N-1)} \tag{4.9}$$

If N is sufficiently large and none of n, M, $N-n$, $N-M$ are too small, then the normalized deviation

$$\frac{\mu - \boldsymbol{E}\mu}{\sqrt{\textbf{Var}\ \mu}}$$

has a nearly standard normal law. Hence, for large N the square of the normalized deviation

$$X^2 = \frac{(\mu - \boldsymbol{E}\mu)^2}{\textbf{Var}\ \mu}$$

is distributed nearly as χ_1^2. The statistic X^2 is often used as a criterion for the random division of N elements into two groups. For small N it is recommended that instead of X^2 one use the statistic

$$Z^2 = \left(|X| - \frac{1}{2\sqrt{\textbf{Var}\ \mu}}\right)^2 \tag{4.10}$$

and follow the rule:

if $P(Z^2; 1) \leq \alpha$ then reject the hypothesis of randomness;

but

if $P(Z^2; 1) \geq \alpha$, then one should conclude that the observed values do not contradict the hypothesis of randomness, where

$$P(x; 1) = \boldsymbol{P}\{\chi_1^2 \geq x\} = \frac{1}{\sqrt{\pi}} \int_x^\infty \frac{1}{\sqrt{z}} e^{-z/2}\, dz, \qquad x > 0.$$

In formula (4.10) the term $1/(2\sqrt{\textbf{Var}\mu})$ is the continuity correction for the approximating distribution function $P(x; 1)$ (see, e.g., Bolshev and Smirnov, 1968).

There are tables of exact critical values of $M - \mu$ (Fisher's criteria) for $n = 3(1)20$ and $N - n = 2(1)n$ which can be used in testing the hypothesis of randomness for small values of N (see Pearson and Hartley, 1956; Latscha, 1953). If $N \geq 20$ then the critical regions of the approximate one-sided tests are defined (with the obvious continuity correction) by the inequalities:

$$\frac{\mu - \boldsymbol{E}\mu}{\sqrt{\mathbf{Var}\ \mu}} \geq \Psi(1-\frac{\alpha}{2}) + \frac{1}{\sqrt{2\mathbf{Var}\ \mu}}$$

$$\frac{\mu - \boldsymbol{E}\mu}{\sqrt{\mathbf{Var}\ \mu}} \leq -\Psi(1-\frac{\alpha}{2}) - \frac{1}{\sqrt{2\mathbf{Var}\ \mu}}. \tag{4.11}$$

This correction differs from the continuity correction proposed by Yates (1934), which should not be used (see Conover, 1974). From Eq. (4.11) it follows that the critical region of the approximate two-sided test with a significance level α is given by the inequality

$$X^2 = \frac{(\mu - \boldsymbol{E}\mu)^2}{\mathbf{Var}\ \mu} \geq \left(\Psi(1-\frac{\alpha}{2}) + \frac{1}{2\sqrt{\mathbf{Var}\ \mu}}\right)^2. \tag{4.12}$$

Now one can see that the approximate formula for critical values of the exact Fisher's test (4.12), leads to results that are different from those based on formula (4.10). In fact, according to Eq. (4.10) the statistic

$$Z^2 = \left(|X| - \frac{1}{2\sqrt{\mathbf{Var}\ \mu}}\right)^2$$

is distributed approximately as χ_1^2, and from Eq. (4.12) it follows that X^2 is distributed approximately as $\left(\xi + 1/(2\sqrt{\mathbf{Var}\ \mu})\right)^2$, where ξ is a standard normal random variable. In other words, the statistic X^2 is distributed approximately as $\chi_1^2(\lambda)$, $\lambda = (4\ \mathbf{Var}\ \mu)^{-1}$. If

$$\mathbf{Var}\ \mu = \frac{nM(N-n)(N-M)}{N^2(N-1)} \to \infty$$

as $n \to \infty$ then the distributions of X^2 and Z^2 are asymptotically the same.

If μ is the number of positive outcomes in a sequence of independent trials with parameters n and p, then to test the hypothesis $H_0 : p = p^{(0)}$ for moderate values of n one could use the same statistic X^2, putting, instead of Eq. (4.9)

$$\boldsymbol{E}\mu = np^{(0)} \quad \text{and} \quad \mathbf{Var}\ \mu = np^{(0)}(1-p^{(0)}).$$

Again the critical region of the approximate test would be given by inequality (4.12). More useful information can be found in the publications of Agresti (1984), Fienberg (1980), Fleiss (1981), Berkson (1978), Grizzle (1967), and Plackett (1964).

Until now we have considered unilateral tests; that is, the hypothesis H_0 is rejected according to the chi-square test if $X^2 > c_\alpha$, where c_α is the critical value corresponding to the level of significance α. In some cases (see Section 5) bilateral tests of chi-square type are of interest: the hypothesis H_0 is accepted if $c_1 < X^2 < c_2$, and otherwise H_0 is rejected, where the critical values $0 < c_1 < c_2 < \infty$ are chosen so that

$$\boldsymbol{P}\{c_1 < X^2 \mid H_0\} = \alpha_1, \qquad \boldsymbol{P}\{X^2 > c_2 \mid H_0\} = \alpha_2,$$

$\alpha_1 + \alpha_2 \leq \alpha$, where α is a given level of significance. Using bilateral criteria of chi-square type is quite reasonable if there are suspicions of "juggled data."

5. CHI-SQUARED TEST FOR RANDOMNESS OF A SAMPLE

Bavarov and Belyaev's Example

As an example, we present a variant of the problem of testing the randomness of selection without replacement studied by Bavarov and Belyaev (1961). Their problem is to construct a test of chi-squared type for *randomness* of a sample, without replacement, from a population of size k consisting of k_i elements of type i, $i = 1, 2, \ldots, s$,

$$k_1 + k_2 + \cdots + k_s = k.$$

We take n independent samplings without replacement, each of size $l \leq k$. For the ith sampling we denote by i_j the type of element appearing at the jth drawing ($i = 1, \ldots, s$; $j = 1, \ldots, l$), and we denote by ν_{ij} the total number of elements of type i at position j over all n samples. *Randomness* of the sample means that at any drawing each of the remaining elements is selected with equal probability. We test the hypothesis H_0 according to which each sample is random and all samples are mutually independent. It is evident that each random variable ν_{ij}

$(i = 1, \dots, s;\ j = 1, \dots, l)$ can be represented as

$$\nu_{ij} = \sum_{x=1}^{n} \nu_{ijx},$$

where

$$\nu_{ijx} = \begin{cases} 1, & \text{if an element of the } i\text{th type at the } j\text{th drawing} \\ & \quad \text{in the sample with number } x, \\ 0, & \text{otherwise.} \end{cases}$$

If H_0 is true then the random variables ν_{ijx} are independent.

Let $p_j^{(i)}$ be the probability that an element of type i is chosen at the jth drawing, that is,

$$p_j^{(i)} = \boldsymbol{P}\{i_j = i\}, \qquad 1 \le i \le s, \quad 1 \le j \le l.$$

In this example, we use, when possible, the notation of Bavarov and Belyaev (1961). If H_0 is true, then

$$p_j^{(i)} = \frac{k_i}{k}, \qquad i = 1, \dots, s; \quad j = 1, \dots, l.$$

As an alternative, we consider the hypothesis H_1, according to which

$$p_j^{(i)} \ne \frac{k_i}{k} \qquad \text{for some pair } (i, j).$$

We denote by $p_{jt}^{(ir)}$ the probability that an element of the ith type appears at the jth drawing and an element of the rth type appears at the tth drawing. If H_0 is true then we have

$$p_{jj}^{(ii)} = p_j^{(i)} = \frac{k_i}{k}, \qquad p_{jj}^{(ir)} = 0 \quad (i \ne r),$$

$$p_{jt}^{(ii)} = \frac{k_i(k_i - 1)}{k(k-1)} \quad (j \ne t), \qquad p_{jt}^{(ir)} = \frac{k_i k_r}{k(k-1)} \quad (j \ne t),$$

$$\boldsymbol{E}\nu_{ijx} = p_j^{(i)}, \qquad \boldsymbol{E}\nu_{ijx}\nu_{rqy} = p_{jq}^{(ir)},$$

$$(x, y = 1, 2, \dots, n; i = 1, \dots, s; j = 1, \dots, l),$$

and hence under H_0

$$\boldsymbol{E}\nu_{ij} = np_j^{(i)}, \qquad \boldsymbol{E}(\nu_{ij} - \boldsymbol{E}\nu_{ij})(\nu_{rq} - \boldsymbol{E}\nu_{rq}) = n(p_{jq}^{(ir)} - p_j^{(i)}p_q^{(r)}).$$

From the central limit theorem, it follows that under H_0 as $n \to \infty$ the vector

$$\begin{aligned}\boldsymbol{\nu} = \Bigg(&\sqrt{\frac{k}{nk_1}}\left(\nu_{11} - \frac{nk_1}{k}\right), \ldots, \sqrt{\frac{k}{nk_1}}\left(\nu_{1l} - \frac{nk_1}{k}\right), \\ &\sqrt{\frac{k}{nk_2}}\left(\nu_{21} - \frac{nk_2}{k}\right), \ldots, \sqrt{\frac{k}{nk_s}}\left(\nu_{sl} - \frac{nk_s}{k}\right)\Bigg)^T,\end{aligned} \tag{5.1}$$

of dimension sl, is $AN_{sl}(\mathbf{0}_{sl}, \boldsymbol{\Lambda})$.

The $sl \times sl$ covariance matrix, $\boldsymbol{\Lambda}$, of the vector $\boldsymbol{\nu}$ is degenerate, with rank $\boldsymbol{\Lambda} = l(s-1)$, and consists of $s \times s$ blocks, $\boldsymbol{\Lambda}_{ij}$, $\boldsymbol{\Lambda} = [\boldsymbol{\Lambda}_{ij}]$, where

$$\begin{aligned}\boldsymbol{\Lambda}_{ij} = \frac{k}{n\sqrt{k_i k_j}} \boldsymbol{E}\Bigg\{&\left(\nu_{i1} - \frac{nk_i}{k}, \nu_{i2} - \frac{nk_i}{k}, \ldots, \nu_{il} - \frac{nk_i}{k}\right)^T \cdot \\ &\left(\nu_{j1} - \frac{nk_j}{k}, \nu_{j2} - \frac{nk_j}{k}, \ldots, \nu_{jl} - \frac{nk_j}{k}\right)\Bigg\},\end{aligned}$$

$i, j = 1, \ldots, s.$

Let $\tilde{\boldsymbol{\Lambda}} = [\boldsymbol{\Lambda}_{ij}]$, the matrix which contains only $(s-1) \times (s-1)$ blocks $(i, j = 1, \ldots, l-1)$. Rank $\tilde{\boldsymbol{\Lambda}} = l(s-1)$, and there exists $\tilde{\boldsymbol{\Lambda}}^{-1}$. The matrix

$$\boldsymbol{\Lambda}^- = \underbrace{\begin{bmatrix} \tilde{\boldsymbol{\Lambda}}^{-1} & \mathbf{0}_{l(s-1)\times l} \\ \mathbf{0}_{l\times l(s-1)} & \mathbf{0}_{l\times l} \end{bmatrix}}_{l(s-1)} \begin{matrix} \}l(s-1) \\ \}l \end{matrix}$$

is a generalized inverse for $\boldsymbol{\Lambda}$, rank $\boldsymbol{\Lambda}^- = l(s-1)$, such that

$$\boldsymbol{\Lambda}\boldsymbol{\Lambda}^- = \boldsymbol{\Lambda}^-\boldsymbol{\Lambda} = \left[\begin{array}{c:c} \boldsymbol{I}_{l(s-1)} & \boldsymbol{0}_{l(s-1)\times l} \\ \hdashline \boldsymbol{0}_{l\times l(s-1)} & \boldsymbol{0}_{l\times l} \end{array}\right], \tag{5.2}$$

$$\text{trace } \boldsymbol{\Lambda}^-\boldsymbol{\Lambda} = l(s-1),$$

and $\boldsymbol{\Lambda}^-$ satisfies Theorem 1.1 with $\boldsymbol{\Sigma}=\boldsymbol{\Lambda}$. In some sense, the matrix $\boldsymbol{\Lambda}^-$ is constructed as $\boldsymbol{\Sigma}^-$ was constructed earlier for the standard Pearson statistic in Section 1. Now by standard methods we can construct two statistics of χ^2 type, the X_n^2 statistic of Pearson, which is the squared length of the vector $\boldsymbol{\nu}$ in Eq. (5.1),

$$X_n^2 = \boldsymbol{\nu}^T\boldsymbol{\nu} = \sum_{ij} \frac{k}{nk_i}\left(\nu_{ij} - \frac{nk_i}{n}\right)^2 = \sum_{ij} \frac{(\nu_{ij} - \boldsymbol{E}\nu_{ij})^2}{\boldsymbol{E}\nu_{ij}} \tag{5.3}$$

and the generalized Pearson statistic Y_n^2 is defined by

$$\begin{aligned} Y_n^2 &= \boldsymbol{\nu}^T\boldsymbol{\Lambda}^-\boldsymbol{\nu} \\ &= \frac{k-1}{n}\left\{\sum_{ij}\frac{1}{k_i}\left(\nu_{ij} - \frac{nk_i}{k}\right)^2 + \frac{1}{k-l}\sum_{i=1}^{s}\left\langle\frac{1}{k_i}\left[\sum_{j=1}^{l}\left(\nu_{ij} - \frac{nk_i}{k}\right)\right]^2\right\rangle\right\} \\ &= \frac{k-1}{k}X_n^2 + \frac{l(k-1)}{k(k-l)}\sum_{i=1}^{s}\frac{(\nu_i - \boldsymbol{E}\nu_i)^2}{\boldsymbol{E}\nu_i}, \end{aligned} \tag{5.4}$$

where $\nu_i = \sum_{j=1}^{l}\nu_{ij}$.

It was shown by Bavarov and Belyaev (1961) that

$$\lim_{n\to\infty} \boldsymbol{P}\{X_n^2 \geq x\} = \boldsymbol{P}\left\{\frac{k}{k-1}\chi^2_{(l-1)(s-1)} + \frac{k-l}{k-1}\chi^2_{s-1} \geq x\right\} \tag{5.5}$$

where $\chi^2_{(l-1)(s-1)}$ and χ^2_{s-1} are independent chi-squared distributed random variables with $(l-1)(s-1) = ls - l - s + 1$ and $s-1$ degrees of freedom respectively.

It follows from Theorem 1.1 and its application to Eqs. (1.15), and (5.2), that

$$\lim_{n\to\infty} \boldsymbol{P}\{Y_n^2 \geq x \mid H_0\} = \boldsymbol{P}\{\chi^2_{l(s-1)} \geq x\}. \tag{5.6}$$

In their paper Bavarov and Belyaev proved that a χ^2-test based on the statistic X_n^2 for testing H_0 is consistent with respect to any alternative of type H_1. It is evident that the test for H_0 based on Y^2 is also consistent with respect to any alternative of type H_1.

Now, we consider the alternative H_1 according to which we have almost deterministic samplings. More exactly, according to H_1 the first element, i, one chooses as before (i.e., the first element drawn is of type i with probability k_i/k), and the remaining elements of all n samples are produced deterministicallly by repeating cycles of the elements $1, \ldots, s$, in order beginning with $i+1$.

Suppose $s = 2$, $k_1/k = k_2/k = 0.5$. Because

$$\nu_{11}+\nu_{21} = \nu_{12}+\nu_{22} = \cdots = \nu_{1l}+\nu_{2l} = n.$$

Then, from Eqs. (5.3) and (5.4), we have

$$X_n^2 = \sum_{j=1}^{l} \left[\frac{2}{\sqrt{n}}(\nu_{ij} - \frac{n}{2})\right]^2, \tag{5.7}$$

$$Y_n^2 = \frac{k-1}{k}\left[\sum_{j=1}^{l}\left[\frac{2}{\sqrt{n}}(\nu_{ij}-\frac{n}{2})\right]^2 + \frac{1}{k-l}\left(\sum_{j=1}^{l}\frac{2}{\sqrt{n}}(\nu_{ij}-\frac{n}{2})\right)^2\right],$$

$$l < k, \tag{5.8}$$

and from Eqs. (5.5) and (5.6)

$$\begin{aligned} \lim_{n\to\infty} \boldsymbol{P}\{Y_n^2 \leq x \mid H_0\} &= \boldsymbol{P}\{\chi_l^2 \leq x\}, \\ \lim_{n\to\infty} \boldsymbol{P}\{\frac{k-1}{k}X_n^2 \leq x \mid H_0\} &= \boldsymbol{P}\{\chi^2_{l-1} + (1-\frac{l}{k})\chi_1^2 \leq x\}, \end{aligned} \tag{5.9}$$

respectively, where χ^2_{l-1} and χ_1^2 are independent.

If H_1 is true then we have

$$\nu_{11} = \nu_{13} = \cdots = \nu_{1,2m+1},$$
$$\nu_{12} = \nu_{14} = \cdots = \nu_{1,2m} = \cdots = n - \nu_{11}.$$

Hence if H_1 is true, then from Eq. (5.7) we obtain

$$X_n^2 = l\left[\frac{2}{\sqrt{n}}(\nu_{11} - \frac{n}{2})\right]^2, \qquad (5.7')$$

and, from Eq. (5.4) we have

$$Y_n^2 = \begin{cases} \frac{k-1}{k}\, l\left[\frac{2}{\sqrt{n}}(\nu_{11} - \frac{n}{2})\right]^2, & \text{if } l \text{ - even,} \\ \frac{k-1}{k}\left(l + \frac{1}{k-l}\right)\left[\frac{2}{\sqrt{n}}(\nu_{11} - \frac{n}{2})\right]^2, & \text{if } l \text{ - odd.} \end{cases} \qquad (5.8')$$

Hence, we obtain

$$\lim_{n\to\infty} \mathbf{P}\{\frac{k-1}{k} X_n^2 \le x \mid H_1\} = \lim_{n\to\infty} \mathbf{P}\{Y_n \le x \mid H_1\} = \mathbf{P}\{\chi_1^2 \le x\}. \quad (5.10)$$

From Eqs. (5.9) and (5.10) it follows that X_n^2 and Y_n^2 distinguish H_0 and H_1, however small the probability of the error of second kind. For this to occur it is necessary that the size of the population, k, be sufficiently large and the test must be *two-sided*.

Two-sided Chi-squared Tests

Two-sided tests sometimes allow us to discover *premeditated sorting*, but of course the idea of a two-sided Pearson's test contradicts the initial intention of Pearson.

Suppose now that $s = l = 2$, $k = 30$, $k_i/k = 0.5$. Then the lower and upper critical values for $[(k-1)/k]X_n^2$ and Y_n^2 (with significance level $\alpha = 0.05$) are (0.0515; 2.997) and (0.0533; 3.099), respectively, and from Eq. (5.10) it follows that as $n \to \infty$

$$\mathbf{P}\{0.0515 < \frac{k-1}{k} X_n^2 < 2.997\} \to 0.7370$$

and

$$\mathbf{P}\{0.0533 < Y_n^2 < 3.099\} \to 0.7389$$

and, hence in this example ($s = l = 2$) the test based on X^2 is asymptotically as powerful as that based on Y^2. But if l increases then the test based on Y^2 will be more powerful. Of course if k is sufficiently large, then asymptotically only the power of both tests will be the same (of course, k must be much bigger than l, $k \gg l$). In this example, and in other such cases, application of a two-sided chi-squared test appears unnatural, but unavoidable, because in practice an alternative hypothesis H_1 is always composite and hence the possibility of constructing a Neyman-Pearson criterion to test H_0 against H_1 is a lucky accident, and not usual.

6. HOW MANY CLASSES?

General Considerations

If the available data are ungrouped, the first questions that arise when one constructs a chi-squared test are:

1. How many classes to use?
2. How to choose these classes?

To evaluate statistical tests, one usually compares their powers, but for this it is necessary to specify an alternative to H_0. As a rule, for chi-squared type testing one takes as alternative, in effect, H_1, the set of all probability distributions distinct from the distribution defined by H_0. Hence, power is, in general, not defined, and there is no apparent criterion for questions 1 and 2. According to the original idea of Pearson, the chi-squared test is a goodness-of-fit test to see whether the data are in concordance with the hypothesis H_0, and we control only the significance level, α, the probability of rejecting H_0, when H_0 is true. If we wish simply to test that the vector of frequencies $\boldsymbol{\nu}=(\nu_1,\ldots,\nu_r)^T$ has the multinomial distribution with parameters n and $\mathbf{p} = (p_1,\ldots,p_r)^T$, then all is clear. The number of classes is given and equal to r. But if we have i.i.d. $X_1,\ldots,X_n$,

$$\mathbf{P}\{X_i < x\} = F(x),$$

where $F(x)$ is unknown and we want to test $H_0 : F(x) = F_0(x)$, $F_0(x)$ is given distribution function, then the statistician must answer questions 1 and 2 in order to choose the grouping classes, $J_1, \ldots, J_r$. If we have no additional information about the distribution of the X_i then we can choose $J_1, \ldots, J_r$ as equiprobable intervals on the real line:

$$\boldsymbol{P}\{X_1 \in J_i \mid H_0\} = 1/r,$$

and we recommend choosing r such that

$$r \leq \min(\frac{1}{\alpha}, \log n)$$

where α is the significance level ($0 < \alpha < 0.5$). Sturges (1926) gives an empirical rule according to which

$$r = [1 + \log n] = 1 + 2.303 \log_{10} n.$$

With this choice of r, the expected number of observations in each class is not small and this is essential for applying the theorem of Tumanyan (Theorem 1.2).

If $r \geq 30$, one can use Tumanyan's theorem (Theorem 1.3), the normal approximation for the distribution of X^2. About the choice of classes see also Schorr (1974), Harrison (1985), and Rayner (1982).

Neyman-Pearson Classes

On the other hand, suppose that we have an alternative, H_1, according to which

$$H_1 : \boldsymbol{P}\{X_1 < x\} = G(x).$$

For simplicity, suppose for a moment that there exist densities $f(x) = F'(x)$, $f_0(x) = F_0'(x)$ and $g(x) = G'(x)$. Consider the case

$$f_0(x) = \begin{cases} 1, & x \in [0,1], \\ 0, & \text{otherwise}. \end{cases}$$

If $g(x)$ is unimodal, one sees that not many classes are needed to test $H_0 : f(x) = f_0(x)$ against $H_1 : f(x) = g(x)$. It is enough to have three, namely $J_1 = [0, x_1)$, $J_2 = [x_1, x_2)$, $J_3 = [x_2, 1]$. Then $\boldsymbol{\nu} = (\nu_1, \nu_2, \nu_3)^T$,

$\boldsymbol{p} = (p_1, p_2, p_3)^T$, $\boldsymbol{q} = (q_1, q_2, q_3)^T$, where

$$p_1 = \int_{J_1} f_0(x)\,dx = x_1, \qquad p_2 = \int_{J_2} f_0(x)\,dx = x_2 - x_1,$$
$$p_3 = \int_{J_3} f_0(x)\,dx = 1 - x_2, \qquad q_i = \int_{J_i} g(x)\,dx, \quad i = 1, 2, 3.$$

We have

$$X_n^2 = \sum_{i=1}^{3} \frac{(\nu_i - np_i)^2}{np_i}$$

and

$$\lim_{n\to\infty} \boldsymbol{P}\{X_n^2 > x \mid H_0\} = \boldsymbol{P}\{\chi_2^2 > x\}.$$

But it is even better to construct Pearson's statistic with only two grouping classes, namely, Neyman-Pearson classes

$$I_1 = \{x : f_0(x) < g(x)\} \text{ and } I_2 = \{x : f_0(x) \geq g(x)\}.$$

It is evident that in our example

$$I_1 = J_2 \text{ and } I_2 = J_1 \cup J_3,$$

and [see Eq. (1.10)]

$$X_n^2 = \frac{(\nu_1 - np_1)^2}{np_1(1-p_1)},$$

where ν_1 is the number of observations $X_1, \ldots, X_n$ in I_1,

$$p_1 = \boldsymbol{P}\{X_i \in I_1 \mid H_0\}$$

(in our example $p_1 = x_2 - x_1$.). Under H_0 , as $n \to \infty$ the statistic X^2 is distributed as χ_1^2. When X_n^2 is constructed we use the rule (3.13) to obtain the χ^2-test of level $\approx \alpha$. By this test, H_0 is rejected if [see Eq. (4.8)]

$$X^2 \geq \left[\Psi(1-\frac{\alpha}{2}) + \frac{1}{2\sqrt{np_1(1-p_1)}}\right]^2 = c_\alpha.$$

We can proceed in the same way if $g(x)$ is multimodal. As before, we can define Neyman-Pearson classes

$$I_1 = \{x : f_0(x) < g(x)\}, \quad I_2 = \{x : f(x) \geq g(x)\},$$

and $\boldsymbol{\nu} = (\nu_1, \nu_2)^T$, $\boldsymbol{p} = (p_1, p_2)^T$, where

$$p_i = \boldsymbol{P}\{X_1 \in I_i \mid H_0\}, \qquad i = 1, 2,$$

and construct the statistic, X^2, which is distributed under H_0 as χ_1^2 as $n \to \infty$. It is clear that to construct I_1 and I_2 we solve the equation $f_0(x) = g(x)$ and unify the corresponding intervals.

Several authors (Kallenberg, Oosterhoff and Schriever, 1985; Drost, 1988, 1989; Borovkov, 1977) have studied the problem of testing $H_0 : f(x) = f_0(x)$ against alternatives

$$H_n : f(x) = g_n(x) = (1 - \eta_n) f_0 + \eta_n g(x),$$

where $g(x)$ is given, $\int g(x) = 1$, and $\eta_n \to 0$ as $n \to \infty$. Since the equation

$$f_0 = g_n(x)$$

is equivalent to

$$f_0(x) = g(x),$$

we have the situation just considered. Asymptotic power comparisons of the chi-squared and likelihood ratio tests are found, for example, in the papers of Sutrick (1986), Quine and Robinson (1985), Drost, Kallenberg, and Oosterhoff (1990).

7. OTHER MEASURES OF FIT

What is a Measure of Fit?

If for two probability distributions F, G, $d(F, G)$ tells us how different G is from F in some sense, then d can be used to define a "measure of fit." The functional $d(\cdot, \cdot)$ should have the properties:

$$d(F, G) \geq 0, \; d(F, F) = 0,$$

and the triangle inequality, for probability distributions F, G, H,

$$d(F,H) \leq d(F,G) + d(G,H).$$

Let $X_1, X_2, \ldots, X_n$ be i.i.d. with distribution F. Then

$$F_n(x) = \frac{1}{n} \sum_{i=1}^{n} I_{(X_i \leq x)}, \ |x| < \infty, \tag{7.1}$$

is called the *empirical distribution function* (EDF) of the sample X_1, $X_2, \ldots, X_n$. For any $d(\cdot,\cdot)$ as above, $d(F_n, F)$ is a *measure of fit* which can be used to construct a goodness-of-fit test as in Section 1 (see, for example, Kac, Kiefer and Wolfowitz, 1955). One example is the Pearson statistic which we have been discussing.

In this section we look at some other measures of fit, give references where they are studied, and say a few words about their relation with the chi-squared family of statistics. There is considerable literature where many useful facts about various measures of fit and large bibliographies can be found, for example, "Theory of Rank Tests" (Hajek and Sidak, 1967), "Order Statistics" (David, 1981), "Approximation Theorems of Mathematical Statistics" (Serfling, 1980), "Omega-Square Criteria" (Martynov, 1977), "The Analysis of Cross-Classified Categorical Data" (Fienberg, 1980), "Analysis of Ordinal Categorical Data" (Agresti, 1984), "Goodness-of-fit Techniques" (D'Agostino and Stephens, Eds., 1986), "Model Selection" (Linhart and Zucchini, 1986), "Theory of Probability and Mathematical Statistics–Selected Works" (Bolshev, 1987), "Asymptotics for Generalized Chi-square Goodness-of-fit Tests" (Drost, 1988), "Goodness-of-fit Statistics for Discrete Multivariate Data" (Read and Cressie, 1988), "Smooth Tests of Goodness of Fit" (Rayner and Best, 1989), "A User's Guide to Principal Components" (Jackson, 1991).

The Kolmogorov Statistic

The most well-known measure of fit based on the EDF is the Kolmogorov statistic

$$K_n = \sup_{|x|<\infty} \sqrt{n}|F_n(x) - F(x)|. \tag{7.2}$$

If F is continuous, the Smirnov transformation $U_i = F(X_i)$, $i = 1, 2, \ldots, n$ lets us assume the data are from the uniform distribution on [0,1], and that K_n is

$$K_n = \sup_{0 \le u \le 1} \sqrt{n}|F_n(u) - u|. \tag{7.3}$$

where the EDF $F_n(x)$ is constructed using $U_1, U_2, \ldots, U_n$. This statistic clearly does not depend on F. Kolmogorov (1933) found its limit distribution

$$\lim_{n\to\infty} \mathbf{P}\{K_n \le z\} = K(z) = \sum_{j=-\infty}^{+\infty} (-1)^j e^{-2j^2z^2}, \qquad 0 < z < \infty. \tag{7.4}$$

Smirnov Transformation for F with Discontinuities

It is not difficult to extend the nonparametric test, based on $\sqrt{n}|F_n(u) - u|$ to right continuous distributions F having discontinuities. Indeed, let $Y_1, Y_2, \ldots, Y_n$ be independent uniformly distributed random variables on [0,1] which are independent of the X_j's, Then the random variables obtained by randomization at the discontinuity points of $F(x) = \mathbf{P}\{X_i \le x\}$,

$$U_i = F(X_i-) + Y_i[F(X_i) - F(X_i-)], \qquad i = 1, \ldots, n, \tag{7.5}$$

are independent and uniformly distributed on [0,1], where

$$F(x-) = \lim_{z\downarrow 0} F(x - z).$$

The proof can be found, for example, in Shorack and Wellner (1986), Huber and Nikulin (1991); see also Nikulin (1992a).

Cramer–von Mises–Smirnov Statistics

The class of quadratic statistics of the form

$$W_n = n \int_{-\infty}^{\infty} [F_n(x) - F(x)]^2 \psi(F(x))\, dF(x), \tag{7.6}$$

or, for F uniform as above,

$$W_n = n \int_0^1 [F_n(u) - u]^2 \psi(u)\, du \tag{7.7}$$

where $\psi(\cdot)$ is some nonnegative function on (0,1), is known as the class of Cramer–von Mises–Smirnov statistics.

The distribution theory of this class of statistics was developed by Smirnov (1936, 1937, 1939, 1970), Cramer (1946), von Mises (1936, 1964), Feller (1948a), Anderson and Darling (1952, 1954), Birnbaum (1953), Darling (1955, 1957), and Doob (1952).

When $\psi(u) = u/(1-u)$, W_n is the Anderson-Darling statistic; when $\psi(u) \equiv 1$, ω_n^2 is the "omega-squared" statistic

$$\omega_n^2 = n \int_0^1 [F_n(u) - u]^2\, du, \tag{7.8}$$

which has the more convenient formula

$$\omega_n^2 = \frac{1}{12n} + \sum_{i=1}^{n} \left[U_{n_i} - \frac{2i-1}{2n} \right]^2, \tag{7.9}$$

where

$$0 \le U_{n0} \le U_{n1} \le \ldots \le U_{nn} \le U_{n,n+1} = 1$$

are the order statistics obtained from the sample $U_1, U_2, \ldots, U_n$ by putting them in nondecreasing order.

Spacings

Let

$$D_{n,i} = U_{n_i} - U_{n,i-i}, \qquad i = 1, 2, \ldots, n. \tag{7.10}$$

The statistics $D_{n,i}$ are called *spacings*. Pyke (1965) studied these statistics and showed that they have some asymptotic optimality properties. Some measures-of-fit to the uniform distribution on [0,1] based on the spacings $D_{n,i}$ are

$$\sum_{i=1}^{n} |D_{n,i} - \frac{1}{n}|, \qquad \sum_{i=1}^{n} \frac{1}{n} \ln(nD_{n,i})$$

and the class

$$\sum_{i=1}^{n} g(D_{n,i})$$

where g is a smooth function. This class contains the well-known *Greenwood-Moran statistic*

$$M_n = \sum_{i=1}^{n} D_{n,i}^2, \tag{7.11}$$

which was introduced by Greenwood (1946) and studied by Moran (1947, 1951) and Weiss (1957). Several authors who have studied statistics based on spacings $D_{n,i}$ compared the first-order spacings $D_{n,i}$ to the multi-order spacings (see, e.g., Kimbal, 1947, 1950; Sherman, 1950; Pyke, 1965; Weiss, 1956, 1974; Hartly and Pfaffenberger, 1972; Rao and Sethuraman, 1975; Cressie, 1976, 1978, 1979; Hegazy and Green, 1975; Del Pino, 1979; Holst, 1979; Rao and Kuo, 1984; Stephens, 1986a,b,c; Hall, 1986; Guttorp and Lockart, 1988). For more references see D'Agostino and Stephens (1986), Read and Cressie (1988), Rayner and Best (1989). Many of the statistics studied in these publications are in the class of generalized omega-squared statistics, as introduced by Hartley and Pfaffenberger (1972); we shall consider them here very briefly.

Generalized Omega-squared Statistics

We describe the omega-squared statistics, which can be viewed as chi-squared statistics with some data point on each interval. Let $\boldsymbol{U} = (U_1, \ldots, U_n)^T$ be an n-dimensional vector, whose components are independent random variables, uniformly distributed on the interval [0,1], and let $\boldsymbol{U}^{(\cdot)} = (U_{n1}, \ldots, U_{nn})^T$ be the order statistic vector, constructed from $\boldsymbol{U}$. It is known (see, e.g., Pyke, 1965; David, 1981) that the order statistic U_{n_i} $(i = 1, 2, \ldots, n)$ has the probability density $p_i(u)$,

$$p_i(u) = \frac{n!}{(i-1)(n-i)!} u^{i-1}(1-u)^{n-i}, \qquad 0 \leq u \leq 1.$$

Let $\boldsymbol{E}\boldsymbol{U}^{(\cdot)}$ and $\boldsymbol{\Sigma} = \|\sigma_{ij}\|$ be the expectation and the covariance matrix of the vector $\boldsymbol{U}^{(\cdot)}$,

$$\begin{aligned} \boldsymbol{E}\boldsymbol{U}^{(\cdot)} &= \left(\frac{1}{n+1}, \frac{2}{n+1}, \dots, \frac{n}{n+1}\right)^T \\ \boldsymbol{\Sigma} &= \boldsymbol{E}(\boldsymbol{U}^{(\cdot)} - \boldsymbol{E}\boldsymbol{U}^{(\cdot)})(\boldsymbol{U}^{(\cdot)} - \boldsymbol{E}\boldsymbol{U}^{(\cdot)})^T, \end{aligned} \tag{7.12}$$

where

$$\sigma_{ij} = \sigma_{ji} = \boldsymbol{E}(U_{n_i} - \boldsymbol{E}U_{n_i})(U_{n_j} - \boldsymbol{E}U_{n_j})$$

$$= \begin{cases} \dfrac{i(n-j-1)}{(n+1)^2(n+2)}, & i \leq j, \\ \dfrac{j(n-i+1)}{(n+1)^2(n+2)}, & i \geq j. \end{cases} \tag{7.13}$$

Let $\boldsymbol{a} = (a_1, a_2, \dots, a_n)^T$ be a vector in R^n, and let $\boldsymbol{C} = \|c_{i_j}\|$ be a positive-definite square matrix of order n. We define (see Hartley and Pfaffenberger, 1972; Nikulin and Osidze, 1987) the *generalized omega-squared statistic* $\Omega_n^2(\boldsymbol{a}, \boldsymbol{C})$, by

$$\Omega_n^2(\boldsymbol{a}, \boldsymbol{C}) = (\boldsymbol{U}^{(\cdot)} - \boldsymbol{a})^T \boldsymbol{C} (\boldsymbol{U}^{(\cdot)} - \boldsymbol{a}). \tag{7.14}$$

Notice that the chi-squared statistics have the same structure. If in formula (7.14) the vector $\boldsymbol{a}$ is

$$\boldsymbol{a} = \left(\frac{1}{2n}, \frac{3}{2n}, \dots, \frac{2i-1}{2n}, \dots, \frac{2n-1}{2n}\right)^T$$

and $\boldsymbol{C}$ is the identity matrix $\boldsymbol{I}_n$, then we obtain the statistic

$$\Omega_n^2(\boldsymbol{a}, \boldsymbol{I}_n) = \omega_n^2 - \frac{1}{12n}, \tag{7.15}$$

where ω_n^2 is the standard omega-squared statistic given by Eq. (7.9).

Omega-squared, Lehmann, Gini, and Downton Statistics

If in formula (7.14) we set

$$\boldsymbol{a} = \boldsymbol{E}\boldsymbol{U}^{(\cdot)} \qquad \text{and} \qquad \boldsymbol{C} = \boldsymbol{I}_n, \tag{7.16}$$

the statistic $\Omega_n^2(\boldsymbol{E}\boldsymbol{U}^{(\cdot)}, \boldsymbol{I}_n)$, which following Lehmann (1973), will be denoted in the sequel by L_n, has the form:

$$L_n = \Omega_n^2(\boldsymbol{E}\boldsymbol{U}^{(\cdot)}, \boldsymbol{I}_n) = \sum_{i=1}^{n}(U_{n_i} - \frac{i}{n+1})^2. \tag{7.17}$$

The statistic ω_n^2 is related to L_n by

$$\omega_n^2 = L_n + \psi_n + \frac{1}{6(n+1)}, \tag{7.18}$$

where ψ_n is the linear order statistic,

$$\psi_n = \sum_{i=1}^{n} \frac{n-2i+1}{n(n+1)} (U_{n_i} - \frac{i}{n+1}).$$

It is known that

$$\boldsymbol{E}\omega_n^2 = \frac{1}{6}, \qquad \textbf{Var}\ \omega_n^2 = \frac{4n-3}{180n}, \tag{7.19}$$

$$\boldsymbol{E}L_n = \frac{n}{6(n+1)}, \qquad \textbf{Var} L_n = \frac{n^2}{45(n+1)^2}. \tag{7.20}$$

From formulas (7.19) and (7.20) and straightforward computation one sees that

$$\boldsymbol{E}\psi_n = 0, \qquad \textbf{Var}\ \psi_n = \frac{(n-1)(n+3)}{180n(n+3)^2},$$

$$\textbf{cov}\ (L_n, \psi_n) = \frac{n-1}{90(n+1)^2}, \qquad \textbf{corr}\ (L_n, \psi_n) = \sqrt{\frac{n-1}{n(n+3)}}.$$

From the definition of ψ_n, it follows that

$$\psi_n = \frac{n-1}{2(n+1)} G + \frac{n-1}{6(n+1)},$$

where

$$G = \frac{2}{n(n-1)} \sum_{i \le i < j \le n} |U_{n_i} - U_{n_j}|$$

is the so-called *Gini's mean difference* statistic (see, e.g., Kendall and Stuart, 1965; Hajek and Sidak, 1967; Gail and Gastworth, 1978; Rayner and Best, 1989). Taking into account that the Gini statistic G differs only by a factor from the well-known *Downton statistic* $\ll \sigma \gg$ (see Kendall and Stuart, 1965), namely

$$\ll \sigma \gg = \frac{\pi}{2} G,$$

we obtain the representation for ω_n^2,

$$\omega_n^2 = L_n + \frac{n-1}{2(n+1)} G + \frac{n}{6(n+1)} = L_n + \sqrt{\pi}\,\frac{n-1}{n+1} \ll \sigma \gg + \frac{n}{6(n+1)}.$$

The Hartley-Pfaffenberger Statistic S_n^2

The covariance matrix $\boldsymbol{\Sigma}$, given by Eq. (7.12) of the order statistic vector $\boldsymbol{U}^{(\cdot)}$ is nonsingular and its inverse $\boldsymbol{\Sigma}^{-1} = \|\sigma^{ij}\|$ has elements σ^{ij}, expressed by

$$\sigma^{ij} = \begin{cases} 2(n+1)(n+2), & \text{if } i = j, \\ -(n+1)(n+2), & \text{if } |i-j| = 1, \\ 0, & \text{if } |i-j| \geq 2. \end{cases} \tag{7.21}$$

If in formula (7.14) we set

$$\boldsymbol{a} = \boldsymbol{E}\boldsymbol{U}^{(\cdot)} \qquad \text{and} \qquad \boldsymbol{C} = \boldsymbol{\Sigma}^{-1}$$

we obtain the Hartley-Pfaffenberger statistic

$$S_n^2 = \Omega_n^2(\boldsymbol{E}\boldsymbol{U}^{(\cdot)}, \boldsymbol{\Sigma}^{-1}) = (\boldsymbol{U}^{(\cdot)} - \boldsymbol{E}\boldsymbol{U}^{(\cdot)})^T \boldsymbol{\Sigma}^{-1} (\boldsymbol{U}^{(\cdot)} - \boldsymbol{E}\boldsymbol{U}^{(\cdot)}). \tag{7.22}$$

It is not difficult to show that

$$\begin{aligned} S_n^2 &= 2(n+1)(n+2)\Big[\sum_{i=1}^{n} U_{ni}^2 - \sum_{i=1}^{n-1} U_{n_i} U_{n,i-1} - U_{nn} - \frac{n}{2(n+1)}\Big] \\ &= 2(n+1)(n+2)\Big[Q_n - 2\sum_{i=1}^{n-1} U_{n_i} U_{n_1 i-1} - U_{nn} - \frac{n}{2(n+1)}\Big], \end{aligned} \tag{7.23}$$

where

$$Q_n = \sum_{i=1}^{n-1} U_{n_i} U_{n,i-1} + nT_n$$

is the statistic studied by Quesenberry and Miller (1977) and Miller and Quesenberry (1979), and

$$T_n = \frac{1}{n} \sum_{i=1}^{n} U_{n_i}^2$$

is the statistic studied by Stephens (1966). See also Quesenberry (1986).

Now we briefly describe some useful properties of the statistic S_n^2. We set $\boldsymbol{V} = (V_1, \ldots, V_{n-1})^T$, where

$$V_i = \frac{U_{n_i}}{U_{nn}}, \qquad i = 1, 2, \ldots, n-1. \tag{7.24}$$

The conditional density of $\boldsymbol{V}$ given U_{nn} is

$$g(v_1, \ldots, v_{n-1} | U_{nn}) = (n-1)!, \qquad 0 \leq v_1 \leq v_2 \leq \cdots \leq v_{n-1} \leq 1. \tag{7.25}$$

Consequently, the statistic $\boldsymbol{V}$ does not depend on V_{nn}, and

$$V_1 \leq V_2 \leq \cdots \leq V_{n-1}, \tag{7.26}$$

that is, $\boldsymbol{V}$ is the order statistic vector, constructed from $n-1$ independent random variables, uniformly distributed on [0,1], (see Pyke, 1965). It follows that

$$2n(n+1) \left[\sum_{i=1}^{n-1} V_i^2 - \sum_{i=1}^{n-1} V_i V_{i+1} - V_{n-1} + \frac{n-1}{2n} \right] = S_{n-1}^2, \tag{7.27}$$

and, consequently

$$S_n^2 = \frac{n+2}{n} U_{nn}^2 S_{n-1}^2 + \frac{(n+1)^2(n+2)}{n} (U_{nn} - \frac{n}{n+1})^2, \tag{7.28}$$

where S_{n-1}^2 and U_{nn} are independent random variables. Since $S_0^2 = 0$

and U_{nn} is distributed as $1-U_{n1}$, we obtain

$$S_n^2 = \frac{n+2}{n}(U_{n1}-1)^2 S_{n-2}^2 + \frac{(n+1)^2(n+2)}{n}(U_{n1}-\frac{1}{n+1})^2. \quad (7.29)$$

Changing n to $(n-1)$ in Eq. (7.29) and noting that $U_{n-1,1}$ is distributed in the same way as $V_1 = U_{n1}/U_{nn}$, we obtain

$$S_{n-1}^2 = \frac{n+1}{n-1}(U_{n1}-U_{nn})^2 S_{n-2}^2 + \frac{(n+1)^2(n+2)}{n}\left(U_{n1}-\frac{1}{n+1}\right)^2. \quad (7.30)$$

Inserting Eq. (7.30) in (7.29), we have the recursion formula:

$$S_n^2 = \frac{(n+1)(n+2)}{n(n-1)}(U_{nn}-U_{n1})^2 S_{n-2}^2 + \frac{n(n+1)(n+2)}{n-1}(U_{n1}-\frac{1}{n}U_{nn})^2$$

$$+\frac{(n+1)^2(n+2)}{2}(U_{nn}-\frac{n}{n+1})^2, \qquad n = 2,3,\ldots \quad (7.31)$$

Let

$$\alpha_n = \boldsymbol{E}S_n^2, \qquad n = 1,2,\ldots. \quad (7.32)$$

From From Eqs. (7.28), (7.11), and (7.13), it follows that $\alpha_n = n$. Now write

$$\mu_n^{(k)} = \boldsymbol{E}(S_n^2 - n)^k. \quad (7.33)$$

Making use of Eq. (7.28), we obtain that

$$\mu_n^2 = \frac{(n+2)^2}{n(n+4)}\mu_{n-1}^{(2)} + \frac{4}{n+4} \sim 4n, \qquad \text{if } n \to \infty \quad (7.34)$$

$$\mu_n^{(3)} = \frac{(n+2)^3}{n^2(n+6)}\mu_{n-1}^{(3)} + \frac{16n(5n^2+3n^2+19n-30)}{(n+3)(n+4)(n+5)(n+6)}$$

$$= \frac{16(n+1)^2(5n-2)}{(n+3)(n+4)(n+5)(n+6)} \sim 80n, \qquad \text{if } n \to \infty, \quad (7.35)$$

and hence the asymmetry coefficient $\sqrt{\beta_1}$ of S_n^2 is

$$\sqrt{\beta_1} = \sqrt{\frac{[\mu^{(3)}]^2}{[\mu_n^{(2)}]^3}} = \frac{2(5n-2)}{(n+5)(n+6)}\sqrt{\frac{(n+3)(n+4)}{n}} \sim \frac{10}{\sqrt{n}}, \qquad n \to \infty. \tag{7.36}$$

The Statistic S_n^2 and the Greenwood-Moran Statistic M_n

From the formula (7.23), it follows that

$$S_n^2 = (n+1)(n+2)M_n - (n+2). \tag{7.37}$$

where

$$M_n = \sum_{i=1}^{n}(U_{n,i+1} - U_{ni})^2 \tag{7.38}$$

is the well-known *Greenwood-Moran statistic*, $U_{n0} = 0$, $U_{nn} = 1$. From Eq. (7.28), it follows that

$$M_n = U_{nn}^2 M_{n-1} + (1 - U_{nn})^2, \qquad M_0 = 1, \tag{7.39}$$

from which, by virtue of the independence of U_n and M_{n-1}, we obtain the moments $\gamma_n^{(k)}$ of M_n:

$$\gamma_n^{(k)} = \mathbf{E}M_n^k = \sum_{m=0}^{k} \frac{k!n(n+2m-1)!(2k-2m)!}{m!(k-m)!(n+2k)!}\gamma_{n-1}^{(m)}. \tag{7.40}$$

For example,

$$\gamma_n^{(1)} = \mathbf{E}M_n = \frac{2}{n+2}, \qquad \gamma_n^{(2)} = \frac{4(n+6)}{(n+2)(n+3)(n+4)}. \tag{7.41}$$

Using Eq. (7.40), we can compute

$$m_n^{(k)} = \mathbf{E}(M_n - \mathbf{E}M_n)^k \tag{7.42}$$

and both the asymmetry coefficient $\sqrt{\beta_1}$ and the excess coefficient p_2.

For example,

$$m_n^{(1)} = 0, \qquad m_2^{(2)} = \mathbf{Var}\ M_n = \frac{4n}{(n+2)^2(n+3)(n+4)}, \tag{7.43}$$

and so on, and

$$\begin{aligned}\sqrt{\beta_1} &= \sqrt{\frac{[m_n^{(3)}]^2}{[m_n^{(2)}]^3}} \\ &= 2\sqrt{\frac{(n+3)(n+4)(5n-2)^2}{n(n+5)(n+6)}} \sim \frac{10}{\sqrt{n}},\end{aligned} \tag{7.44}$$

as $n \to \infty$; then,

$$\begin{aligned}\beta_2 = \frac{m_n^{(4)}}{[m_n^{(2)}]^2} &= 3\frac{(n+3)(n+4)(n^3+101n^2+14n-8)^2}{n(n+5)(n+6)(n+7)(n+8)} \\ &\sim 3(1+\frac{82}{n}),\end{aligned} \tag{7.45}$$

as $n \to \infty$.

Furthermore, from Eq. (7.42) one can easily obtain the central moment $\mu_n^{(k)}$ of the statistic S_n^2, making use of the fact that

$$\mu_n^{(k)} = (n+1)^k(n+2)^k m_n^{(k)}. \tag{7.46}$$

Using Eq. (7.46), we can compute, for example, the excess coefficient β_2 of the statistic S_n^2:

$$\beta_2 = \frac{\mu_n^{(4)}}{[\mu_n^{(2)}]^2} = 3\frac{(n+3(n+4)(n^3+101n^2+14n-8)}{n(n+5)(n+6)(n+7)(n+8)} \sim 3(1+\frac{82}{n}), \tag{7.47}$$

as $n \to \infty$.

Finally we remark that the limit distribution of S_n^2 as $n \to \infty$ was obtained by both Moran (1947) and Weiss (1956):

$$\lim_{n\to\infty} \boldsymbol{P}\left\{\sqrt{(1+\frac{3}{n})(1+\frac{3}{n+1})}\,\frac{S_n^2-n}{2\sqrt{n+1}} \le x\right\} = \Phi(x). \tag{7.48}$$

The Kulldorf Statistic

Now let us suppose that $n = km+m-1 = (k+1)m-1$, and let us consider the statistic

$$\boldsymbol{U} = (U_{nm}, U_{n,2m}, \ldots, U_{n,km})^T, \tag{7.49}$$

which is obtained from the order vector statistic $\boldsymbol{U}^{(\cdot)}=(U_{n1},\ldots,U_{nn})^T$ by uniform thinning. We say that the elements of $\boldsymbol{U}$ form nonoverlapping m-step spacings

$$U_{n,jm} - U_{n,(j-1)m}, \qquad j = 1,2,\ldots,k+1, \tag{7.50}$$

on the interval [0,1]. Using Eqs. (7.12) and (7.13), we find

$$\boldsymbol{EU} = \left(\frac{1}{k+1}, \frac{2}{k+1}, \ldots, \frac{k}{k+1}\right)^T \qquad \text{and} \qquad \boldsymbol{\Sigma} = \|\sigma_{ij}\|,$$

where

$$\sigma_{ij} = \boldsymbol{E}(U_{n,im} - \frac{i}{k+1})(U_{n,jm} - \frac{j}{k+1})$$

$$= \begin{cases} \dfrac{i(k-j+1)}{(k+1)^2[m(k+1)+1]}, & \text{if } i \le j, \\ \dfrac{j(k-i+1)}{(k+1)^2[m(k+1)+1]}, & \text{if } i \ge j, \end{cases}$$

and

$$\boldsymbol{\Sigma}^{-1} = \|\sigma^{ij}\|,$$

where

$$\sigma_{ij} = \begin{cases} 2(k+1)[m(k+1)+1], & \text{if } i = j, \\ -(k+1)[m(k+1)+1], & \text{if } |i-j| = 1, \\ 0, & \text{if } |i-j| \ge 2. \end{cases}$$

Now let us define the *statistic of Kulldorf* (see Hartley and Pfaffenberger, 1972)

$$Y_n^2 = (\boldsymbol{U} - \boldsymbol{E}\boldsymbol{U})^T \Sigma^{-1} (\boldsymbol{U} - \boldsymbol{E}\boldsymbol{U}). \tag{7.51}$$

This can be regarded as a quantile chi-squared statistic. It can be represented as

$$Y_n^2 = (k+1)[m(k+2)+1]\{M_n^{(m)} - \frac{1}{k+1}\}, \tag{7.52}$$

where

$$M_n^{(m)} = \sum_{i=1}^{k} (U_{n,(i+1)m} - U_{n,im})^2$$

is the Greenwood-Moran statistic. Using the properties of $M_n^{(k)}$ and formula (7.40), we find that

$$\boldsymbol{E} M_n^{(m)} = \frac{m+1}{m(k+1)+1}$$

and (7.53)

$$\mathbf{Var}\ M_n^{(m)} = \frac{2m(m+1)k}{[m(k+1)+1]^2[m(k+1)+2][m(k+1)+3]}$$

and hence

$$\boldsymbol{E} Y_n^2 = k \quad \text{and} \quad \mathbf{Var} Y_n^2 = \frac{2m(m+1)k(k+1)^2}{[m(k+1)+2][m(k+1)+3]} \tag{7.54}$$

(see, e.g., Nikulin, 1992a). If k is fixed and $m \to \infty$, then, from the theorem of Mosteller (1946), it follows that

$$\lim_{n \to \infty} \boldsymbol{P}\{Y_n^2 \leq x\} = \boldsymbol{P}\{\chi_k^2 \leq x\}.$$

Asymptotic Properties of the Statistic Y_n^2

Here we give some interesting results from J.S. Rao and Sethuraman (1975), Del Pino (1979), Jammalamadaka (J.S. Rao) and Tiwari (1985,

1987). They study the asymptotic properties of the statistic

$$V_n^{(m)} = \frac{[m(k+1)]^2}{k} \sum_{i=1}^{k} (U_{n,(i+1)m} - U_{n,im})^2 = \frac{m^2(k+1)^2}{k} M_n^{(m)}, \quad (7.55)$$

which is a linear transformation of the statistic Y_n^2, since we have Eq. (7.52). Hence the asymptotic optimality properties of $V_n^{(m)}$ are valid also for Y_n^2 and we can focus our attention on the results of Jammalamadaka, Tiwari, and Del Pino about $V_n^{(m)}$. At first we remark that, from Eq. (7.40), it follows that

$$\boldsymbol{E} V_n^{(m)} = \frac{(m+1)[m(k+1)]^2}{k[m(k+1)+1]},$$

$$\textbf{Var } V_n^{(m)} = \frac{2m(m+1)[m(k+1)]^4}{k[m(k+1)+1]^2[m(k+1)+2][m(k+1)+3]}. \quad (7.56)$$

Now, following the paper of Jammalamadaka and Tiwari (1987), we assume that m is fixed and $k \to \infty$ as $n \to \infty$, so that $n/k \to m$. From Eq. (7.56) we obtain

$$\boldsymbol{E} V_n^{(m)} = m(m+1) + o(1), \quad \textbf{Var } V_n^{(m)} = \frac{2m(m+1)}{k} + o(\frac{1}{k}), \quad (7.57)$$

and the statistic $V^{(m)}$ is asymptotically normal,

$$\lim_{k\to\infty} \boldsymbol{P} \left\{ \sqrt{n} \frac{V_n^{(m)} - m(m+1)}{\sqrt{2m(m+1)}} \le x \right\} = \Phi(x). \quad (7.58)$$

This is an analogue of Tumanyan's theorem for the quantile chi-squared statistic. Furthermore, Jammalamadaka and Tiwari show that a test based on $V_n^{(m)}$ is asymptotically unable to detect alternatives approaching the uniform at a faster rate than $n^{-1/4}$. Namely, against the alternative H_1:

$$X_i \sim f_n(x) = 1 + \frac{\ell(x)}{n^{1/4}}, \qquad 0 \le x \le 1, \quad (7.59)$$

where $\ell(x)$ is a square integrable and continuously differentiable function on [0,1], such that

$$\int_0^1 \ell(x)\,dx = 0.$$

The sequence of alternatives shown in Eq. (7.59) has been considered also by Weiss (1956, 1957) and Chibisov (1961b).

As was shown by Jammalamadaka and Tiwari

$$\lim_{k\to\infty} \boldsymbol{P}\left\{\sqrt{k}\,\frac{V_n^{(m)} - m(m+1)}{\sqrt{2m}m+1)} \le x|H_1\right\} = \Phi(x-\mu), \tag{7.60}$$

where

$$\mu = \sqrt{m}(m+1)\int_0^1 \ell^2(t)\,dt.$$

It is evident that if $\ell(x) \equiv 0$, then we have Eq. (7.50).

Gihman's Statistic

There are also a number of measures of fit, including Pearson's statistic, that are based on grouped data. In this paragraph, we show that the Pearson statistic X^2 is also of the form $d(F_n, F)$ where F_n is the EDF, as described at the beginning of this section. Let $U_1, U_2, \ldots, U_n$ denote independent uniform [0,1] random variables, and let $\boldsymbol{p} = (p_1, \ldots, p_{k+1})^T$ be a vector of positive probabilities

$$p_1 + p_2 + \cdots + p_k + p_{k+1} = 1, \qquad p_i > 0,\ k \ge 1.$$

We set $x_0 = 0$, $x_{k+1} = 1$,

$$x_i = p_1 + p_2 + \cdots + p_i, \qquad i = 1, \ldots, k.$$

The intervals

$$[0, x_1], (x_1, x_2], \ldots, (x_{k-1}, x_k], (x_k, x_{k+1}]$$

partition [0,1].

Denote by $\nu = (\nu_1, \nu_2, \ldots, \nu_k, \nu_{k+1})^T$ the vector of frequencies

$$\nu_i = \#U_j'\text{s} \in (x_{i-1}, x_i], \qquad i = 1, \ldots, k+1;\ j = 1, \ldots, n.$$

The empirical distribution function $F_n^*(x)$, $x \in [0,1]$ of the grouped data ν is:

$$F_n^*(x) = \begin{cases} 0, & x = x_0 \\ \frac{1}{n}(\nu_1 + \nu_2 + \cdots + \nu_i), & x_{i-1} < x \leq x_i,\ i = 1, \ldots, k+1, \end{cases} \tag{7.61}$$

and the *statistic of Gihman* is $\boldsymbol{Z}_n = (Z_{n1}, \ldots, Z_{nk})^T$, where

$$Z_{n_i} = \sqrt{n}[F_n^*(x) - x_i], \qquad i = 1, \ldots, k. \tag{7.62}$$

It is evident that

$$\boldsymbol{E Z}_n = (0, 0, \ldots, 0)^T = \mathbf{0}_k \qquad \text{and} \qquad \boldsymbol{E Z}_n \boldsymbol{Z}_n^T = \boldsymbol{\Sigma},$$

where

$$\boldsymbol{\Sigma} = \begin{Vmatrix} x_1 & x_1 & x_1 & \cdots & x_1 \\ x_1 & x_2 & x_2 & \cdots & x_2 \\ x_1 & x_2 & x_3 & \cdots & x_3 \\ \vdots & \vdots & \vdots & & \\ x_1 & x_2 & x_3 & \cdots & x_k \end{Vmatrix} - \begin{Vmatrix} x_1 \\ x_2 \\ x_3 \\ \vdots \\ x_k \end{Vmatrix} \|x_1, x_2, x_3, \ldots, x_k\|. \tag{7.63}$$

Suppose that $k = k(n) \to \infty$ as $n \to \infty$ and that the length of the grouping intervals, p_i, tend to zero uniformly and sufficiently fast that the condition of Gihman holds:

$$\max_{1 \leq i \leq k+1} np_i \to 0 \text{ as } n \to \infty. \tag{7.64}$$

Denote by K_n^* the analogue of the Kolmogorov statistic K_n, given by Eq. (7.3), for grouped data,

$$K_n^* = \max_{1 \leq i \leq k} |Z_{n_i}|. \tag{7.65}$$

Theorem (Gihman, 1961). *If $k \to \infty$ and the condition of Gihman shown in Eq. (7.64) is fulfilled, then the statistic K_n^* and the Kolmogorov* *statistic K_n are asymptotically equivalent.*

$$\lim_{n\to\infty} \mathbf{P}\{K_n^* \le z\} = \lim_{n\to\infty} \mathbf{P}\{K_n \le z\} = K(z), \tag{7.66}$$

where $K(z)$ is the Kolmogorov distribution function, given by Eq. (7.4).

Hence, under the conditions of Gihman, for large values of n, $\boldsymbol{Z}_n$ can be used to construct an approximation to the statistic ω^2 of Cramer–von Mises–Smirnov, W_n^2 of Anderson and Darling, and so on. If the k grouping intervals are fixed, the Pearson statistic X_n^2 is a quadratic form of the Gihman statistic $\boldsymbol{Z}_n$, with the appropriate matrix $\boldsymbol{\Sigma}^{-1}$ (see below), and this is also based explicitly on the EDF. From the multivariate central limit theorem it follows (see, e.g., Mosteller, 1946, Wilks, 1962) that as $n \to \infty$ the law of $\boldsymbol{Z}_n$ converges to the k-dimensional normal distribution $N(\mathbf{0}_r, \boldsymbol{\Sigma})$, where $\boldsymbol{\Sigma}$ is given by Eq. (7.63). The covariance matrix $\boldsymbol{\Sigma}$ is nonsingular and its inverse is $\boldsymbol{\Sigma}^{-1} = \|\sigma^{ij}\|$, where

$$\begin{aligned}
\sigma^{ij} &= 0, && \text{if } |i-j| \ge 2,\\
\sigma^{i,i+1}, &= -\frac{1}{x_{i+1}-x_i} = -\frac{1}{p_{i+1}}, && i = 1,2,\dots,k-1,\\
\sigma^{i,i-1} &= -\frac{1}{x_i - x_{i-1}} = -\frac{1}{p_i}, && i = 2,3,\dots,k,\\
\sigma^{ii} &= -(\sigma^{i,i+1} + \sigma^{i,i-1}) && \text{if } i = j\\
&= \frac{1}{x_{i+1}-x_i} + \frac{1}{x_i - x_{i-1}},
\end{aligned} \tag{7.67}$$

Let us define the statistic

$$Y_n^2 = \boldsymbol{Z}_n^T \boldsymbol{\Sigma}^{-1} \boldsymbol{Z}_n.$$

By direct calculation, Y_n^2 is the Pearson statistic

$$Y_n^2 = \sum_{i=1}^{k+1} \frac{(\nu_i - np_i)}{np_i} = X^2.$$

From the asymptotic normality of $\boldsymbol{Z}_n$ we obtain that

$$\lim_{n\to\infty} \mathbf{P}\{Y_n^2 \le x\} = \mathbf{P}\{\chi_k^2 \le x\}.$$

Recall from Section 1 that if, instead of Gihman's condition, we have $k = k(n) \to \infty$ and the condition of Tumanyan's theorem (Theorem 1.3),

$$\min_{1 \le i \le k+1} np_i \to \infty,$$

is satisfied, then

$$\sup_{|x|<\infty} \left| \boldsymbol{P}\{Y_n^2 \ge x\} - 1 + \Phi\left(\frac{x-k}{\sqrt{2k}}\right) \right| \to 0,$$

where $\Phi(x)$ is the distribution function of the standard normal law.

It is evident that under certain conditions one obtains the Cochran Yarnold limit (see Section 1).

The Hellinger or Freeman-Tukey Statistic

Now we consider some statistics which like Pearson's statistic are based on grouped data and which have asymptotics similar to the chi-squared type statistics.

In Theorems 1.2 and 1.3, the standard statistic of Pearson, X_n^2, can be replaced by

$$F_n^2 := 4H_n^2 = 4\sum_{i=1}^{r} (\sqrt{\nu_i} - \sqrt{np_i})^2 .$$

This can be seen by writing

$$\begin{aligned} X_n^2 &= \sum_{i=1}^{r} \frac{(\nu_i - np_i)^2}{np_i} = \sum_{i=1}^{r} \frac{(\nu_i - np_i)^2}{(\sqrt{\nu_i} + \sqrt{np_i})^2} \, \frac{(\sqrt{\nu_i} + \sqrt{np_i})^2}{np_i} \\ &= 4\sum_{i=1}^{r} (\sqrt{\nu_i} - \sqrt{np_i})^2 \, \frac{(\nu_i + \sqrt{np_i})^2}{4np_i}. \end{aligned}$$

The second factor under the sum can be expressed as

$$\left[\frac{4np_i}{(\nu_i + \sqrt{np_i})^2}\right]^{-1} = (1 - \alpha_n)^{-1} ,$$

where

$$\alpha_n = \frac{\left(\sqrt{\frac{\nu_i}{n}} - \sqrt{p_i}\right)\left(\sqrt{\frac{\nu_i}{n}} + 3\sqrt{p_i}\right)}{\left(\sqrt{\frac{\nu_i}{n}} + \sqrt{p_i}\right)^2}.$$

Since ν_i/n goes to p_i in probability, by the law of large numbers, α_n goes to 0 in probability, and hence $4H_n^2 = (1+\beta_n)X_n^2$, where $\beta_n \to 0$ in probability. It follows that as $n \to \infty$ the limit distribution of $F_n^2 = 4H_n^2$ is the same as that of X_n^2, that is,

$$\lim_{n\to\infty} \boldsymbol{P}\{F_n^2 > x\} = \boldsymbol{P}\{\chi_{r-1}^2 > x\}$$

(see Fraser, 1976). The statistic H_n^2 is called *the Hellinger* or *Freeman-Tukey statistic*. Sometimes the statistic F_n^2 is preferable to X_n^2, for example when some p_i are small. Fraser (1976) gives some interesting examples concerning F_n^2.

Pearson's statistic may be regarded as a distance between the probability vectors

$$\boldsymbol{p} = (p_1, \ldots, p_r)^T \quad \text{and} \quad \frac{1}{n}\boldsymbol{\nu} = \left(\frac{\nu_1}{n}, \cdots, \frac{\nu_r}{n}\right)^T.$$

In fact, if $\boldsymbol{P}$ and $\widetilde{\boldsymbol{P}}$ are probability distributions with densities f and $\tilde{f}$ with respect to some measure $\boldsymbol{Q}$, the chi-squared distance between $\boldsymbol{P}$ and $\widetilde{\boldsymbol{P}}$ can be defined as $\rho_{\chi^2}(\boldsymbol{P}, \widetilde{\boldsymbol{P}})$, where

$$\rho_{\chi^2}(\boldsymbol{P}, \widetilde{\boldsymbol{P}}) = \boldsymbol{E}_{\boldsymbol{Q}}(f - \tilde{f})^2/f.$$

If $\boldsymbol{P}$ and $\widetilde{\boldsymbol{P}}$ are $\boldsymbol{p}$ and $(1/n)\boldsymbol{\nu}$ respectively, and $\boldsymbol{Q}$ is counting measure, then

$$\rho_{\chi^2}(\boldsymbol{p}, \frac{1}{n}\boldsymbol{\nu}) = \sum_{i=1}^{r} \left(\frac{\nu_i}{n} - p_i\right)^2 / p_i = \frac{1}{n}X_n^2.$$

As we know

$$\boldsymbol{E}\frac{\nu_i}{n} = p_i \quad \text{and} \quad \mathbf{Var}\ \frac{\nu_i}{n} = \frac{1}{n}p_i(1-p_i) \leq \frac{1}{4n},$$

so that, from the law of large numbers, if n is large then $\nu_i/n \approx p_i$. Hence $\rho_{\chi^2}[\boldsymbol{p}, 1/(n)\boldsymbol{\nu}]$ can be considered as a measure of discrepancy between a given multinomial distribution with parameters n and $\boldsymbol{p}$

and the empirical multinomial distribution with parameters n and $(1/n)\boldsymbol{\nu}$. Kaufman, Mathai, and Rathie (1972) gave two axiomatic characterizations for the statistic X^2, regarded as the chi-squared measure between $\boldsymbol{p}$ and $(1/n)\boldsymbol{\nu}$.

Similarly, the Hellinger statistic, introduced above, can be regarded as a multiple of the Hellinger distance between $\boldsymbol{p}$ and $\boldsymbol{\nu}/n$. Hellinger distance $\rho_H(\boldsymbol{P}, \widetilde{\boldsymbol{P}})$ is defined by

$$\rho_H(\boldsymbol{P}, \widetilde{\boldsymbol{P}}) = \boldsymbol{E}_{\boldsymbol{Q}} \left(\sqrt{f} - \sqrt{\tilde{f}}\right)^2,$$

where $\boldsymbol{Q}, f$, and $\tilde{f}$ are as above. The Hellinger distance between $(1/n)\boldsymbol{\nu}$ and $\boldsymbol{p}$ is

$$\rho_H(\frac{1}{n}\boldsymbol{\nu}, \boldsymbol{p}) = \sum_{i=1}^{r} \left(\sqrt{\frac{\nu_i}{n}} - \sqrt{p_i}\right)^2 = \frac{1}{n} H_n^2,$$

where H_n^2 is the Hellinger statistic.

The Kullback-Leibler or Log-Likelihood Ratio Statistic

Another member of the same family of distances between probability measures $\boldsymbol{P}, \widetilde{\boldsymbol{P}}$ is the Kullback-Leibler distance, $\rho_K(\boldsymbol{P}, \widetilde{\boldsymbol{P}})$, defined by

$$\rho_K(\boldsymbol{P}, \widetilde{\boldsymbol{P}}) = \boldsymbol{E}_{\boldsymbol{Q}} z \ln \frac{z}{\tilde{z}},$$

where

$$z = \frac{d\boldsymbol{P}}{d\boldsymbol{Q}}, \quad \tilde{z} = \frac{d\widetilde{\boldsymbol{P}}}{d\boldsymbol{Q}}, \quad \boldsymbol{Q} = \frac{1}{2}(\boldsymbol{P} + \widetilde{\boldsymbol{P}}).$$

The corresponding *Kullback-Leibler* or *log likelihood ratio statistic*, called G^2, is

$$G^2 = 2 \log \frac{l(\boldsymbol{\nu})}{l(n\boldsymbol{p})} = 2 \sum_{i=1}^{r} \nu_i \log \left(\frac{\nu_i}{np_i}\right),$$

where $l(\boldsymbol{p})$ is the likelihood function (1.3). The statistics X^2 and G^2, which Neyman (1949) showed to be equivalent, are members of the family

$$\frac{2}{\lambda(\lambda+1)}\sum_{i=1}^{r}\nu_i\left[\left(\frac{\nu_i}{np_i}\right)^{\lambda}-1\right] \tag{7.68}$$

of power divergence statistics, studied in the book of Read and Cressie (1988). For $\lambda = 1$, Eq. (7.68) is X^2, and the limit as $\lambda \to 0$ of Eq. (7.68) is G^2. The reader interested in analysis of discrete multivariate data will find several issues addressed by Read and Cressie. For example, when n is small, they find reason to choose $\lambda = 2/3$ in Eq. (7.68) to get good χ^2_{r-1} approximation. The power of X^2 and G^2 against various composite alternatives was studied also by Larntz (1973, 1978). See also Rao (1973), Koehler (1977), Haberman (1974, 1977), Linhart and Zucchini (1986), Rayner and Best (1989), and Cressie and Read (1989).

Morris (1975) proved a central limit theorem for multinomial sums and used it as a basis for comparing the power of Pearson's chi-squared statistic with that of the likelihood ratio or Kullback-Leibler test based on the same grouping. He proved central limit theorems for Pearson's statistic, improving Tumanyan's theorem (Theorem 1.3), and for the Kullback-Leibler statistic. More information about the Kullback-Leibler statistic can be found in Borovkov (1984). A comparison of Hellinger, Kullback-Leibler, and chi-squared distances between probability measuures is found in the monograph by Greenwood and Shiryaev (1986).

Remark. To test $H_0 : p_i = p_i^{(0)} = \frac{1}{r}$. Watanabe (1952) proposed the statistic

$$w_{nr}^2 = \frac{1}{nr}\sum_{m=1}^{r-1}\sum_{i=1}^{m}\left(\nu_i - np_i^{(0)}\right)^2 = \frac{1}{nr}\sum_{m=1}^{r-1}\sum_{i=1}^{m}(\nu_i - \frac{n}{r})^2.$$

Watanabe proved that *if H_0 is true then*

$$\lim_{n\to\infty} \boldsymbol{P}\{w_{nr}^2 \le x\} = \boldsymbol{P}\{w^2 \le x\}.$$

Walsh (1960) has given a table and a rule which allows us to compare the approximate and the exact significance levels for some r and for some critical values of the statistic w_{nr}^2.

Composite Null Hypothesis:
Obtaining the Distribution of the Test Statistic

A statistical test based on any measure of fit, $d(F_n, F)$ can be applied to a composite hypothesis H_0:

$$X_i \sim F(x; \boldsymbol{\theta}), \qquad \boldsymbol{\theta} \in \Theta_0 \subset \Theta \subset R^3.$$

For instance, we can base a test on $d(F_n, F(\cdot; \boldsymbol{\theta}_n^*))$ where $\boldsymbol{\theta}_n^*$ is a consistent sequence of estimators of $\boldsymbol{\theta}$. Such a test may depend on the method of estimation, in particular on the asymptotic properties of $\boldsymbol{\theta}_n^*$.

Asymptotic properties of the *empirical process*

$$\sqrt{n}(F_n(x) - F(x; \boldsymbol{\theta}_n^*)),$$

which underlie the study of goodness-of-fit statistics, have been discussed in the works of Anderson and Darling (1952, 1954), Darling (1955, 1957), Gihman (1953, 1961), Durbin (1973b, 1975). Properties of the Kolmogorov, Cramer–von Mises–Smirnov, Anderson–Darling and other statistics described in this section have been considered, for example, by Chibisov (1961a, 1962, 1965), Bolshev (1987), Martynov (1976, 1977), and Dzhaparidze and Nikulin (1982a).

Subsequently there have appeared many interesting papers both in the general theory of goodness-of-fit testing and its applications when H_0 is a composite hypothesis. Surveys and studies on this topic include D'Agostino and Stephens (1986), Drost (1988), Bernstein (1982), and Drost, Kallenberg, and Oosterhoff (1990), where one can find many additional references.

The theory of goodness-of-fit tests, and in particular of chi-squared type tests, developed as part of the general theories of hypothesis testing and estimation. Practical difficulties associated with obtaining the distribution of test statistics, and of evaluation of statistics explain why people continue to be interested in chi-squared tests even if they are not so efficient as other tests (e.g., likelihood ratio tests). One way to obtain an approximation to the distribution of a test statistic is by simulation. An advantage of chi-squared type tests is that the asymptotic distribution of the test statistic is known, even for complex hypotheses. This brings us to the final topic of this section.

Considerations in Choosing a Measure of Fit

There are two main considerations in choosing a measure of fit. One is do we know the distribution, or if n is large, the limiting distribution of $d(F_n, F)$ so that we can compute c_α? The limiting distribution of $d(F_n, F)$, which is known for many of the measures-of-fit listed here becomes more problematic when H_0 is complex and one wants the limiting distribution of $d(F_n, F(\cdot; \boldsymbol{\theta}_n^*))$ for some estimator sequence $\boldsymbol{\theta}_n^*$. Chapter 2 of this guide is devoted to this question in the case of chi-squared testing. For most other measures of fit less has been done about finding limit laws for complex H_0. Good sources of information and further references are the book edited by D'Agostino and Stephens (1986) and the book of Read and Cressie (1988).

A second consideration in choosing a measure of fit is the power of the corresponding test. We want the probability of rejecting, or not accepting, H_0 when H_0 is false, $\boldsymbol{P}\{d(F_n, G) > c_\alpha\}$, $G \neq F$, to be large. This question cannot be pursued rigorously without specifying what G or class of G one has in mind. The best test of a simple H_0 or F against a simple alternative G is the likelihood ratio test, given that it can be implemented. Other less precise statements can be made: in chi-squared testing we lose power by grouping data, but asymptotically as n and the number of grouping intervals increases we regain some or most of this power, depending on exactly how the asymptotics go and what we say about alternatives. We know of no satisfactory general study of this problem. However, if certain alternatives to H_0 are seen as particularly hazardous or likely to occur, one wants to think about how to choose the grouping intervals to protect against them when using a chi-squared test. An analogous problem exists when one uses any goodness-of-fit test.

8. PEARSON'S STATISTIC AND THE THEORY OF ESTIMATION

Suppose the random vector $\boldsymbol{\nu} = (\nu_1, \ldots, \nu_r)^T$ has the multinomial distribution with parameters n and $\boldsymbol{p} = (p_1, \ldots, p_r)^T$, $\boldsymbol{\nu}^T\mathbf{1}_r = n$, $\nu_i \geq 0$, $\boldsymbol{p}^T\mathbf{1}_r = 1$, $p_i > 0$, where $\boldsymbol{p}$ is unknown. The likelihood function

$\ell(\boldsymbol{p}) = \ell(\boldsymbol{p}, \boldsymbol{\nu})$ is given by

$$\ell(\boldsymbol{p}) = \frac{n!}{\nu_1! \dots \nu_r!} p_1^{\nu_1} p_2^{\nu_2} \dots p_r^{\nu_r} = \tilde{\ell}(\tilde{\boldsymbol{p}}), \tag{8.1}$$

where

$$\tilde{\ell}(\tilde{\boldsymbol{p}}) = \frac{n!}{\nu_1! \dots \nu_{r-1}!(n - \nu_1 - \nu_2 - \dots - \nu_{r-1})!}$$

$$p_1^{\nu_1} p_2^{\nu_2} \dots p_{r-1}^{\nu_{r-1}} (1 - p_2 - \dots - p_{r-1})^{n - \nu_1 - \dots - \nu_{r-1}}, \tag{8.2}$$

because $p_1 + \dots + p_r = 1$, $\nu_1 + \dots + \nu_r = n$. It follows that $\tilde{\boldsymbol{\nu}} = (\nu_1, \dots, \nu_{r-1})^T$ is a sufficient statistic for $\boldsymbol{p}$. To find the maximum likelihood estimator $\hat{\boldsymbol{p}} = (\hat{p}_1, \dots, \hat{p}_r)^T$ for $\boldsymbol{p}$ it suffices to find the maximum likelihood estimator $\widehat{\tilde{\boldsymbol{p}}} = (\hat{p}_1, \dots, \hat{p}_{r-1})^T$ for $\tilde{\boldsymbol{p}} = (p_1, \dots, p_{r-1})^T$, because $\hat{p}_r = 1 - \hat{p}_1 - \dots - \hat{p}_{r-1}$; we need only estimate $r-1$ parameters, $p_1, \dots, p_{r-1}$. To find $\widehat{\tilde{\boldsymbol{p}}}$, which maximizes $\tilde{\ell}(\tilde{\boldsymbol{p}})$, we must solve the likelihood equation

$$\mathbf{grad}_{\tilde{\boldsymbol{p}}} \log \tilde{\ell}(\tilde{\boldsymbol{p}}) = \mathbf{0}_{r-1}. \tag{8.3}$$

From Eqs. (8.2) and (8.3) we obtain the system

$$\frac{\nu_i}{p_i} - \frac{\nu_r}{p_r} = 0, \qquad i = 1, \dots, r-1, \tag{8.4}$$

which is equivalent to Eq. (8.3), and from Eq. (8.4) we find that

$$\hat{p}_1 = \frac{\nu_1}{n}, \quad \hat{p}_2 = \frac{\nu_2}{n}, \dots, \hat{p}_{r-1} = \frac{\nu_{r-1}}{n}; \quad \hat{p}_r = \frac{\nu_r}{n}. \tag{8.5}$$

Accordingly $\hat{\boldsymbol{p}} = (1/n)\boldsymbol{\nu}$ and $\widehat{\tilde{\boldsymbol{p}}} = (1/n)\tilde{\boldsymbol{\nu}}$ are the maximum likelihood estimators for $\boldsymbol{p}$ and $\tilde{\boldsymbol{p}}$ respectively. From Eqs. (8.1), (8.2) and (1.15), we have

$$\boldsymbol{E}\hat{\boldsymbol{p}} = \boldsymbol{p}, \qquad \mathbf{Var}\,\hat{\boldsymbol{p}} = \frac{1}{n}\boldsymbol{\Sigma} = \frac{1}{n}[\boldsymbol{P} - \boldsymbol{p}\boldsymbol{p}^T], \tag{8.6}$$

$$\boldsymbol{E}\widehat{\tilde{\boldsymbol{p}}} = \tilde{\boldsymbol{p}}, \qquad \mathbf{Var}\,\widehat{\tilde{\boldsymbol{p}}} = \frac{1}{n}\tilde{\boldsymbol{\Sigma}} = \frac{1}{n}[\tilde{\boldsymbol{P}} - \tilde{\boldsymbol{p}}\tilde{\boldsymbol{p}}^T]. \tag{8.7}$$

We see that $\hat{\boldsymbol{p}}$ and $\widehat{\tilde{\boldsymbol{p}}}$ are the best unbiased estimators for $\boldsymbol{p}$ and $\tilde{\boldsymbol{p}}$ respectively, because $\tilde{\boldsymbol{\nu}}$ is the minimal sufficient statistic. In Eqs. (8.6) and (8.7), $\boldsymbol{P}$ and $\tilde{\boldsymbol{P}}$ are diagonal matrices with elements $p_1, \dots, p_{r-1}, p_r$

and $p_1, p_2, \ldots, p_{r-1}$ on the main diagonals respectively. As was shown in Section 1 the matrix $\boldsymbol{\Sigma}$ is degenerate, rank $\boldsymbol{\Sigma} = r-1$, the matrix $\widetilde{\boldsymbol{\Sigma}}$ is positive definite, rank $\widetilde{\boldsymbol{\Sigma}} = r-1$, and

$$\widetilde{\boldsymbol{\Sigma}}^{-1} = \widetilde{\boldsymbol{P}}^{-1} + \frac{1}{p_r} \mathbf{1}_{r-1} \times \mathbf{1}_{r-1}^T. \tag{8.8}$$

Furthermore, from the multivariate central limit theorem (see the appendix), if $n \to \infty$, then

$$\sqrt{n}(\hat{\boldsymbol{p}} - \boldsymbol{p}) \quad \text{is} \quad AN(\mathbf{0}_r, \boldsymbol{\Sigma}) \tag{8.9}$$

$$\sqrt{n}(\widehat{\tilde{\boldsymbol{p}}} - \tilde{\boldsymbol{p}}) \quad \text{is} \quad AN(\mathbf{0}_{r-1}, \widetilde{\boldsymbol{\Sigma}}) \tag{8.10}$$

and the distribution of the vector $\widehat{\tilde{\boldsymbol{p}}}$ is not degenerate, rank $\widetilde{\boldsymbol{\Sigma}} = r-1$. The vector of frequencies can be represented as the sum

$$\boldsymbol{\nu} = \sum_{i=1}^{n} \boldsymbol{\mu}_i = (\nu_1, \ldots, \nu_r)^T \tag{8.11}$$

of independent random vectors, $\boldsymbol{\mu}_i = (\mu_{i1}, \ldots, \mu_{ir})^T$, where μ_{ix}, $x = 1, \ldots, r$ are independent random variables,

$$\begin{aligned} \boldsymbol{P}\{\mu_{ix} = 1\} &= p_x, \qquad & \boldsymbol{P}\{\mu_{ix} = 0\} &= 1 - p_x, \\ \boldsymbol{E}\mu_{ix} &= p_x, \qquad & \mathbf{Var}\,\mu_{ix} &= p_x(1-p_x) \end{aligned} \tag{8.12}$$

Now we compute the information matrix $\boldsymbol{J}(\tilde{\boldsymbol{p}})$ of μ_i, at $\tilde{\boldsymbol{p}}$. From Eq. (8.11) and from the definition of $\boldsymbol{J}(\tilde{\boldsymbol{p}})$, we have

$$\boldsymbol{J}(\tilde{\boldsymbol{p}}) = \frac{1}{n} \boldsymbol{E} \left\{ (\mathbf{grad}_{\tilde{\boldsymbol{p}}} \tilde{\ell}(\tilde{\boldsymbol{p}}))(\mathbf{grad}_{\tilde{\boldsymbol{p}}} \tilde{\ell}(\tilde{\boldsymbol{p}}))^T \right\}, \tag{8.13}$$

and hence, from Eqs. (8.3), (8.4), (8.7), and (8.12), we obtain

$$\boldsymbol{J}(\tilde{\boldsymbol{p}}) = \widetilde{\boldsymbol{\Sigma}}^{-1} = \widetilde{\boldsymbol{P}}^{-1} + \frac{1}{p_r} \mathbf{1}_{r-1} \times \mathbf{1}_{r-1}^T. \tag{8.14}$$

The matrix $n\boldsymbol{J}(\tilde{\boldsymbol{p}})$ is the information matrix of $\tilde{\boldsymbol{\nu}}$ (or $\boldsymbol{\nu}$). From Eqs. (8.14) and (8.2) we obtain that $\widehat{\tilde{\boldsymbol{p}}}$ is an efficient MVUE (minimum variance unbiassed estimator) for $\tilde{\boldsymbol{p}}$. Furthermore from Eqs. (8.10) and (8.14), it follows that $\widehat{\tilde{\boldsymbol{p}}}$ is the *best asymptotically normal* (BAN) estimator for $\tilde{\boldsymbol{p}}$.

A sequence of estimators $\{\hat{\boldsymbol{\theta}}_n\}$ of the parameter $\boldsymbol{\theta}$, $\boldsymbol{\theta} \in \boldsymbol{\Theta} \subset \boldsymbol{R}^s$, is said to be BAN if and only if $\{\hat{\boldsymbol{\theta}}_n\}$ is consistent and $\sqrt{n}(\hat{\boldsymbol{\theta}}_n - \boldsymbol{\theta})$ is $AN(\mathbf{0}_s, \boldsymbol{J}^{-1}(\boldsymbol{\theta}))$ as $n \to \infty$, where $\boldsymbol{J}^{-1}(\boldsymbol{\theta})$ is the inverse of the information matrix of Fisher $\boldsymbol{J}(\boldsymbol{\theta})$. In our case, we have

$$\sqrt{n}(\widehat{\boldsymbol{p}}) \quad \text{is} \quad AN(\mathbf{0}_{r-1}, \boldsymbol{J}^{-1}(\tilde{\boldsymbol{p}})). \tag{8.15}$$

As was shown in Section 1, Pearson's statistic is

$$\begin{aligned} X^2 &= \sum_{i=1}^{r} \frac{(\nu_i - np_i)^2}{np_i} = \frac{1}{n}(\boldsymbol{\nu} - n\boldsymbol{p})^T \boldsymbol{\Sigma}^- (\boldsymbol{\nu} - n\boldsymbol{p}) \\ &= \frac{1}{n}(\tilde{\boldsymbol{\nu}} - n\tilde{\boldsymbol{p}})^T \widetilde{\boldsymbol{\Sigma}}^{-1} (\tilde{\boldsymbol{\nu}} - n\tilde{\boldsymbol{p}}) \end{aligned} \tag{8.16}$$

Now, using Eqs. (8.7) and (8.14)–(8.16), we can write

$$X^2 = \frac{1}{n}(\tilde{\boldsymbol{\nu}} - n\tilde{\boldsymbol{p}})^T \boldsymbol{J}(\tilde{\boldsymbol{p}})(\tilde{\boldsymbol{\nu}} - n\tilde{\boldsymbol{p}}) = n(\widehat{\boldsymbol{p}} - \tilde{\boldsymbol{p}})^T \boldsymbol{J}(\tilde{\boldsymbol{p}})(\widehat{\boldsymbol{p}} - \hat{\boldsymbol{p}}). \tag{8.17}$$

Of course, as in Section 1, we have

$$\boldsymbol{\Sigma}^- = \left[\begin{array}{c:c} \boldsymbol{J}(\tilde{\boldsymbol{p}}) & \mathbf{0}_{r-1} \\ \hdashline \mathbf{0}_{r-1}^T & 0 \end{array}\right]_{r \times r} \tag{8.18}$$

From the properties of the estimator $\widehat{\boldsymbol{p}}$, it follows that if c_α is the α-upper quantile of the chi-squared distribution with $r-1$ degrees of freedom,

$$\boldsymbol{P}\{\chi^2_{r-1} \geq c_\alpha\} = \alpha, \tag{8.19}$$

then

$$\Delta(\boldsymbol{p}) = \{\boldsymbol{p} : X^2 < c_\alpha\} \tag{8.20}$$

represents the best smallest ellipsoidal confidence region for $\boldsymbol{p}$ having, as $n \to \infty$, limiting confidence coefficient $1-\alpha$. The chi-squared test of Pearson rejects H_0 , according to which $\boldsymbol{\nu}$ has the multinomial distribution (8.1), if $X^2 \geq c_\alpha$, and accepts H_0 if $X^2 < c_\alpha$, which is entirely in accordance with Eq. (8.20).

If in Eq. (8.17) instead of $\boldsymbol{J}(\tilde{\boldsymbol{p}})$ we put $\boldsymbol{J}(\widehat{\tilde{\boldsymbol{p}}})$, replacing the parameter $\tilde{\boldsymbol{p}}$ by the MVUE $\widehat{\tilde{\boldsymbol{p}}}$, then the corresponding statistic has the form

$$\begin{aligned}\widetilde{X}^2 &= \frac{1}{n}(\tilde{\boldsymbol{\nu}} - n\tilde{\boldsymbol{p}})^T \boldsymbol{J}(\widehat{\tilde{\boldsymbol{p}}})(\tilde{\boldsymbol{\nu}} - n\boldsymbol{p}) = \frac{1}{n}(\widehat{\tilde{\boldsymbol{p}}} - \tilde{\boldsymbol{p}})^T \boldsymbol{J}(\widehat{\tilde{\boldsymbol{p}}})(\widehat{\tilde{\boldsymbol{p}}} - \tilde{\boldsymbol{p}}) \\ &= \sum_{i=1}^{r} \frac{(\nu_i - np_i)^2}{\nu_i}.\end{aligned}$$

Clearly

$$\lim_{n\to\infty} \boldsymbol{P}\{\widetilde{X}^2 > x \mid H_0\} = \boldsymbol{P}\{\chi^2_{r-1} > x\}, \tag{8.21}$$

and one can use $\widetilde{X}^2$ to construct a χ^2-test in the standard way.

CHAPTER 2

The Chi-Squared Test for a Composite Hypothesis

9. ESTIMATING AND THEN TESTING: FISHER'S THEOREM

Composite Hypotheses

Let $X_1, \ldots, X_n$ be i.i.d. random variables with unknown distribution. Suppose we have in mind a family of distributions, $F(x, \boldsymbol{\theta})$, $\boldsymbol{\theta} = (\theta_1, \ldots, \theta_s)^T \in \boldsymbol{\Theta}$, $\boldsymbol{\Theta}$ open in $\boldsymbol{R}^s$, and we wish to test the hypothesis H_0 that the distribution of the X_i is a member of this family,

$$H_0 : \boldsymbol{P}_{\boldsymbol{\theta}}\{X_i < x\} = F(x; \boldsymbol{\theta}), \quad \boldsymbol{\theta} = (\theta_1, \ldots, \theta_s)^T \in \boldsymbol{\Theta} \subset \boldsymbol{R}^s.$$

We assume that different values of $\boldsymbol{\theta}$ correspond to distinct distributions: if $\boldsymbol{\theta}_1, \boldsymbol{\theta}_2 \in \boldsymbol{\Theta}$, $\boldsymbol{\theta}_1 \neq \boldsymbol{\theta}_2$, then $F(x; \boldsymbol{\theta}_1) \not\equiv F(x; \boldsymbol{\theta}_2)$. If $\boldsymbol{\Theta}$ has only one element then we have the problem considered in Section 1, and we call H_0 a *simple hypothesis*. If $\boldsymbol{\Theta}$ has more than one element we say that H_0 is a *composite hypothesis*. Now we consider how to test H_0 in the composite case. This will involve estimation of θ, but first let us write Pearson's X^2 statistic for each $\boldsymbol{\theta}$.

Grouping Intervals and Multinomial Distribution

As in Section 1 we choose $-\infty = a_0 < a_1 < \cdots < a_r = \infty$, boundary points of grouping intervals. This time we must have $r > s+1$, where $\boldsymbol{\Theta} \subset \boldsymbol{R}^s$. Let

$$p_j(\boldsymbol{\theta}) = \int\limits_{a_{j-1}}^{a_j} dF(x;\boldsymbol{\theta}) = \boldsymbol{P}_{\boldsymbol{\theta}}\{a_{j-1} \leq X_1 < a_j\} \tag{9.1}$$

where $\boldsymbol{P}_{\boldsymbol{\theta}}$ denotes the probability with respect to $F(x;\boldsymbol{\theta})$. Again, $\boldsymbol{\nu} = (\nu_1, \ldots, \nu_r)^T$ denotes, the vector of frequencies

$$\nu_j := \int\limits_{a_{j-1}}^{a_j} dF_n(x),$$

obtained from the data by counting the number of $X_1, \ldots, X_n$ which lie in each of the successive intervals

$$(a_0, a_1), [a_1, a_2), \ldots, [a_{r-1}, a_r).$$

Under H_0 the distribution of $\boldsymbol{\nu}$ is multinomial, with parameters r and $\boldsymbol{p}(\boldsymbol{\theta}) = (p_1(\boldsymbol{\theta}), \ldots, p_r(\boldsymbol{\theta}))^T$:

$$\boldsymbol{P}_{\boldsymbol{\theta}}(\nu_1 = n_1, \ldots, \nu_r = n_r) = \frac{n!}{n_1! \ldots n_k!} p_1^{n_1}(\boldsymbol{\theta}) \ldots p_r^{n_r}(\boldsymbol{\theta}), \tag{9.2}$$

where $n_1 + \cdots + n_r = n$, $n_i \in \{0, \ldots, n\}$. For each $\boldsymbol{\theta} \in \boldsymbol{\Theta}$ we have the random vector

$$\boldsymbol{X}_n(\boldsymbol{\theta}) = \left(\frac{\nu_1 - np_1(\boldsymbol{\theta})}{\sqrt{np_1(\boldsymbol{\theta})}}, \ldots, \frac{\nu_r - np_r(\boldsymbol{\theta})}{\sqrt{np_r(\boldsymbol{\theta})}} \right)^T \tag{9.3}$$

and the random variable of Pearson's type

$$X_n^2(\boldsymbol{\theta}) = \boldsymbol{X}_n^T(\boldsymbol{\theta})\boldsymbol{X}_n(\boldsymbol{\theta}) = \sum_{i=1}^{r} \frac{(\nu_i - np_i(\boldsymbol{\theta}))^2}{np_i(\boldsymbol{\theta})}. \tag{9.4}$$

If H_0 is true, then $\boldsymbol{P}_{\boldsymbol{\theta}}$ is in effect for some $\boldsymbol{\theta} \in \boldsymbol{\Theta}$ and for this value of $\boldsymbol{\theta}$,

$$\lim_{n\to\infty} \boldsymbol{P}_{\boldsymbol{\theta}}\{X_n^2(\boldsymbol{\theta}) \leq x\} = \boldsymbol{P}(\chi_{r-1}^2 \leq x). \tag{9.5}$$

We can write Eq. (9.3) as $\boldsymbol{X}_n(\boldsymbol{\theta}) = (1/\sqrt{n})\boldsymbol{P}^{-1/2}(\boldsymbol{\nu} - n\boldsymbol{p})$, where

$$\boldsymbol{P} = \boldsymbol{P}(\boldsymbol{\theta}) = \begin{bmatrix} p_1(\boldsymbol{\theta}) & 0 & \dots & 0 \\ 0 & p_2(\boldsymbol{\theta}) & \dots & 0 \\ \vdots & \vdots & \ddots & \vdots \\ 0 & 0 & \dots & p_r(\boldsymbol{\theta}) \end{bmatrix},$$

$$\boldsymbol{p} = \boldsymbol{p}(\boldsymbol{\theta}) = (p_1(\boldsymbol{\theta}), \dots, p_r(\boldsymbol{\theta}))^T,$$

and Eq. (9.4) as

$$X_n^2(\boldsymbol{\theta}) = \frac{1}{n}(\boldsymbol{\nu} - n\boldsymbol{p})^T \boldsymbol{P}^{-1}(\boldsymbol{\nu} - n\boldsymbol{p}).$$

The Estimation Problem

We cannot immediately define a test of H_0 as we did in Section 1 because we do not know which $\boldsymbol{\theta} \in \boldsymbol{\Theta}$ is in effect. This is the essential difference between simple and composite hypotheses in applying the Pearson test. It is necessary to choose *one* distribution, $F(x; \boldsymbol{\theta}^*)$, from the family $\{F(x; \boldsymbol{\theta}), \boldsymbol{\theta} \in \Theta\}$ in the best possible way, using the data, so that the composite hypothesis H_0 is reduced to a simple hypothesis,

$$H_0^* : \boldsymbol{P}(X_i < x) = F(x; \boldsymbol{\theta}^*).$$

There are various methods that we might use to compute an estimate $\boldsymbol{\theta}^*$ for the unknown true value of $\boldsymbol{\theta}$, which we denote by $\boldsymbol{\theta}_o$. For each such method, we must then consider the distribution $\boldsymbol{P}_{\boldsymbol{\theta}_o}(X_n^2(\boldsymbol{\theta}^*) \leq x)$, where $\boldsymbol{P}_{\boldsymbol{\theta}_o} = \boldsymbol{P}_0$ is the true unknown distribution and $\boldsymbol{\theta}^*$ is our estimate, substituted for $\boldsymbol{\theta}$ in formula (9.4) and, in turn, into formula (9.1). Notice that the distribution $F(x; \boldsymbol{\theta}^*)$ is used to define the random variable $X_n^2(\boldsymbol{\theta}^*)$, after which the law of this random variable under the distribution $F(x; \boldsymbol{\theta}_o)$ is considered. It may be that there is a limit distribution of $\boldsymbol{P}_{\boldsymbol{\theta}_o}(X_n^2(\boldsymbol{\theta}^*) \leq x)$ when $n \to \infty$, but this does *not* follow immediately from Eq. (9.5). Moreover, if a limit distribution exists, its identity may depend on the method used to construct the estimate $\boldsymbol{\theta}^*$.

From a publication of Fisher (1928) (see also Cramer, 1946, Sverdrup, 1967), we have the following three methods of estimating $\boldsymbol{\theta}_o$, which are asymptotically equivalent in a sense defined below.

The Minimum Chi-Squared Estimator

Fisher's first idea was to make $X_n^2(\boldsymbol{\theta}^*)$ as small as possible by defining $\boldsymbol{\theta}^* = \tilde{\boldsymbol{\theta}}_n$ according to

$$X_n^2(\tilde{\boldsymbol{\theta}}_n) = \min_{\boldsymbol{\theta}\in\Theta} X_n^2(\boldsymbol{\theta}). \tag{9.6}$$

This is called, quite naturally, *the minimum chi-squared method of estimating* $\boldsymbol{\theta}_o$, and $\{\tilde{\boldsymbol{\theta}}_n\}$ is called *the sequence of minimum chi-squared estimators*. Under some regularity conditions, $\tilde{\boldsymbol{\theta}}_n$ can be found by solving the equations

$$\begin{aligned} -\frac{1}{2}\frac{\partial X_n^2(\boldsymbol{\theta})}{\partial\theta_j} &= \sum_{i=1}^{r}\left[\frac{(\nu_i - np_i(\boldsymbol{\theta}))}{p_i(\boldsymbol{\theta})} + \frac{(\nu_i - np_i(\boldsymbol{\theta}))^2}{2np_i^2(\boldsymbol{\theta})}\right]\frac{\partial p_i(\boldsymbol{\theta})}{\partial\theta_j} \\ &= 0, \quad j = 1,\ldots,s. \end{aligned} \tag{9.7}$$

In general, the system of equations (9.7) is difficult to solve. However, the term

$$\frac{(\nu_i - np_i(\boldsymbol{\theta}))^2}{2np_i^2(\boldsymbol{\theta})}$$

becomes negligible as $n \to \infty$. Fisher's second method was to replace system (9.7) by

$$\sum_{i=1}^{r}\frac{\nu_i - np_i(\boldsymbol{\theta})}{p_i(\boldsymbol{\theta})}\frac{\partial p_i(\boldsymbol{\theta})}{\partial\theta_j} = 0, \quad j = 1,\ldots,s. \tag{9.8}$$

Since $\sum_{i=1}^{r} p_i(\boldsymbol{\theta}) \equiv 1$, we can rewrite Eq. (9.8) as

$$\sum_{i=1}^{r}\frac{\nu_i}{p_i(\boldsymbol{\theta})}\frac{\partial p_i(\boldsymbol{\theta})}{\partial\theta_j} = 0, \quad j = 1,\ldots,s. \tag{9.9}$$

We denote by $\boldsymbol{\lambda}(\boldsymbol{\theta})$ the vector $\boldsymbol{\lambda}(\boldsymbol{\theta}) = (\lambda_1(\boldsymbol{\theta}),\ldots,\lambda_s(\boldsymbol{\theta}))^T$, where

$$\lambda_j(\boldsymbol{\theta}) = \sum_{i=1}^{r}\frac{\nu_i}{p_i(\boldsymbol{\theta})}\frac{\partial p_i(\boldsymbol{\theta})}{\partial\theta_j}, \quad j = 1,\ldots,s. \tag{9.10}$$

The vector $\boldsymbol{\lambda}(\boldsymbol{\theta})$ is called the *informant vector of the grouped data*, $\boldsymbol{\nu} = (\nu_1,\ldots,\nu_r)^T$. In terms of $\boldsymbol{\lambda}(\boldsymbol{\theta})$ the system of equations (9.9) is

$$\boldsymbol{\lambda}(\boldsymbol{\theta}) = \mathbf{0}_s, \quad \text{the zero vector in } \boldsymbol{R}^s. \tag{9.11}$$

For a reason explained below, Fisher called Eq. (9.9) or (9.11), using the frequency vector $\boldsymbol{\nu}$ (the grouped data!), *the maximum-likelihood equation.*

The Multinomial Maximum-Likelihood Estimator

As we have seen, the vector $\boldsymbol{\nu} = (\nu_1, \ldots, \nu_r)^T$ has the multinomial distribution with parameters n and $\boldsymbol{p}(\boldsymbol{\theta}) = (p_1(\boldsymbol{\theta}), \ldots, p_r(\boldsymbol{\theta}))^T$,

$$\boldsymbol{P}_{\boldsymbol{\theta}}(\nu_1 = n_1, \ldots, \nu_r = n_r) = \frac{n!}{n_1! \ldots n_r!} p_1^{n_1}(\boldsymbol{\theta}) \ldots p_r^{n_r}(\boldsymbol{\theta}),$$

some $\boldsymbol{\theta} \in \boldsymbol{\Theta}$, under hypothesis H_0 . *The likelihood function*, $\ell(\boldsymbol{\theta})$, of $\boldsymbol{\nu}$ is

$$\ell(\boldsymbol{\theta}) = \frac{n!}{\prod_{i=1}^r \nu_i!} \prod_{i=1}^r p_i^{\nu_i}(\boldsymbol{\theta}).$$

It is evident that Eq. (9.11) is the same as

$$\mathbf{grad}\ \log \ell(\boldsymbol{\theta}) = \mathbf{0}_s,$$

where **grad** denotes the gradient, the vector of the partial derivatives. Another notation is Δ. The method of maximum likelihood is to use any estimator $\hat{\boldsymbol{\theta}}_n$ which maximizes the likelihood function $\ell(\boldsymbol{\theta})$, and this is why Fisher called Eq. (9.11) the maximum-likelihood equation and the estimator $\hat{\boldsymbol{\theta}}_n$ *the multinomial maximum-likelihood estimator.*

The Modified Minimum Chi-Squared Estimator

Fisher's third method is to consider the so-called *modified minimum chi-squared estimator* $\bar{\boldsymbol{\theta}}_n$ which minimizes the quadratic form (studied by Neyman, 1949),

$$\bar{X}_n^2(\boldsymbol{\theta}) := \sum_{i=1}^r \frac{(\nu_i - np_i(\boldsymbol{\theta}))^2}{\nu_i} \tag{9.6$'$}$$

with respect to $\boldsymbol{\theta}$. Notice that here ν_i replaces $np_i(\boldsymbol{\theta})$ in the denominator, that is, we use the estimator ν_i/n for $p_i(\boldsymbol{\theta})$ in the denominator of $X_n^2(\boldsymbol{\theta})$. (See also Section 8.)

We now have three methods to construct a sequence $\{\boldsymbol{\theta}_n^*\}$ of estimators of $\boldsymbol{\theta}_o$, called $\{\tilde{\boldsymbol{\theta}}_n\}$, $\{\hat{\boldsymbol{\theta}}_n\}$, $\{\bar{\boldsymbol{\theta}}_n\}$, the minimum chi-squared, maximum likelihood, and minimum modified chi-squared methods. (These constructions use grouped data. In Section 10, we consider what happens if we use ungrouped data.) We will see now that under rather general conditions the three constructions have similar asymptotic properties, namely, the statistics $X_n^2(\boldsymbol{\theta}_n^*)$ have the same limit distribution for $\boldsymbol{\theta}_n^* = \tilde{\boldsymbol{\theta}}_n$, $\hat{\boldsymbol{\theta}}_n$, or $\bar{\boldsymbol{\theta}}_n$, and the random vectors $\sqrt{n}(\boldsymbol{\theta}_n^* - \boldsymbol{\theta}_o)$ have the same limit distribution in all three cases, with respect to the measure $\boldsymbol{P}_{\boldsymbol{\theta}_o}$. We will say that such sequences of estimates are *asymptotically equivalent* and sometimes we use the notation $\tilde{\boldsymbol{\theta}}_n$ to denote any of them. We now formulate these results more exactly. First some definitions.

Cramer's Conditions and Some Limit Theorems

Definition 9.1. A sequence of estimators $\{\boldsymbol{\theta}_n^*\}$ is called *consistent* for the parameter $\boldsymbol{\theta}$ if for any $\epsilon > 0$,

$$\boldsymbol{P}_{\boldsymbol{\theta}}\{|\boldsymbol{\theta}_n^* - \boldsymbol{\theta}| > \epsilon\} \to 0 \quad \text{as} \quad n \to \infty.$$

Here

$$|\boldsymbol{\theta}_n^* - \boldsymbol{\theta}\,| = \sqrt{(\boldsymbol{\theta}_n^* - \boldsymbol{\theta})^T(\boldsymbol{\theta}_n^* - \boldsymbol{\theta})} = \sqrt{\sum_{j=1}^{s}(\theta_{n_j}^* - \theta_j)^2}, \tag{9.12}$$

$$\boldsymbol{\theta}_n^* = (\theta_{n1}^*, \ldots, \theta_{ns}^*)^T, \qquad \boldsymbol{\theta} = (\theta_1, \ldots, \theta_s)^T \in \Theta \subset \boldsymbol{R}^s.$$

Instead of Eq. (9.12) we often write

$$\boldsymbol{\theta}_n^* - \boldsymbol{\theta} \longrightarrow \boldsymbol{0}_s \text{ in } \boldsymbol{P}_{\boldsymbol{\theta}} \text{ probability}$$

or

$$\boldsymbol{\theta}_n^* \longrightarrow \boldsymbol{\theta} \text{ in } \boldsymbol{P}_{\boldsymbol{\theta}} \text{ probability, as } n \to \infty.$$

Definition 9.2. A sequence of estimators $\{\boldsymbol{\theta}_n^*\}$ is called $\sqrt{n}$-consistent for the parameter $\boldsymbol{\theta}$ if for any $\epsilon > 0$ there exist C_ϵ and n_ϵ such that

$$\boldsymbol{P}_{\boldsymbol{\theta}}(\sqrt{n}|\boldsymbol{\theta}_n^* - \boldsymbol{\theta}| > C_\epsilon) < \epsilon \text{ for all } n > n_\epsilon.$$

Now we look at some properties of $\{\tilde{\boldsymbol{\theta}}_n\}$, $\{\hat{\boldsymbol{\theta}}_n\}$, $\{\bar{\boldsymbol{\theta}}_n\}$. For the remainder of this section we suppose that the following four conditions of Cramer hold (see Cramer, 1946; Birch, 1964; Chibisov, 1971),

1. $\sum_{i=1}^{r} p_i(\boldsymbol{\theta}) = 1,\ \boldsymbol{\theta} \in \boldsymbol{\Theta}$.
2. $p_i(\boldsymbol{\theta}) > c > 0,\ i = 1, \ldots, r$.
3. $\partial^2 p_i(\boldsymbol{\theta})/\partial\theta_\ell \partial\theta_j$ are continuous functions $\ell, j = 1, \ldots, s$.
4. The rank of the matrix $\boldsymbol{D} = \boldsymbol{D}(\boldsymbol{\theta}) = \|\partial p_i(\theta)/\partial\theta_j\|$ is s. (9.13)
   From condition 4, it follows that
5. The rank of the matrix $\boldsymbol{B} = \boldsymbol{B}(\boldsymbol{\theta})$ is s, where

$$\boldsymbol{B} := \left\| \begin{matrix} \frac{1}{\sqrt{p_1(\boldsymbol{\theta})}} \frac{\partial p_1(\boldsymbol{\theta})}{\partial \theta_1} & \cdots & \frac{1}{\sqrt{p_1(\boldsymbol{\theta})}} \frac{\partial p_1(\boldsymbol{\theta})}{\partial \theta_s} \\ \vdots & \vdots & \vdots \\ \frac{1}{\sqrt{p_r(\boldsymbol{\theta})}} \frac{\partial p_r(\boldsymbol{\theta})}{\partial \theta_1} & \cdots & \frac{1}{\sqrt{p_r(\boldsymbol{\theta})}} \frac{\partial p_r(\boldsymbol{\theta})}{\partial \theta_s} \end{matrix} \right\|_{r \times s}$$

Definition 9.3. The matrix

$$\boldsymbol{J} = \boldsymbol{J}(\boldsymbol{\theta}) := \boldsymbol{B}(\boldsymbol{\theta})^T \boldsymbol{B}(\boldsymbol{\theta})$$

is called the *Fisher information matrix* of the multinomial distribution with parameters $\boldsymbol{p}(\boldsymbol{\theta}) = (p_1(\boldsymbol{\theta}), \ldots, \boldsymbol{p}_r(\boldsymbol{\theta}))^T$. The Fisher information in the vector of frequencies, $\boldsymbol{\nu} = (\nu_1, \ldots, \nu_r)^T$, about $\boldsymbol{\theta}$ is $n\boldsymbol{J}(\boldsymbol{\theta})$.

Remark 9.1. We observe that Eq. (9.9) or Eq. (9.11) is equivalent to the equation

$$\boldsymbol{B}^T(\boldsymbol{\theta})\boldsymbol{X}_n(\boldsymbol{\theta}) = \mathbf{0}_s. \tag{9.14}$$

Let us prove that $\{\tilde{\boldsymbol{\theta}}_n\}$ is a consistent sequence for $\boldsymbol{\theta}_o$, that is, that $\tilde{\boldsymbol{\theta}}_n \to \boldsymbol{\theta}_o$ in $\boldsymbol{P}_{\boldsymbol{\theta}_o}$ probability as $n \to \infty$. For any $\boldsymbol{\theta} \in \Theta$ we have

$$X_n^2(\boldsymbol{\theta}) = \sum_{i=1}^{r} \frac{(\nu_i - np_i(\boldsymbol{\theta}))^2}{np_i(\boldsymbol{\theta})} = n \sum_{i=1}^{r} \frac{(\nu_i/n - p_i(\boldsymbol{\theta}))^2}{p_i(\boldsymbol{\theta})}.$$

If $\boldsymbol{\theta}_o$ is the true value of the parameter $\boldsymbol{\theta}$, then the law of large numbers says

$$\nu_i/n \to p_i(\boldsymbol{\theta}_o) \text{ in } \boldsymbol{P}_o \text{ probability}, \quad i = 1, \ldots, s, \tag{9.15}$$

and therefore by the continuity theorem, as $n \to \infty$,

$$\frac{1}{n} X_n^2(\boldsymbol{\theta}_o) = \sum_{i=1}^{r} \frac{(\nu_i/n - p_i(\boldsymbol{\theta}_0))^2}{p_i(\boldsymbol{\theta}_o)} \to 0 \text{ in } \boldsymbol{P}_o \text{ probability.} \tag{9.16}$$

But with $\boldsymbol{P}_o$ probability 1,

$$0 \le X_n^2(\tilde{\boldsymbol{\theta}}_n) \le X_n^2(\boldsymbol{\theta}_o)$$

and hence from Eq. (9.16)

$$\frac{1}{n} X_n^2(\boldsymbol{\theta}_o) \to 0 \text{ in } \boldsymbol{P}_o \text{ probability as } n \to \infty,$$

which is equivalent, under our hypotheses, to

$$\frac{\nu_i}{n} - p_i(\tilde{\boldsymbol{\theta}}_n) \to 0, \quad i = 1, \ldots, k, \text{ in } \boldsymbol{P}_o \text{ probability, as } n \to \infty. \tag{9.17}$$

The relations (9.15) and (9.17) give us

$$p_i(\tilde{\boldsymbol{\theta}}_n) \to p_i(\boldsymbol{\theta}_o), \quad i = 1, \ldots, s, \quad \text{in } \boldsymbol{P}_o \text{ probability as } n \to \infty, \tag{9.18}$$

so that $\{\tilde{\boldsymbol{\theta}}_n\}$ is a consistent sequence of estimators for $\boldsymbol{\theta}_o$. One can show in the same way that the sequences $\{\bar{\boldsymbol{\theta}}_n\}$ and $\hat{\boldsymbol{\theta}}_n\}$ are consistent for $\boldsymbol{\theta}_o$.

Theorem 9.1. *Let $\{\bar{\boldsymbol{\theta}}_n\}$ be the sequence of minimum modified chi-squared estimators of the parameter $\boldsymbol{\theta}_o$, defined by Eq. (9.6′). Then*

$$\sqrt{n}(\bar{\boldsymbol{\theta}}_n - \boldsymbol{\theta}_o) = \boldsymbol{J}^{-1}(\boldsymbol{\theta}_o)\boldsymbol{B}^T(\boldsymbol{\theta}_o)\boldsymbol{X}_n(\boldsymbol{\theta}_o) + o_{\boldsymbol{P}_o}(\mathbf{1}_s), \text{ as } n \to \infty, \tag{9.19}$$

where $\boldsymbol{J}(\boldsymbol{\theta}) = \boldsymbol{B}^T(\boldsymbol{\theta})\boldsymbol{B}(\boldsymbol{\theta})$ is the information matrix, and $o_{\boldsymbol{P}_o}(\mathbf{1}_s)$ is a random vector converging to $\mathbf{0}_s = (0, \ldots, 0)^T \in \boldsymbol{R}^s$, in $\boldsymbol{P}_o$ probability.

Remark 9.2. It follows from Eq. (9.19) that $\{\bar{\boldsymbol{\theta}}_n\}$ is an $\sqrt{n}$-consistent sequence of estimators of $\boldsymbol{\theta}_o$, as in Definition 9.2. Section 15 shows a numerical method for improving any $\sqrt{n}$-consistent sequence of estimators to a sequence with property (9.19).

Remark 9.3. From the discussion above we see that since Eq. (9.19) holds for the sequence $\{\bar{\boldsymbol{\theta}}_n\}$, the same condition holds also for the sequence of minimum chi-squared estimators and for the sequence of multinomial maximum likelihood estimators. Remark 9.2 also applies to these sequences.

Proof of Theorem 9.1. See Cramer (1946) and Rao (1973).

By its definition, the minimum modified chi-squared estimator $\{\bar{\boldsymbol{\theta}}_n\}$ satisfies

$$\mathbf{grad}\, \bar{X}_n^2(\boldsymbol{\theta})|_{\boldsymbol{\theta}=\bar{\boldsymbol{\theta}}_n} = \mathbf{0}_s, \quad \text{where} \quad \bar{X}_n^2(\boldsymbol{\theta}) = \sum_{i=1}^{r} \frac{(\nu_i - np_i(\boldsymbol{\theta}))^2}{\nu_i},$$

or, equivalently,

$$\sum_{i=1}^{r} \frac{\nu_i - np_i(\bar{\boldsymbol{\theta}}_n)}{\sqrt{\nu_i}} \sqrt{\frac{n}{\nu_i}} \left. \frac{\partial p_i(\boldsymbol{\theta})}{\partial \theta_j} \right|_{\boldsymbol{\theta}=\bar{\boldsymbol{\theta}}_n} = 0, \quad j = 1, \ldots, s. \tag{9.20}$$

Because $\{\bar{\boldsymbol{\theta}}_n\}$ is a consistent sequence of estimators for $\boldsymbol{\theta}_o$, we have for $i = 1, \ldots, r$,

$$\begin{aligned} \nu_i - np_i(\bar{\boldsymbol{\theta}}_n) &= \nu_i - np_i(\boldsymbol{\theta}_o) - n(p_i(\bar{\boldsymbol{\theta}}_n) - p_i(\boldsymbol{\theta}_o)) \\ &= \nu_i - np_i(\boldsymbol{\theta}_o) - n \sum_{j=1}^{s} \left[\left. \frac{\partial p_i(\boldsymbol{\theta})}{\partial \theta_j} \right|_{\boldsymbol{\theta}=\boldsymbol{\theta}_o} + o_{\boldsymbol{P}_o}(1) \right] (\bar{\theta}_{nj} - \theta_{oj}), \end{aligned} \tag{9.21}$$

where $\bar{\theta}_{nj}$ and θ_{oj} are the components of $\bar{\boldsymbol{\theta}}_n, \boldsymbol{\theta}_o$. For each $i = 1, \ldots, r$, also,

$$\begin{aligned} (np_i/\nu_i)^{1/2} &= 1 + o_{\boldsymbol{P}_o}(1), \\ (n/\nu_i)^{1/2} \left. \frac{\partial p_i(\boldsymbol{\theta})}{\partial \theta_j} \right|_{\boldsymbol{\theta}=\bar{\boldsymbol{\theta}}_n} &= \frac{1}{\sqrt{p_i(\boldsymbol{\theta}_o)}} \left. \frac{\partial p_i(\boldsymbol{\theta})}{\partial \theta_j} \right|_{\boldsymbol{\theta}=\boldsymbol{\theta}_o} + o_{\boldsymbol{P}_o}(1). \end{aligned} \tag{9.22}$$

Using Eq. (9.22) we can write Eq. (9.21) in the form

$$\boldsymbol{B}^T(\boldsymbol{\theta}_o)\boldsymbol{X}_n(\boldsymbol{\theta}_o) - \sqrt{n} \left[\boldsymbol{B}^T(\boldsymbol{\theta}_o)\boldsymbol{B}(\boldsymbol{\theta}_o) + o_{\boldsymbol{P}_o}(\mathbf{1}_{s\times s}) \right] (\bar{\boldsymbol{\theta}}_n - \boldsymbol{\theta}_o) = o_{\boldsymbol{P}_o}(\mathbf{1}_s). \tag{9.23}$$

The matrix $\boldsymbol{J}(\boldsymbol{\theta}_o)$ is nonsingular, and has rank s, by condition 4. Because

$$\lim_{n\to\infty} \det\left(\boldsymbol{J}(\boldsymbol{\theta}_o)+o_{\boldsymbol{P}_o}(\mathbf{1}_{s\times s})\right) = \det \boldsymbol{J}(\boldsymbol{\theta}_o) \text{ in } \boldsymbol{P}_o \text{ probability}$$

and $\det \boldsymbol{J}(\boldsymbol{\theta}_o) \neq 0$, we have from Eq. (9.23)

$$\left[\boldsymbol{J}(\boldsymbol{\theta}_o)+o_{\boldsymbol{P}_o}(\mathbf{1}_{s\times s})\right]^{-1} \boldsymbol{B}^T(\boldsymbol{\theta}_o)\boldsymbol{X}_n(\boldsymbol{\theta}_o)-\sqrt{n}(\bar{\boldsymbol{\theta}}_n-\boldsymbol{\theta}_o) = o_{\boldsymbol{P}_o}(\mathbf{1}_s),$$

and hence

$$\sqrt{n}(\bar{\boldsymbol{\theta}}_n-\boldsymbol{\theta}_o) = \boldsymbol{J}^{-1}(\boldsymbol{\theta}_0)\boldsymbol{B}^T(\boldsymbol{\theta}_o)\boldsymbol{X}_n(\boldsymbol{\theta}_o)+o_{\boldsymbol{P}_o}(\mathbf{1}_s). \tag{9.24}$$

Remark 9.4. From Eqs. (9.21) and (9.22), we have that for any $\sqrt{n}$-consistent sequence $\{\boldsymbol{\theta}_n^*\}$ of estimators for $\boldsymbol{\theta}_o$,

$$\begin{aligned}\boldsymbol{X}_n(\boldsymbol{\theta}_n^*) = \boldsymbol{X}_n(\boldsymbol{\theta}_o)&-\sqrt{n}\boldsymbol{B}(\boldsymbol{\theta}_o)(\boldsymbol{\theta}_n^*-\boldsymbol{\theta}_o)\\ &+o_{\boldsymbol{P}_o}(\mathbf{1}_{r\times s})\sqrt{n}(\boldsymbol{\theta}_n^*-\boldsymbol{\theta}_o)+o_{\boldsymbol{P}_o}(\mathbf{1}_r).\end{aligned} \tag{9.25}$$

The term $o_{\boldsymbol{P}_o}(\mathbf{1}_{r\times s})\sqrt{n}(\boldsymbol{\theta}_n^*-\boldsymbol{\theta}_o)$ is also $o_{\boldsymbol{P}_o}(\mathbf{1}_r)$ if $\sqrt{n}(\boldsymbol{\theta}_n^*-\boldsymbol{\theta}_o)$ has a tight family of distributions with respect to $\boldsymbol{P}_o$ probability. Applying this remark to the sequence $\{\tilde{\boldsymbol{\theta}}_n\}$, we obtain

Corollary 9.1.

$$\boldsymbol{X}_n(\tilde{\boldsymbol{\theta}}_n) = (\boldsymbol{I}_r-\boldsymbol{Q}(\boldsymbol{\theta}_o))\boldsymbol{X}_n(\boldsymbol{\theta}_o)+o_{\boldsymbol{P}_o}(\mathbf{1}_r), \tag{9.26}$$

where

$$\boldsymbol{Q}(\boldsymbol{\theta}) = \boldsymbol{B}(\boldsymbol{\theta})\boldsymbol{J}^{-1}(\boldsymbol{\theta})\boldsymbol{B}^T(\boldsymbol{\theta})$$

is a projection matrix and

$$\boldsymbol{J}(\boldsymbol{\theta}) = \boldsymbol{B}^T(\boldsymbol{\theta})\boldsymbol{B}(\boldsymbol{\theta}).$$

For any vector $\boldsymbol{z}$ from $\boldsymbol{R}^r$ the vector $\boldsymbol{Q}(\boldsymbol{\theta})\boldsymbol{z}$ is the projection of $\boldsymbol{z}$ on the space spanned by the columns of $\boldsymbol{B}(\boldsymbol{\theta})$, which has dimension s because rank $\boldsymbol{B}(\boldsymbol{\theta}) =$ rank $\boldsymbol{J}(\boldsymbol{\theta}) = s$. Similarly

$$\boldsymbol{I}_r-\boldsymbol{Q}(\boldsymbol{\theta})$$

is the projection matrix onto the complimentary $r-s$ dimensional subspace. Hence under H_0 $\boldsymbol{X}_n(\boldsymbol{\theta}_o)$ is $AN(\mathbf{0}_r, \boldsymbol{\Sigma})$, where

$$\boldsymbol{\Sigma} = \boldsymbol{\Sigma}(\boldsymbol{\theta}_o) = \boldsymbol{I}_r - \boldsymbol{q}(\boldsymbol{\theta}_o)\boldsymbol{q}^T(\boldsymbol{\theta}_o),$$
$$\boldsymbol{q}(\boldsymbol{\theta}) = \left(\sqrt{p_1(\boldsymbol{\theta})}, \ldots, \sqrt{p_r(\boldsymbol{\theta})}\right)^T, \tag{9.27}$$

and rank $\boldsymbol{\Sigma} = r - 1$. From Corollary 9.1 we obtain that under H_0

$$\boldsymbol{X}_n(\tilde{\boldsymbol{\theta}}_n) \text{ is } AN(\mathbf{0}_r, \boldsymbol{W}(\boldsymbol{\theta}_0)), \tag{9.28}$$

where

$$\boldsymbol{W}(\boldsymbol{\theta}) = (\boldsymbol{I}_r - \boldsymbol{Q}(\boldsymbol{\theta}))\boldsymbol{\Sigma}(\boldsymbol{\theta})(\boldsymbol{I}_r - \boldsymbol{Q}(\boldsymbol{\theta}))^T. \tag{9.29}$$

The definitions of $\boldsymbol{B}(\boldsymbol{\theta})$ and $\boldsymbol{q}(\boldsymbol{\theta})$, given by Eqs. (9.13) and (9.27), and the condition $\boldsymbol{p}^T(\boldsymbol{\theta})\mathbf{1}_r = 1$ give us

$$\boldsymbol{B}^T(\boldsymbol{\theta})\boldsymbol{q}(\boldsymbol{\theta}) = \mathbf{0}_s. \tag{9.30}$$

Notice that the matrices $\boldsymbol{q}(\boldsymbol{\theta})\boldsymbol{q}^T(\boldsymbol{\theta})$, $\boldsymbol{\Sigma}(\boldsymbol{\theta})$, and $\boldsymbol{Q}(\boldsymbol{\theta})$ are idempotent, and that

$$\text{rank } \mathbf{1}_r = r, \quad \text{rank } \boldsymbol{q}(\boldsymbol{\theta})\boldsymbol{q}^T(\boldsymbol{\theta}) = 1, \quad \text{rank } \boldsymbol{\Sigma}(\boldsymbol{\theta}) = r - 1, \quad \text{rank } \boldsymbol{Q}(\boldsymbol{\theta}) = s.$$

By straightforward calculation, using these properties, one obtains

$$\boldsymbol{W}(\boldsymbol{\theta}) = \boldsymbol{I}_r - \boldsymbol{q}(\boldsymbol{\theta})\boldsymbol{q}^T(\boldsymbol{\theta}) - \boldsymbol{Q}(\boldsymbol{\theta}) \tag{9.31}$$

and $\boldsymbol{W}^2(\boldsymbol{\theta}) = \boldsymbol{W}(\boldsymbol{\theta})$, that is, $\boldsymbol{W}(\boldsymbol{\theta})$ is also an idempotent matrix and rank $\boldsymbol{W}(\boldsymbol{\theta}) = r - s - 1$ because $\boldsymbol{W}(\boldsymbol{\theta})$, $\boldsymbol{q}(\boldsymbol{\theta})\boldsymbol{q}^T(\boldsymbol{\theta})$ and $\boldsymbol{Q}(\boldsymbol{\theta})$ are projections orthogonal to each other.

We now summarize these results:

Theorem 9.2. *Under H_0 as $n \to \infty$*

$$\begin{pmatrix} \sqrt{n}(\tilde{\boldsymbol{\theta}}_n - \boldsymbol{\theta}_0) \\ \boldsymbol{X}_n(\tilde{\boldsymbol{\theta}}_n) \end{pmatrix} \textit{ is } AN\left(\mathbf{0}_{s+r}, \boldsymbol{G}(\boldsymbol{\theta}_0)\right), \tag{9.32}$$

where

$$\boldsymbol{G}(\theta_0) = \left[\begin{array}{c:c} \boldsymbol{J}^{-1}(\boldsymbol{\theta}_0) & \mathbf{0}_{s\times r} \\ \hdashline \mathbf{0}_{r\times s} & \boldsymbol{W}(\boldsymbol{\theta}_0) \end{array}\right]_{(r+s)(r+s)}, \tag{9.33}$$

$rank\ \boldsymbol{J}^{-1}(\boldsymbol{\theta}_0) = s,\ rank\ \boldsymbol{W}(\theta_0) = r - s - 1.$

Theorem 9.3 follows from Theorem 9.2.

Theorem 9.3. *Under H_0 the limiting joint distribution of*

$$\sqrt{n}(\tilde{\boldsymbol{\theta}}_n - \boldsymbol{\theta}_0) \text{ and } X_n^2(\tilde{\boldsymbol{\theta}}_n)$$

as $n \to \infty$ is the joint distribution of a normal $N(\mathbf{0}_s, \boldsymbol{J}^{-1}(\boldsymbol{\theta}_0))$ random vector and a random variable χ^2_{r-s-1}, which are mutually independent.

Proof. From (9.33) we obtain that $\boldsymbol{X}_n(\tilde{\boldsymbol{\theta}}_n)$ and $\sqrt{n}(\tilde{\boldsymbol{\theta}}_n - \boldsymbol{\theta}_0)$ are asymptotically independent, and hence the statistic of Pearson $X_n^2(\tilde{\boldsymbol{\theta}}_n) = \boldsymbol{X}_n^T(\tilde{\boldsymbol{\theta}}_n)\boldsymbol{X}_n(\tilde{\boldsymbol{\theta}}_n)$ is also asymptotically independent of $\sqrt{n}(\tilde{\boldsymbol{\theta}}_n - \boldsymbol{\theta}_0)$ as $n \to \infty$. To show that $X_n^2(\tilde{\boldsymbol{\theta}}_n)$ is distributed as χ^2_{r-s-1} as $n \to \infty$ we need

Lemma 9.1 (Chernoff and Lehmann, 1954). *Let*

$$\boldsymbol{Y} = (Y_1, \ldots, Y_r)^T \sim N_r(\mathbf{0}_r, \boldsymbol{A}).$$

Then the random variable

$$\boldsymbol{Y}^T \boldsymbol{Y} = \sum_{i=1}^{r} Y_i^2 \tag{9.34}$$

is distributed as

$$\lambda_1 \xi_1^2 + \lambda_2 \xi_2^2 + \cdots + \lambda_r \xi_r^2, \tag{9.35}$$

where $\xi_1, \xi_2, \ldots, \xi_r$ are independent standard normal, $N_1(0,1)$, random variables and $\lambda_1, \lambda_2, \ldots, \lambda_r$ are the eigenvalues of the matrix $\boldsymbol{A}$.

If $\boldsymbol{A}$ is an idempotent matrix, that is,

$$\boldsymbol{A}^2 = \boldsymbol{A}^T \boldsymbol{A} = \boldsymbol{A}\boldsymbol{A}^T = \boldsymbol{A},$$

then all its eigenvalues, λ_j, are 0 or 1 and hence rank $\boldsymbol{A}$ = $tr\ \boldsymbol{A}$ = m, where m is the number of λ_j which are equal to 1 ($m \leq r$). In this case from Eqs. (9.34) and (9.35) we obtain

$$\boldsymbol{P}\{\boldsymbol{Y}^T \boldsymbol{Y} < x\} = \boldsymbol{P}\{\chi^2_m < x\}. \tag{9.36}$$

Indeed, the matrix $\boldsymbol{W}(\boldsymbol{\theta}_0)$ is idempotent (see Eq. 9.31) and rank $\boldsymbol{W}(\boldsymbol{\theta}_0)= r-s-1$. Hence, from Lemma 9.1 and Eq. (9.36) it follows that

$$\lim_{n\to\infty} \boldsymbol{P}\{X_n^2(\tilde{\boldsymbol{\theta}}_n) < x\} = \boldsymbol{P}\{\chi^2_{r-s-1} < x\}. \tag{9.37}$$

Theorem 9.3 is proved.

Fisher's Theorem

Theorem 9.4, of Fisher, consolidates our information about the limit distribution of $X_n^2(\boldsymbol{\theta})$, defined by Eq. (9.4), when $\boldsymbol{\theta}$ is estimated, and readies us to construct χ^2 tests for some examples. Suppose we observe the random vector $\boldsymbol{\nu}=(\nu_1,\ldots,\nu_r)^T$, which has a multinomial distribution (9.2) with parameters n and $\boldsymbol{p}=(p_1,\ldots,p_r)^T$, $\boldsymbol{p}^T\mathbf{1}_r=1$, $p_i>0$. The true value $\boldsymbol{p}^{(0)}=(p_1^{(0)},\ldots,p_r^{(0)})^T$ of the vector of probabilities $\boldsymbol{p}$ is unknown. Consider the hypothesis H_0 according to which

$$\boldsymbol{p}=\boldsymbol{p}(\boldsymbol{\theta})=(\boldsymbol{p}_1(\boldsymbol{\theta}),\ldots,\boldsymbol{p}_r(\boldsymbol{\theta}))^T,$$

where $\boldsymbol{\theta}=(\theta_1,\ldots,\theta_s)^T\in\boldsymbol{\Theta}\subset\boldsymbol{R}^s$, $r>s+1$, and $\boldsymbol{p}(\boldsymbol{\theta})$ satisfies Cramer's conditions 1–4. If H_0 is true, there exists $\boldsymbol{\theta}_0\in\boldsymbol{\Theta}$ such that $\boldsymbol{p}(\boldsymbol{\theta}_0)=\boldsymbol{p}^{(0)}$. And conversely, if H_0 is not true, then we have

$$\boldsymbol{p}(\boldsymbol{\theta})\neq\boldsymbol{p}^{(0)} \quad \text{for all} \quad \boldsymbol{\theta}\in\boldsymbol{\Theta}. \tag{9.38}$$

The following theorem is attributed to Fisher.

Theorem 9.4. *If H_0 is true and $\tilde{\theta}_n$ is an estimator that satisfies Eq. (9.19), then*

1. $$\lim_{n\to\infty} \boldsymbol{P}\{X_n^2(\boldsymbol{\theta}_0) > x\} = \boldsymbol{P}\{\chi^2_{r-1} > x\}, \tag{9.39}$$

2. $$\lim_{n\to\infty} \boldsymbol{P}\{X_n^2(\tilde{\boldsymbol{\theta}}_n) > x\} = \boldsymbol{P}\{\chi^2_{r-s-1} > x\}, \tag{9.40}$$

If H_0 is not true then

3. $$\lim_{n\to\infty} \boldsymbol{P}\{X_n^2(\tilde{\boldsymbol{\theta}}_n) > x\} - \boldsymbol{P}\{\chi^2_{r-s-1}(\lambda) > x\} = 0, \tag{9.41}$$

where

$$\lambda = n \cdot \min_{\boldsymbol{\theta} \in \boldsymbol{\Theta}} \sum_{i=1}^{r} \frac{[p_i^{(0)} - p_i(\boldsymbol{\theta})]^2}{p_i^{(0)}}. \tag{9.42}$$

According to Pearson's test based on $X_n^2(\tilde{\boldsymbol{\theta}}_n)$, we should reject H_0 if $X_n^2(\tilde{\boldsymbol{\theta}}_n) > c_\alpha$, where c_α is the α-upper quantile of χ^2_{r-s-1}.

Example: A χ^2 Test of Homogeneity for Grouped Data. Suppose we have I independent vectors

$$\boldsymbol{\nu}_1 = (\nu_{11}, \ldots, \nu_{1r})^T, \ldots, \boldsymbol{\nu}_I = (\nu_{I1}, \ldots, \nu_{Ir})^T,$$

each $\boldsymbol{\nu}_i$ having a multinomial distribution with parameters $(n_i, \boldsymbol{p}_i)$,

$$\begin{aligned} \boldsymbol{p}_i &= (p_{i1}, \ldots, p_{ir})^T, \quad i = 1, \ldots, I, \\ \boldsymbol{p}_i^T \mathbf{1}_r &= 1, \quad p_{ij} > 0, \qquad i = 1, \ldots, I; \quad j = 1, \ldots, r. \end{aligned}$$

Then

$$\boldsymbol{P}\{\nu_{i1} = x_{i1}, \ldots, \nu_{ir} = x_{ir}\} = \frac{n_i!}{x_{i1}! \ldots x_{ir}!} p_{i1}^{x_{i1}} \ldots p_{ir}^{x_{ir}}, \tag{9.43}$$

for any $\boldsymbol{x}_i = (x_{i1}, \ldots, x_{ir})^T$, $\boldsymbol{x}_i^T \mathbf{1}_r = 1$, $x_{ij} > 0$, $i = 1, \ldots, I$, $j = 1, \ldots, r$.

Consider the hypothesis H_0, according to which

$$H_0 : \boldsymbol{p}_1 = \boldsymbol{p}_2 = \ldots = \boldsymbol{p}_I = \boldsymbol{p},$$

where $\boldsymbol{p} = (p_1, \cdots, p_r)^T$ is such that $\boldsymbol{p}^T \mathbf{1}_r = 1$, $p_i > 0$. We write

$$\boldsymbol{\nu} = \sum_{i=1}^{I} \boldsymbol{\nu}_i = (\nu_1, \ldots, \nu_r)^T, \qquad \nu_j = \sum_{i=1}^{I} \nu_{ij}, \quad j = 1, \ldots, r,$$

$n = n_1 + n_2 + \cdots + n_I$. Then under H_0, from Theorem 9.4 of Fisher and from the independence of $\boldsymbol{\nu}_1, \ldots, \boldsymbol{\nu}_r$, it follows that the statistic

$$X^2(\boldsymbol{p}) = \sum_{i=1}^{I} \frac{(\nu_{ij} - n_i p_j)^2}{n_i p_j} \tag{9.44}$$

has in the limit as $\min_i n_i \to \infty$ a chi-squared distribution with $I(r-1)$ degrees of freedom; that is,

$$\lim_{\min n_i \to \infty} \boldsymbol{P}\{X^2 > x \mid H_0\} = \boldsymbol{P}\{\chi^2_{I(r-1)} > x\}. \tag{9.45}$$

Now, suppose that $\boldsymbol{p}_1, \ldots, \boldsymbol{p}_I$ are unknown and we want to test

$$H_0 : \boldsymbol{p}_1 = \boldsymbol{p}_2 = \cdots = \boldsymbol{p}_r = \boldsymbol{p},$$

where $\boldsymbol{p}$ is unknown also. The likelihood function $\ell(\boldsymbol{p})$ of the vector $\boldsymbol{\nu}$ is a product of likelihoods of the form shown in Eq. (2.4) and under H_0 is

$$L(\boldsymbol{p}) = c p_1^{\nu_1} p_2^{\nu_2} \ldots p_r^{\nu_r}, \tag{9.46}$$

where $c = \prod_{i=1}^{I} c_i$, and c_i is the coefficient in Eq. (9.43).

From Eq. (9.46) we obtain that the maximum-likelihood estimator $\hat{\boldsymbol{p}}$ of $\boldsymbol{p} = (p_1, \ldots, p_r)^T$ is

$$\hat{\boldsymbol{p}} = \frac{1}{n}\boldsymbol{\nu} = (\frac{1}{n}\nu_1, \ldots, \frac{1}{n}\nu_r)^T. \tag{9.47}$$

From Fisher's Theorem 9.4, it follows that the statistic

$$X_n^2(\hat{\boldsymbol{p}}) = \sum_{i=1}^{I}\sum_{j=1}^{r} \frac{(\nu_{ij} - n_i \frac{\nu_j}{n})^2}{n_i \frac{\nu_j}{n}} \tag{9.48}$$

has, in the limit as $\min n_i \to \infty$, the chi-squared distribution with $(I-1)$ $(r-1)$ degrees of freedom, that is,

$$\lim_{\min n_i \to \infty} \boldsymbol{P}\{X_n^2(\hat{\boldsymbol{p}}) > x \mid H_0\} = \boldsymbol{P}\{\chi^2_{(I-1)(r-1)} > x\}. \tag{9.49}$$

Pearson's test based on the statistic $X_n^2(\hat{\boldsymbol{p}})$ rejects the hypothesis of homogeneity if

$$X_n^2(\hat{\boldsymbol{p}}) > c_\alpha,$$

where c_α is the α upper quantile of the chi-squared distribution with $(I-1)(r-1)$ degrees of freedom.

For example suppose that two groups of 300 students each had the same exam. In the first group, 144 students received an A, 80 a B, 43 a C, and 33 failed. In the second group the distribution of the marks was 154 – A, 72 – B, 35 – C, 39 – failed. In this case homogeneity would mean that the two groups are realizations of random vectors having the same multinomial distribution.

Here, the data is in table form

i	ν_{i1}	ν_{i2}	ν_{i3}	ν_{i4}
1	144	80	43	33
2	154	72	35	39

Under H_0 the maximum-likelihood estimator of $\boldsymbol{p}$ is $\hat{\boldsymbol{p}} = (\hat{p}_1, \hat{p}_2, \hat{p}_3, \hat{p}_4)^T$, where

$$\hat{p}_1 = \frac{298}{600}, \qquad \hat{p}_2 = \frac{152}{600}, \qquad \hat{p}_3 = \frac{78}{600}, \qquad \hat{p}_4 = \frac{72}{600},$$

since, in this example,

$$\begin{gathered} N = n_1 + n_2 = 300 + 300 = 600, \\ \nu_1 = \nu_{11} + \nu_{21} = 298, \qquad \nu_2 = \nu_{12} + \nu_{22} = 152, \\ \nu_3 = \nu_{13} + \nu_{23} = 78, \qquad \nu_4 = \nu_{14} + \nu_{24} = 72, \end{gathered}$$

and $\hat{p}_i = \nu_i / N$. To test H_0 we construct a chi-squared test, based on the statistic (9.48). With our data we have

$$\begin{aligned} X^2(\hat{\boldsymbol{p}}) &= 2\left\{\frac{(33 - 300 \cdot \frac{72}{600})^2}{36} + \frac{(43 - 300 \cdot \frac{78}{600})^2}{39} \right. \\ &\qquad \left. + \frac{(80 - 300 \cdot \frac{152}{600})^2}{76} + \frac{(144 - 300 \cdot \frac{298}{600})^2}{149}\right\} \\ &= 2\left(\frac{9}{36} + \frac{16}{39} + \frac{16}{76} + \frac{25}{149}\right) < 2\left(\frac{1}{4} + \frac{1}{2} + \frac{1}{4} + \frac{1}{5}\right) \\ &= 2.4 < \chi^2_{3,0.95} = 7.815, \end{aligned}$$

where $\chi^2_{3,0.95}$ is the upper 0.05-quantile of the chi-squared distribution with three degrees of freedom. Since

$$X^2(\hat{\boldsymbol{p}}) < \chi^2_{3,0.95} = 7.815,$$

we accept H_0 if $\alpha = 0.05$.

Analysis of Contingency Tables – Simple Examples

Suppose that the results of multinomial trials can be classified according to two different criteria, A, and B. One may wish to test the hypothesis, H_0, of independence of these criteria. Suppose we have n independent multinomial trials where each trial results in one of IJ outcomes. In Table 9.1, ν_{ij} represents the frequency of trials whose outcomes fall in the category "i" of criterion A, and category "j" of criterion B ($i=1,\ldots,I$; $j=1,\ldots,J$). Denote by p_{ij} the corresponding probability (see Table 9.2). Of course $\sum_{ij} p_{ij} = 1$. The hypothesis H_0 of independence is equivalent to the hypothesis that

$$p_{ij} = p_{i\cdot}p_{\cdot j}$$

where

$$\sum_{j=1}^{J} p_{ij} = p_{i\cdot}, \quad \sum_{i=1}^{I} p_{ij} = p_{\cdot j} \tag{9.50}$$

The probability $p_{i\cdot}$ is the probability of an outcome in category i (criterion A), and $p_{\cdot j}$ is the probability of an outcome in category j (criterion B) in each trial. Assume now that all probabilities p_{ij}, $p_{i\cdot}$, $p_{\cdot j}$ are unknown. Under H_0, we have only $I+J-2$ unknown parameters to estimate, namely

$$p_{i\cdot} \quad \text{for} \quad i = 1,\ldots,I-1,$$
$$p_{\cdot j} \quad \text{for} \quad j = 1,\ldots,J-1,$$

because

$$p_{I\cdot} = 1-\sum_{i=1}^{I-1} p_{i\cdot}, \quad p_{\cdot J} = 1-\sum_{j=1}^{J-1} p_{\cdot j}. \tag{9.51}$$

The likelihood function $\ell(\boldsymbol{p})$ of the vector of frequencies $\boldsymbol{\nu} = (\nu_{11}, \ldots, \nu_{IJ})^T$ is

$$\ell(\boldsymbol{p}) = \prod_{i,j} p_{ij}^{\nu_{ij}}, \qquad \boldsymbol{p} = (p_{11},\ldots,p_{IJ})^T. \tag{9.52}$$

Under H_0

Table 9.1

Criterion A Category	Criterion B Category				
	1	2	...	J	Total
1	ν_{11}	ν_{12}	...	ν_{1J}	$\nu_{1\cdot}$
2	ν_{21}	ν_{22}	...	ν_{2J}	$\nu_{2\cdot}$
.	.	.		.	.
.	.	.		.	.
.	.	.		.	.
I	ν_{I1}	ν_{I2}	...	ν_{IJ}	$\nu_{I\cdot}$
Total	$\nu_{\cdot 1}$	$\nu_{\cdot 2}$	...	$\nu_{\cdot J}$	n

Table 9.2

Criterion A Category	Criterion B Category				
	1	2	...	J	Total
1	p_{11}	p_{12}	...	p_{1J}	$p_{1\cdot}$
2	p_{21}	p_{22}	...	p_{2J}	$p_{2\cdot}$
.	.	.		.	.
.	.	.		.	.
.	.	.		.	.
I	p_{I1}	p_{I2}	...	p_{IJ}	$p_{I\cdot}$
Total	$p_{\cdot 1}$	$p_{\cdot 2}$	...	$p_{\cdot J}$	1

Table 9.3

ν_{11}	ν_{12}	$\nu_{1\cdot}$
ν_{21}	ν_{22}	$\nu_{2\cdot}$
$\nu_{\cdot 1}$	$\nu_{\cdot 2}$	n

$$\ell(\boldsymbol{p}) = \prod_{i,j} p_{i\cdot}^{\nu_{ij}} p_{\cdot j}^{\nu_{ij}}$$

and hence

$$\ln \ell(\boldsymbol{p}) = \sum_{i=1}^{I} \sum_{j=1}^{J} \nu_{ij}(\ln p_{i\cdot} + \ln p_{\cdot j}) = \sum_{i=1}^{I} \nu_{i\cdot} \ln p_{i\cdot} + \sum_{j=1}^{J} \nu_{\cdot j} \ln p_{ij}, \quad (9.53)$$

where

$$\nu_{i\cdot} = \sum_{j=1}^{J} \nu_{ij}, \qquad \nu_{\cdot j} = \sum_{i=1}^{I} \nu_{ij}.$$

We find from Eq. (9.53) that the maximum likelihood estimators $\hat{p}_{i\cdot}$, $\hat{p}_{\cdot j}$ for $p_{i\cdot}$, $p_{\cdot j}$ are

$$\hat{p}_{i\cdot} = \frac{\nu_{i\cdot}}{n}, \quad \hat{p}_{\cdot j} = \frac{\nu_{\cdot j}}{n}, \quad i = 1, \ldots, I; \; j = 1, \ldots, J. \quad (9.54)$$

From Eq. (9.54), it follows that

$$\hat{p}_{ij} = \hat{p}_{i\cdot}\hat{p}_{\cdot j} = \frac{\nu_{i\cdot}}{n} \cdot \frac{\nu_{\cdot j}}{n} \quad (9.55)$$

is the maximum likelihood estimator for p_{ij}. Let

$$X_n^2(\boldsymbol{p}) = \sum_{i=1}^{I} \sum_{j=1}^{J} \frac{(\nu_{ij} - np_{ij})^2}{np_{ij}}, \quad (9.56)$$

and

$$X_n^2(\hat{\boldsymbol{p}}) = \sum_{i=1}^{I} \sum_{j=1}^{J} \frac{(\nu_{ij} - n\hat{p}_{ij})^2}{n\hat{p}_{ij}}. \quad (9.57)$$

From Theorem 9.4 of Fisher we obtain

$$\lim_{n\to\infty} \boldsymbol{P}\{X_n^2(\boldsymbol{p}) > x\} = \boldsymbol{P}\{\chi_{IJ-1}^2 > x\}, \quad (9.58)$$

$$\lim_{n\to\infty} \boldsymbol{P}\{X_n^2(\hat{\boldsymbol{p}}) > x \mid H_0\} = \boldsymbol{P}\{\chi_{(I-1)(J-1)}^2 > x\}. \quad (9.59)$$

The hypothesis H_0 should be rejected if $X_n^2(\hat{\boldsymbol{p}}) > c_\alpha$, where c_α is the α upper quantile of the chi-squared distribution with $(I-1)(J-1) = IJ - 1 - (I + J - 2)$ degrees of freedom.

Table 9.4

A	B		Total
	1	2	
1	pP	qP	P
2	pQ	qQ	Q
Total	p	q	1

Table 9.5

Final Exam	Midterm		Total
	1	2	
1	18	30	48
2	79	173	252
Total	97	203	300

More about the analysis of contingency tables can be found in Freeman and Malton (1951), Cox (1970), Goodman and Kruskal (1979), Feinberg (1980), Fleiss (1981), Linhart and Zucchini (1986). The work of Diaconis and Efron (1985) has a fresh look at 2-way or contingency tables, and is followed by a lively discussion and good bibliography. More recent are the works of Mirvaliev (1988) on contingency tables for Poisson distributions and Oluyede (1994) on $r \times c$ contingency tables.

The following are simple illustrative examples.

Let us consider the case when $I = J = 2$. Table 9.1 is a 2×2 table. Under H_0 Table 9.2 is Table 9.4, where

$$P = p_{1\cdot}, \qquad Q = p_{2\cdot} = 1 - p_{1\cdot}, \qquad p = p_{\cdot 1}, \qquad q = p_{\cdot 2} = 1 - p_{\cdot 1}.$$

One can verify that, after some manipulation, Pearson's statistic (9.57) is

$$X_n^2(\hat{\boldsymbol{p}}) = n\frac{(\nu_{11}\nu_{22} - \nu_{12}\nu_{21})^2}{\nu_{1\cdot}\nu_{2\cdot}\nu_{\cdot 1}\nu_{\cdot 2}}. \tag{9.60}$$

Under H_0, the statistic (9.60) is distributed in the limit, when $n \to \infty$, as χ_1^2.

For example, suppose that 300 students had a midterm exam in math and 97 of them received a pass (1) and the other 203 students received a fail (2). One month later these students had a final exam in math and this time 48 received a pass (1) and of these only 18 had passed the first exam (i.e., only 18 students passed both). On the basis of this data, we construct a chi-squared test at the significance level $\alpha = 0.1$ for testing the hypothesis H_0 of independence of passing the two exams. Table 9.3 for this example is Table 9.5.

The maximum-likelihood estimators of p and P are

$$\hat{p} = \frac{\nu_{\cdot 1}}{n} = \frac{97}{300} \qquad \text{and} \qquad \hat{P} = \frac{\nu_{1\cdot}}{n} = \frac{48}{300}.$$

The value of Pearson's statistic X^2, given by Eq. (9.54), can be evaluated using formula (9.60):

$$X^2 = \frac{\left(18 - \frac{48\cdot 97}{300}\right)^2}{\frac{48\cdot 97}{300}} + \frac{\left(30 - \frac{48\cdot 203}{300}\right)^2}{\frac{48\cdot 203}{300}}$$

$$+ \frac{\left(79 - \frac{252\cdot 97}{300}\right)^2}{\frac{252\cdot 97}{300}} + \frac{\left(173 - \frac{252\cdot 203}{300}\right)^2}{\frac{252\cdot 203}{300}}$$

$$= \frac{300(18\cdot 173 - 30\cdot 79)^2}{97\cdot 203\cdot 48\cdot 252} = \frac{100(248)^2}{97\cdot 203\cdot 64\cdot 7}$$

$$= \frac{200}{203}\cdot\frac{93}{97}\cdot\frac{31}{42} < 1.$$

Under H_0 Pearson's statistic X^2 is distributed approximately as χ_1^2 and we accept H_0, since the observed value of X^2 is less than $\chi_{1,0.9}^2 = 2.707$.

Estimation with Constraints

Diamond, Mitra, and Roy (1960) have given some useful generalizations of Fisher's theorem. Here is one of them. In Fisher's theorem, the vector of frequencies $\boldsymbol{\nu} = (\nu_1, \ldots, \nu_r)^T$ satisfies $s+1$ linear conditions $(r \geq s+2)$

$$\begin{aligned} &\nu_1 + \cdots + \nu_r = n \\ &\sum_{i=1}^{r} \frac{\nu_i - np_i(\tilde{\boldsymbol{\theta}}_n)}{np_i(\tilde{\boldsymbol{\theta}}_n)} \frac{\partial p_i(\boldsymbol{\theta})}{\partial \theta_j}\bigg|_{\boldsymbol{\theta}=\tilde{\boldsymbol{\theta}}_n} = 0, \quad j = 1, \ldots, s. \end{aligned} \tag{9.61}$$

Following the work of Diamond et al., suppose that we have u $(u < s)$ functions $f_1(\boldsymbol{\theta}), \ldots, f_u(\boldsymbol{\theta})$ which constrain the parameter $\boldsymbol{\theta}$:

$$f_m(\boldsymbol{\theta}) = 0, \quad m = 1, 2, \ldots, u,$$

and such that

a. Every $f_m(\boldsymbol{\theta})$ has continuous second partial derivatives,

$$\frac{\partial^2 f_m(\boldsymbol{\theta})}{\partial \theta_i \partial \theta_j} \quad (i, j = 1, \ldots, s; \quad m = 1, \ldots, u);$$

b. The matrix

$$\left[\frac{\partial f_m(\boldsymbol{\theta})}{\partial \theta_i}\right] \quad (m = 1, \ldots, u; \quad i = 1, \ldots, s)$$

is of rank u.

Consider the hypothesis H_0: $p_j = p_j(\boldsymbol{\theta})$ defined by Eq. (9.1) $(j = 1, \ldots, r)$ subject to $f_m(\boldsymbol{\theta}) = 0$ for $m = 1, \ldots, u$, and the alternative H_n:

$$H_n : p_j = p_{jn} = p_j(\boldsymbol{\theta}_0) + \frac{\alpha_j}{\sqrt{n}},$$

also subject to $f_m(\boldsymbol{\theta}) = 0$, where

$$\sum_{j=1}^{r} \alpha_j = 0, \qquad \sum_{j=1}^{r} \alpha_j^2 > 0,$$

$p_{jn} \neq p_j(\boldsymbol{\theta})$ for any $\boldsymbol{\theta} \in \boldsymbol{\Theta}$ and $\boldsymbol{\theta}_0 \in \boldsymbol{\Theta}$. Then,

1. the equations

$$\sum_{i=1}^{r} \frac{\nu_i - np_i(\boldsymbol{\theta})}{p_i(\boldsymbol{\theta})} \frac{\partial p_i(\boldsymbol{\theta})}{\partial \theta_j} + \sum_{m=1}^{u} \lambda_m \frac{\partial f_m(\boldsymbol{\theta})}{\partial \theta_j} = 0, \quad j = 1, \dots, s,$$

$$f_m(\boldsymbol{\theta}) = 0, \quad m = 1, \dots, u$$

for minimizing $X_n^2(\boldsymbol{\theta})$ subject to the constraints and subject to H_0, have exactly one solution

$$\boldsymbol{\theta}_n^* = (\theta_1^*, \dots, \theta_s^*)^T, \qquad \lambda_n^* = (\lambda_1^*, \dots, \lambda_u^*)^T$$

such that $\boldsymbol{\theta}_n^* \to \boldsymbol{\theta}_0$ in probability $\boldsymbol{P}_{\boldsymbol{\theta}_0}$ as $n \to \infty$;

2. $\lim_{n\to\infty} \boldsymbol{P}\{X_n^2(\boldsymbol{\theta}_n^*) > x \mid H_0\} = \boldsymbol{P}\{\chi^2_{r-s+u-1} > x\}$

3. $\lim_{n\to\infty} \boldsymbol{P}\{X_n^2(\boldsymbol{\theta}_n^*) > x \mid H_n\} = \boldsymbol{P}\{\chi^2_{r-s+u-1}(\lambda) > x\}$

for any $x > 0$, where

$$\lambda = \boldsymbol{d}^T[\boldsymbol{I}_r - \boldsymbol{B}_*(\boldsymbol{B}_*^T)^{-1}\boldsymbol{B}_*^T]\boldsymbol{d},$$

and

$$\boldsymbol{d} = \left(\frac{d_1}{\sqrt{p_1(\boldsymbol{\theta}_0)}}, \dots, \frac{d_r}{\sqrt{p_r(\boldsymbol{\theta}_0)}} \right)^T,$$

$$\boldsymbol{B}_* = [b_{ij}^*]_{r\times(s-u)}, \qquad b_{ij}^* = \frac{1}{\sqrt{p_i(\boldsymbol{\theta}_0)}} \left. \frac{\partial p_j^*(\boldsymbol{\vartheta})}{\partial \vartheta_j} \right|_{\boldsymbol{\theta}(\boldsymbol{\vartheta})=\boldsymbol{\theta}_0},$$

where p_j^* is $p_j(\boldsymbol{\theta})$ expressed in terms of the $t-u$ independent θ_j which are designated by ϑ_j.

Remark 9.5. To test H_0, we can use the Hellinger statistic

$$F_n^2(\tilde{\boldsymbol{\theta}}_n) = 4\sum_{i=1}^{r} \left(\sqrt{\nu_i} - \sqrt{np_i(\tilde{\boldsymbol{\theta}}_n)} \right)^2,$$

which was defined in Section 7. As was shown by Fraser (1976),

$$\lim_{n \to \infty} \boldsymbol{P}\{F_n^2(\tilde{\boldsymbol{\theta}}_n) > x \mid H_0\} = P\{\chi^2_{r-s-1} > x\}.$$

where $\tilde{\boldsymbol{\theta}}_n$ is the minimum chi-squared estimator of $\boldsymbol{\theta}$. A study of the power of the chi-squared test, when one uses the minimum chi-squared estimator, was made by Gvanceladze and Chibisov (1979).

Remark 9.6. We have considered χ^2 criteria for testing a hypothesis H_0 when we supposed the data $X_1, \ldots, X_n$ to be independent. But this independence is not essential so long as the vector of frequencies $\boldsymbol{\nu} = (\nu_1, \ldots, \nu_r)^T$ is asymptotically normally distributed, which is true, for example, when $X_1, \ldots, X_n$ is a Markov sequence. This question is considered, e.g. by Rajarshi (1986), Gleser and Moore (1983), Chanda (1981).

We conclude this section with some examples.

Example: Simple Genetic Model. Characteristics of animals or plants depend on genes, which are passed from parents to their offspring. In the usual model, genes appear in pairs and each gene of a particular pair can be either of two types called A and a. The possible genotypes are

$$AA, Aa, aa,$$

since, biologically, the genotypes Aa and aA are equivalent. We assume that each of two parents independently contributes to the offspring one of its two genes with equal probability. Suppose that the equilibrium frequencies of genotypes AA, Aa, and aa are p_1, p_2, p_3, that is, the probabilities p_1, p_2, p_3 for AA, Aa, and aa in successive generations are the same. Suppose we have data from a population of size $n = 8,000$ with frequency vector

$$\nu_1 = 2014, \qquad \nu_2 = 5012, \qquad \nu_3 = 979,$$

$\nu_1 + \nu_2 + \nu_3 = 8000$. We wish to test the composite hypothesis H_0 according to which genotype probabilities p_1, p_2, p_3 are of the form

$$p_1 = p_1(\theta) = \frac{1 - 4\theta}{2}, \qquad p_2 = \frac{1 + 2\theta}{2}, \qquad p_3 = p_3(\theta) = \theta$$

for some $\theta \in \boldsymbol{\Theta} = (0, 0.25)$. Under H_0 the vector $\boldsymbol{\nu}$ has a trinomial distribution

$$\boldsymbol{P}\{\nu_1 = x_1, \nu_2 = x_2, \nu_3 = x_3 | H_0\}$$
$$= \frac{n!}{\nu_1!\nu_2!\nu_3!} p_1^{\nu_1}(\theta) p_2^{\nu_2}(\theta) p_3^{\nu_3}(\theta), \qquad \theta \in \Theta.$$

To test H_0 we construct a chi-squared test. We choose a significance level α; for example, we take $\alpha = 0.05$. According to Fisher's theorem under H_0 the statistic

$$X_n^2(\tilde{\theta}_n) = \min_{\theta \in \Theta} X_n^2(\theta)$$

has in the limit as $n \to \infty$ the chi-squared distribution with $3-1-1=1$ degrees of freedom. We reject H_0 if

$$X_n^2(\tilde{\theta}_n) > \chi^2_{1,1-\alpha} = \chi^2_{1,0.95} = 3.841.$$

Instead of the minimum chi-squared estimator $\tilde{\theta}_n$, we may wish to use the unbiased estimator

$$\theta_n^* = \frac{\nu_3}{n}.$$

The statistic $X_n^2(\theta_n^*)$ is stochastically larger than $\chi_n^2(\tilde{\theta}_n)$. Since

$$\theta_n^* = \frac{\nu_3}{n} = \frac{974}{8000} \approx \frac{1}{8} = 0.125,$$

let us use $\theta_n^* \approx 0.125$ as an approximation. Then

$$X_n^2(\tilde{\theta}_n) = \min_{\theta \in \Theta} X_n^2(\theta) \leq X_n^2(0.125)$$

$$= \frac{[\nu_1 - np_1(0.125)]^2}{np_1(0.125)} + \frac{[\nu_2 - np_2(0.125)]^2}{np_2(0.125)} + \frac{[\nu_3 - np_3(0.125)]^2}{np_3(0.125)}$$

$$= \frac{(2014-2000)^2}{2000} + \frac{(5012-5000)^2}{5000} + \frac{(974-1000)^2}{1000}$$

$$= 0.7336 < \chi^2_{1,0.95} = 3.841.$$

Hence, the hypothesis H_0 is accepted at the level of significance $\alpha = 0.05$.

Example: Another Genetic Model. Each individual in a given population of size n has one of four possible genotypes

$$AB, Ab, aB, ab$$

with probabilities p_1, p_2, p_3, p_4, respectively. We want to test the hypothesis H_0 that the values of p_i are in accordance with the laws of Mendelian inheritance, which predict that the probabilities p_i are of the form

$$\begin{aligned} p_1 &= p_1(\theta) = \frac{2+\theta}{4}, \quad p_2 = p_2(\theta) = \frac{1-\theta}{4}, \\ p_3 &= p_3(\theta) = \frac{1-\theta}{4}, \quad p_4 = p_4(\theta) = \frac{\theta}{4}, \end{aligned}$$

$0<\theta<1$. The vector of observed frequencies is $\boldsymbol{\nu}=(\nu_1, \nu_2, \nu_3, \nu_4)^T$ where again $\nu_1+\nu_2+\nu_3+\nu_4 = n$.

Consider a probability model according to which the vector $\boldsymbol{\nu}$ has a four-dimensional multinomial distribution $M_4(n, \boldsymbol{p})$, where $\boldsymbol{p}=\boldsymbol{p}(\theta)= (p_1, p_2, p_3, p_4)^T$. In this case, the likelihood function of $\boldsymbol{\nu}$ is

$$\begin{aligned} \ell(\theta) &= \frac{n!}{\nu_1!\nu_2!\nu_3!\nu_4!} p_1^{\nu_1}(\theta) p_2^{\nu_2}(\theta) p_3^{\nu_3}(\theta) p_4^{\nu_4}(\theta) \\ &= \frac{n!4^{-n}}{\nu_1!\nu_2!\nu_3!\nu_4!} (2+\theta)^{\nu_1}(1-\theta)^{\nu_2+\nu_3}\theta^{\nu_4}, \end{aligned}$$

which implies that the statistic $T = (\nu_1, \nu_2+\nu_3, \nu_4)^T$ is sufficient for θ. Since $\nu_1+\nu_2+\nu_3+\nu_4 = n$, $U = (\nu_1, \nu_4)^T$ is a minimal sufficient statistic for θ. To estimate θ we use the maximum likelihood method. We have

$$\ln \ell(\theta) = \ln \frac{n!4^{-n}}{\nu_1!\nu_2!\nu_3!\nu_4!} \nu_1 \ln(2+\theta) + (\nu_2+\nu_3)\ln(1-\theta) + \nu_3 \ln \theta,$$

and hence the informant $\lambda(\theta)$ is

$$\lambda(\theta) = \frac{d}{d\theta} \ln \ell(\theta) = \frac{\nu_1}{2+\theta} - \frac{\nu_1+\nu_2}{1-\theta} + \frac{\nu_4}{\theta}.$$

To find the multinomial maximum-likelihood estimator $\tilde{\theta}_n$, which is asymptotically efficient, we need to resolve the liklihood equation $\lambda(\theta) = 0,$, that is, to find the solution of the equation

$$\frac{\nu_1}{2+\theta} - \frac{\nu_1+\nu_2}{1-\theta} + \frac{\nu_4}{\theta} = 0,$$

which reduces to

$$2\nu_4 + (3\nu_1 + \nu_4 - 2n)\theta - n\theta^2 = 0, \qquad 0 < \theta < 1.$$

The positive root is

$$\tilde{\theta}_n = \frac{3\nu_1 + \nu_4 - 2n + \sqrt{(3\nu_1 + \nu_4 - 2n)^2 + 8n\nu_4}}{2n}.$$

Now we construct a chi-squared test for H_0 . Since $r = 4$, $s = 1$,

$$\lim_{n\to\infty} \boldsymbol{P}\{X_n^2(\tilde{\theta}_n) \geq x | H_0\} = \boldsymbol{P}\{\chi_2^2 \geq x\}.$$

We reject H_0 if

$$X_n^2(\tilde{\theta}_n) \geq \chi^2_{2,1-\alpha},$$

where $\chi^2_{2,1-\alpha}$ is the upper α quantile of the chi-squared distribution with two degrees of freedom. Suppose that with $n = 290$ we obtain the frequency vector

$$\nu_1 = 187, \qquad \nu_2 = 35, \qquad \nu_3 = 37, \qquad \nu_4 = 31,$$

Then $\tilde{\theta}_n = 0.4835$ and Pearson's statistic is

$$X_n^2(\tilde{\theta}_n) = \sum_{i=1}^{4} \frac{[\nu_i - np_i(\tilde{\theta}_n)]^2}{np_i(\tilde{\theta}_n)} = 0.91.$$

If we take significance level $\alpha = 0.1$, then $\chi^2_{2,0.9} = 4.605$. Since

$$X_n^2(\tilde{\theta}_n) = 0.91 < \chi^2_{2,0.9} = 4.605,$$

we accept H_0 .

Example: A Markov Chain. In this example from Huber and Nikulin (1991; see also Nikulin and Huber, 1993), we construct a Pearson-type statistic using conditional distributions. Let $X_1, X_2, \ldots, X_n$ be observations from a Markov chain taking values 0 and 1. We wish to test the hypothesis H_0 that the X_i are independent with some law

$\boldsymbol{P}\{X_i = 1\} = \boldsymbol{p}$, $\boldsymbol{P}\{X_i = 0\} = q = 1-p$, $0 < p < 1$. In any case, the X_i are observations from a Markov chain with some transition matrix

$$\boldsymbol{P} = \begin{pmatrix} p_{00} & p_{01} \\ p_{10} & p_{11} \end{pmatrix}, p_{01}+p_{00} = p_{10}+p_{11} = 1. \tag{9.62}$$

The conditional distribution of X_{r+1} given X_r is

$$\boldsymbol{P}\{X_{r+1} = j | X_r = i\} = p_{ij}, \qquad i,j \in \{0,1\}, \qquad r = 1,\ldots,n-1.$$

We regard H_0 as the independent case in which the transition probabilities do not depend on i, that is the transition matrix is

$$\boldsymbol{P} = \boldsymbol{P}_0 = \begin{pmatrix} q & p \\ q & p \end{pmatrix}, \qquad p+q = 1. \tag{9.63}$$

Let ν_0 be the number of 0s in the sequence $X_1,\ldots,X_{n-1}$ and $\nu_1 = n-1-\nu_0$ the number of 1s. Consider the pairs (X_r, X_{r+1}), $r = 1,\ldots,n-1$. Let ν_{ij} denote the number of these pairs which are (i,j), $i,j \in \{0,1\}$.

A Pearson statistic for testing H_0 can be based on the transitions to 1, that is, those (X_r, X_{r+1}) with $X_{r+1} = 1$. Under H_0, with ν_0 and ν_1 fixed, ν_{01} and ν_{11} are binomial with means $\nu_0 p$ and $\nu_1 p$. We construct a vector of deviations $(\nu_{01} - \nu_0 p, \nu_{11} - \nu_1 p)$ and the X^2 random variable

$$X^2 = (\nu_{01} - \nu_0 p)^2/\nu_0 pq + (\nu_{11} - \nu_1 p)^2/\nu_1 pq. \tag{9.64}$$

If the values of ν_0 and ν_1 go to infinity with n, then, under H_0,

$$\lim P\{X^2 > x | \nu_{01}\nu_{11}H_0\} = P\{\chi_2^2 > x\}.$$

The conditioning on the values of ν_0, ν_1 can be removed since they go almost surely to ∞, that is,

$$\lim P\{X^2 > x | H_0\} = P\{\chi_2^2 > x\}.$$

With the maximum-likelihood estimator $\hat{p} = \nu_1/(n-1)$ in place of the unknown p in Eq. (9.64) we have the statistic

$$
\begin{aligned}
X^2 &= \frac{(n-1)^2}{\nu_1[(n-1)-\nu_1]}\left\{\frac{\left[\nu_{01}-\frac{\nu_1(n-1+\nu_1)}{n-1}\right]^2}{n-1-\nu_1}+\frac{\left[\nu_{11}-\frac{\nu_1^2}{n-1}\right]^2}{\nu_1}\right\} \\
&= \frac{1}{\nu_1(n-1-\nu_1)}\left\{\frac{[(n-1)\nu_{01}-\nu_1(n-1-\nu_1)]^2}{n-1-\nu_1}+\frac{[(n-1)\nu_{11}-\nu_1^2]^2}{\nu_1}\right\} \\
&= \frac{1}{\nu_1^2(n-1-\nu_1)^2}\{(n-1)^2\nu_{01}^2\nu_1-2(n-1)(n-1-\nu_1)\nu_1^2\nu_{01} \\
&\qquad +\nu_1^3(n-1-\nu_1)^2+(n-1)^2\nu_{11}^2(n-1-\nu_1) \\
&\qquad -2(n-1)\nu_{11}\nu_1^2(n-1-\nu_1)+\nu_1^4(n-1-\nu_1)\},
\end{aligned}
\tag{9.65}
$$

which when $n \to \infty$ has limit law χ_1^2 by Fisher's Theorem 9.4. Using the law of large numbers we can reduce expression (9.65) to

$$X^2 \approx \frac{n^3}{\nu_1^2}\left[\frac{\nu_{01}}{\nu_0}-\frac{\nu_1}{n}\right]^2. \tag{9.66}$$

This statistic can be used to construct a test of the hypothesis H_0 that the observed Markov chain is in fact an independent sequence.

10. ESTIMATING AND THEN TESTING: DISTRIBUTION OF TEST STATISTIC DEPENDS ON METHOD OF ESTIMATION

Using Ungrouped Data

We continue to consider testing of a composite hypothesis, H_0, for which the distribution of our data is in the class $F(x;\boldsymbol{\theta})$, $\boldsymbol{\theta} \in \Theta \subset \boldsymbol{R}^s$. We choose grouping intervals and define the vector of probabilities $\boldsymbol{p}(\theta)$ and the vector of frequencies $\boldsymbol{\nu}$ as in Section 9. Again, we assume that Cramer's conditions 1–4 hold. In Section 9, we constructed estimators of $\boldsymbol{\theta}$ in various ways using the frequency vector $\boldsymbol{\nu}$. But $\boldsymbol{\nu}$ is not a sufficient statistic for $\boldsymbol{\theta}$, that is, some of the information about $\boldsymbol{\theta}$ in the original data is lost by grouping. In this paragraph, we look at what happens if we use for an estimator of $\boldsymbol{\theta}$ the maximum-likelihood estimator based on the ungrouped data, $X_1, \ldots, X_n$, which we denote here by $\hat{\boldsymbol{\theta}}_n$. Theorem 9.4 of Fisher tells us nothing about the behavior of Pearson's statistic when we use $\hat{\boldsymbol{\theta}}_n$ to estimate $\boldsymbol{\theta}$. The limiting distribution of the statistic

$$X_n^2(\hat{\boldsymbol{\theta}}_n) = \sum_{i=1}^{r} \frac{(\nu_i - np_i(\hat{\boldsymbol{\theta}}_n))^2}{np_i(\hat{\boldsymbol{\theta}})}, \tag{10.1}$$

as $n \to \infty$, was found by Chernoff and Lehmann (1954), and some generalizations were obtained by Roy (1956), Watson (1957, 1958, 1959), Dahiya and Gurland (1970a,b, 1972a,b, 1973), Chibisov (1971), Moore (1971), Durbin (1973a), and Pollard (1979).

Regularity Conditions

The following assumptions and computations lead up to the Chernoff and Lehmann result (Theorem 10.1), which the reader might wish to look at immediately.

Let us suppose, in addition to Cramer's conditions 1–4, these conditions:

5. The distribution function $F(x;\boldsymbol{\theta})$ has a density

$$f(x;\boldsymbol{\theta}) = \frac{d}{dx} F(x;\boldsymbol{\theta}), \quad |x| < \infty, \quad \boldsymbol{\theta} \in \boldsymbol{\Theta},$$

such that the derivatives

$$\frac{\partial^2 f(x;\boldsymbol{\theta})}{\partial\theta_i \partial\theta_j}, \quad i = 1, j, \ldots, s,$$

exist and are continuous,

6. The *Fisher information matrix*

$$\boldsymbol{i}(\boldsymbol{\theta}) = \boldsymbol{E}_{\boldsymbol{\theta}} \boldsymbol{\Lambda}_j \boldsymbol{\Lambda}_j^T, \quad j = 1, \ldots, n, \tag{10.2}$$

corresponding to any one observation X_j is finite and positive definite for each $\boldsymbol{\theta} \in \boldsymbol{\Theta}$, where

$$\boldsymbol{\Lambda}_j = \textbf{grad} \ \log f(X_j;\boldsymbol{\theta}) = (\Lambda_{j1}, \ldots, \Lambda_{js})^T \tag{10.3}$$

is *the vector-informant*, corresponding to the observation X_j,

$$\Lambda_{ji} = \Lambda_{ji}(\boldsymbol{\theta}) = \frac{\partial \log f(X_j, \boldsymbol{\theta})}{\partial \theta_i}, \quad i = 1, \ldots, s. \tag{10.4}$$

7. Differentiation with respect to parameters under the integral sign in

$$\int_{-\infty}^{\infty} f(x; \boldsymbol{\theta})\, dx = 1 \tag{10.5}$$

is permissible, that is,

$$\frac{\partial}{\partial \theta_i} \int_{-\infty}^{\infty} f(x; \boldsymbol{\theta})\, dx = \int_{-\infty}^{\infty} \frac{\partial}{\partial \theta_i} f(x; \boldsymbol{\theta})\, dx = 0. \tag{10.6}$$

Maximum Likelihood, Sufficiency, and Asymptotic Efficiency

Let us denote by

$$L(\boldsymbol{\theta}) = \prod_{i=1}^{n} f(x_i; \boldsymbol{\theta}) \tag{10.7}$$

the likelihood function of the sample $X_1, \ldots, X_n$, and

$$\boldsymbol{\Lambda}(\boldsymbol{\theta}) = \textbf{grad} \ \log L(\boldsymbol{\theta}) = \sum_{j=1}^{n} \textbf{grad} \ \log f(X_j; \boldsymbol{\theta}) = \sum_{j=1}^{n} \boldsymbol{\Lambda}_j(\boldsymbol{\theta}), \tag{10.8}$$

where $\boldsymbol{\Lambda}_j(\boldsymbol{\theta})$ is defined by Eq. (10.3). The vector $\boldsymbol{\Lambda}(\boldsymbol{\theta})$ is called the *vector-informant* of the sample $X_1, \ldots, X_n$. From the definitions of $\boldsymbol{\Lambda}(\boldsymbol{\theta})$ and $\boldsymbol{I}(\theta)$, and because $X_1, \ldots, X_n$ are i.i.d. we have

$$\boldsymbol{E}_{\boldsymbol{\theta}} \boldsymbol{\Lambda}(\boldsymbol{\theta}) \boldsymbol{\Lambda}^T(\boldsymbol{\theta}) = n\boldsymbol{I}(\boldsymbol{\theta}). \tag{10.9}$$

The matrix $n\boldsymbol{I}(\boldsymbol{\theta})$ is called *the Fisher information matrix* of the sample $X_1, X_2, \ldots, X_n$. From Eqs. (10.4) and (10.6) and condition 7 it follows that

$$\boldsymbol{E}_{\boldsymbol{\theta}} \boldsymbol{\Lambda}_j = \mathbf{0}_s, \qquad \textbf{Var}_{\boldsymbol{\theta}} \boldsymbol{\Lambda}_j = \boldsymbol{I}(\boldsymbol{\theta}), \tag{10.10}$$

$$\boldsymbol{E}_{\boldsymbol{\theta}} \boldsymbol{\Lambda} = \mathbf{0}_s, \qquad \textbf{Var}_{\boldsymbol{\theta}} \boldsymbol{\Lambda} = n\boldsymbol{I}(\boldsymbol{\theta}). \tag{10.11}$$

Let us denote by $\{\hat{\boldsymbol{\theta}}_n\}$ a sequence of maximum-likelihood estimators of the parameter $\boldsymbol{\theta}$. For each n,

$$\boldsymbol{\Lambda}(\hat{\boldsymbol{\theta}}_n) = \mathbf{0}_s, \tag{10.12}$$

where $\boldsymbol{\Lambda}(\boldsymbol{\theta})$ is defined by Eq. (10.8).

That this sequence of estimators is asymptotically efficient is well-known. We will not go into the formal definition of asymptotic efficiency but use as a working definition relation (10.17), which is also commonly termed BAN, standing for "best asymptotically normal." If T is a sufficient statistic, then $\hat{\theta}_n = \hat{\theta}_n(T)$ (see, e.g., Ibragimov and Hasminskii, 1981).

Example. The Normal Family. Let $X_1, \ldots, X_n$ be independent $N_1(\mu, \sigma^2)$-distributed random variables,

$$\boldsymbol{E}X_1 = \mu, \qquad \mathbf{Var}\ X_1 = \sigma^2.$$

Then

$$\boldsymbol{\theta} = (\mu, \sigma^2)^T, \quad \boldsymbol{\theta} \in \boldsymbol{\Theta} = \{(\mu, \sigma^2)^T : |\mu| < \infty, \sigma^2 > 0\},$$

and the likelihood function $L(\boldsymbol{\theta})$ has the form

$$L(\boldsymbol{\theta}) = \frac{1}{(2\pi)^{n/2}\sigma^n} \exp\left\{ -\frac{1}{2\sigma^2} \sum_{i=1}^{n} (X_i - \mu)^2 \right\}.$$

It is easy to verify that

$$\hat{\boldsymbol{\theta}}_n = (\bar{X}_n, s_n^2)^T,$$

where

$$\bar{X}_n = \frac{1}{n} \sum_{i=1}^{n} X_i, \quad s_n^2 = \frac{1}{n} \sum_{i=1}^{n} (X_i - \bar{X}_n)^2. \tag{10.13}$$

A minimal sufficient statistic is

$$\boldsymbol{T} = \left(\sum_{i=1}^{n} X_i, \sum_{i=1}^{n} X_i^2 \right)^T, \tag{10.14}$$

and from Eq. (10.13), one can easily see that $\hat{\boldsymbol{\theta}}_n$ is a function of $\boldsymbol{T}$, $\hat{\boldsymbol{\theta}}_n = \hat{\boldsymbol{\theta}}_n(\boldsymbol{T})$ (see, e.g., Rao, 1973).

Chernoff and Lehmann's Theorem

Now we begin the climb toward the theorem of Chernoff and Lehmann. Under conditions 5–7, the maximum-likelihood estimator $\hat{\boldsymbol{\theta}}_n$, calculated relative to the ungrouped observations $X_1, \ldots, X_n$, is a solution of the *likelihood equation*

$$\Lambda(\boldsymbol{\theta}) = \mathbf{0}_s \tag{10.15}$$

and as $n \to \infty$ the sequence $\{\hat{\boldsymbol{\theta}}_n\}$ satisfies the asymptotic relation

$$\sqrt{n}(\hat{\boldsymbol{\theta}}_n - \boldsymbol{\theta}_0) = \frac{1}{\sqrt{n}} \boldsymbol{I}^{-1}(\boldsymbol{\theta}_0)\boldsymbol{\Lambda}(\boldsymbol{\theta}_0) + o_{\boldsymbol{P}_0}(\mathbf{1}_s), \tag{10.16}$$

where $\boldsymbol{\theta}_0$ is the true value of the parameter $\boldsymbol{\theta}$. One can verify this as was done for Eq. (1.24) (see, e.g., Ibragimov and Hasminskii, 1981). Moreover, from Eqs. (10.16) and (10.8) under conditions 5–7, by the central limit theorem,

$$\sqrt{n}(\hat{\boldsymbol{\theta}}_n - \boldsymbol{\theta}_0) \quad \text{is} \quad AN(\mathbf{0}_s, \boldsymbol{I}^{-1}(\boldsymbol{\theta}_0)), \tag{10.17}$$

so $\hat{\boldsymbol{\theta}}_n$ is BAN for $\boldsymbol{\theta}$ (see Section 8). In fact we can improve any $\sqrt{n}$-consistent estimator sequence to a sequence having this property (see Section 15).

Furthermore, because for any $i = 1, \ldots, r$

$$\nu_i = \sum_{j=1}^{n} \mu_{ij}, \qquad \boldsymbol{\nu} = \sum_{i=1}^{n} \mu_i, \tag{10.18}$$

where $\mu_i = (\mu_{i1}, \ldots, \mu_{ir})^T$,

$$\mu_{ij} = \begin{cases} 1, & \text{if } X_j \in [a_{i-1}, a_i), \qquad j = 1, \ldots, n, \\ 0, & \text{otherwise,} \end{cases}$$

from condition 7 and formulas (10.4), (10.6), and (10.8), we have

$$\boldsymbol{E}_{\boldsymbol{\theta}}\left\{\nu_i \frac{\partial \log L(\boldsymbol{\theta})}{\partial \theta_j}\right\} = n\boldsymbol{E}_{\boldsymbol{\theta}}\{\mu_{i1}\Lambda_{1j}\} = n \int_{a_{i-1}}^{a_i} \frac{\partial f(x;\boldsymbol{\theta})}{\partial \theta_j}\, dx, \tag{10.19}$$

and hence we obtain from Eqs. (10.2, (10.6), (10.7), and (10.19) that

$$\boldsymbol{E}_{\boldsymbol{\theta}}\{\boldsymbol{X}_n(\boldsymbol{\theta})\boldsymbol{\Lambda}^T(\boldsymbol{\theta})\} = \sqrt{n}\boldsymbol{B} \tag{10.20}$$

where, as defined in Eq. (9.3),

$$\boldsymbol{X}_n(\boldsymbol{\theta}) = \left(\frac{\nu_1 - np_1(\boldsymbol{\theta})}{\sqrt{np_1(\boldsymbol{\theta})}}, \ldots, \frac{\nu_r - np_r(\boldsymbol{\theta})}{\sqrt{np_r(\boldsymbol{\theta})}}\right)^T = \frac{1}{\sqrt{n}}\boldsymbol{P}^{-1/2}(\boldsymbol{\nu} - n\boldsymbol{p}), \tag{10.21}$$

$\boldsymbol{P} = \boldsymbol{P}(\boldsymbol{\theta})$ is the diagonal matrix with the elements $p_1(\boldsymbol{\theta}), p_2(\boldsymbol{\theta}), \ldots, p_r(\boldsymbol{\theta})$ on the main diagonal,

$$\boldsymbol{B} = \boldsymbol{B}(\boldsymbol{\theta}) = [b_{ij}]_{r\times s}, \qquad b_{ij} = \frac{1}{\sqrt{p_i(\boldsymbol{\theta})}}\int_{a_{i-1}}^{a_i} \frac{\partial f(x; (\boldsymbol{\theta})}{\partial \theta_j}\, dx. \tag{10.22}$$

As we know (see Section 9) $\boldsymbol{B}^T\boldsymbol{B} = \boldsymbol{J}(\boldsymbol{\theta})$ is Fisher's information matrix for the multinomial distribution with paramcters n and $\boldsymbol{p}$. From the theorem of Fisher we have

$$\mathbf{Var}_{\boldsymbol{\theta}}\boldsymbol{X}_n(\boldsymbol{\theta}) = \boldsymbol{I}_r - \boldsymbol{q}\boldsymbol{q}^T, \tag{10.23}$$

where $\boldsymbol{q} = \boldsymbol{q}(\boldsymbol{\theta}) = \boldsymbol{P}^{1/2}\mathbf{1}_r$, and because

$$\boldsymbol{B}^T\boldsymbol{P}^{1/2}\mathbf{1}_r = \boldsymbol{B}^T\boldsymbol{q} = \mathbf{0}_s \tag{10.24}$$

we have from Eq. (10.20)

$$\boldsymbol{E}_{\boldsymbol{\theta}}\left\{\left[\frac{1}{\sqrt{n}}\boldsymbol{\Lambda} - \boldsymbol{B}^T\boldsymbol{X}_n(\boldsymbol{\theta})\right]\left[\frac{1}{\sqrt{n}}\boldsymbol{\Lambda} - \boldsymbol{B}^T\boldsymbol{X}_n(\boldsymbol{\theta})\right]^T\right\} = \boldsymbol{I}(\boldsymbol{\theta}) - \boldsymbol{J}(\boldsymbol{\theta}), \tag{10.25}$$

and hence one can see from Eq. (10.25) that the matrix $\boldsymbol{I}(\boldsymbol{\theta}) - \boldsymbol{J}(\boldsymbol{\theta})$ is nonnegative (positive semidefinite). But because the statistic $\boldsymbol{\nu} = (\nu_1, \ldots, \nu_r)^T$ is not sufficient, from Eq. (10.25) we obtain that $\boldsymbol{I}(\boldsymbol{\theta}) - \boldsymbol{J}(\boldsymbol{\theta})$ is positive definite, rank $[\boldsymbol{I}(\boldsymbol{\theta}) - \boldsymbol{J}(\boldsymbol{\theta})]{=}s$, and all s eigenvalues, $\lambda_1, \ldots, \lambda_s$, of $\boldsymbol{I}(\boldsymbol{\theta}) - \boldsymbol{J}(\boldsymbol{\theta})$ are positive. Comparing Eq. (9.24) and Eq. (10.16), we see that $\hat{\boldsymbol{\theta}}_n$ is more concentrated about the true value, $\boldsymbol{\theta}_0$, of the parameter $\boldsymbol{\theta}$, than is $\tilde{\boldsymbol{\theta}}_n$.

From Eqs.(10.16) to (10.25), it follows that under H_0 as $n \to \infty$ the joint limit distribution of

$$\begin{pmatrix} \boldsymbol{X}_n(\boldsymbol{\theta}_0) \\ \sqrt{n}(\hat{\boldsymbol{\theta}}_n - \boldsymbol{\theta}_0) \end{pmatrix} \text{ is } AN\Big(\mathbf{0}_{r+s}, \boldsymbol{G}(\boldsymbol{\theta}_0)\Big), \tag{10.26}$$

where

$$\boldsymbol{G}(\boldsymbol{\theta}) = \left[\begin{array}{c:c} \boldsymbol{I}_r - \boldsymbol{q}(\boldsymbol{\theta})\boldsymbol{q}^T(\boldsymbol{\theta}) & \boldsymbol{B}(\boldsymbol{\theta})\boldsymbol{I}^{-1}(\boldsymbol{\theta}) \\ \hdashline \boldsymbol{I}^{-1}(\boldsymbol{\theta})\boldsymbol{B}^T(\boldsymbol{\theta}) & \boldsymbol{I}^{-1}(\boldsymbol{\theta}) \end{array}\right]_{(r+s)(r+s)} \tag{10.27}$$

In the preceding paragraph we saw that under our regularity conditions for any $\sqrt{n}$-consistent sequence $\{\boldsymbol{\theta}_n^*\}$ of estimators $\boldsymbol{\theta}_n^*$ for $\boldsymbol{\theta}_0$ we have

$$\boldsymbol{X}_n(\boldsymbol{\theta}_n^*) = \boldsymbol{X}_n(\boldsymbol{\theta}_0) - \sqrt{n}\boldsymbol{B}(\boldsymbol{\theta}_0)(\boldsymbol{\theta}_n^* - \boldsymbol{\theta}_0) + o_{\boldsymbol{P}_0}(\mathbf{1}_r), \tag{10.28}$$

and hence using Eqs. (10.16), (10.17), and (10.23) to (10.26), we obtain that under H_0 as $n \to \infty$

$$\boldsymbol{X}_n(\hat{\boldsymbol{\theta}}_n) \text{ is } AN(\mathbf{0}_r, \boldsymbol{W}(\boldsymbol{\theta}_0)), \tag{10.29}$$

where

$$\boldsymbol{W}(\boldsymbol{\theta}_0) = \boldsymbol{I}_r - \boldsymbol{q}\boldsymbol{q}^T - \boldsymbol{B}\boldsymbol{I}^{-1}\boldsymbol{B}^T, \tag{10.30}$$

$\boldsymbol{B} = \boldsymbol{B}(\boldsymbol{\theta})$, $\boldsymbol{I} = \boldsymbol{I}(\boldsymbol{\theta})$, $\boldsymbol{q} = \boldsymbol{q}(\boldsymbol{\theta})$ are defined by Eqs. (10.10), (10.22), and (10.23). Rank $W(\theta) = r-1$, because the matrix $\boldsymbol{I}(\boldsymbol{\theta}) - \boldsymbol{J}(\boldsymbol{\theta})$ is positive definite. Furthermore, $\boldsymbol{q}\boldsymbol{q}^T$ is a projection of rank 1, orthogonal to $\boldsymbol{B}$ and hence to $\boldsymbol{B}\boldsymbol{I}^{-1}\boldsymbol{B}^T$, where rank $\boldsymbol{B}\boldsymbol{I}^{-1}\boldsymbol{B}^T = s$. It follows that $r-s-1$ of the characteristic roots of $\boldsymbol{W}$ are equal to 1, one is equal to 0, and the other s positive characteristic roots $\lambda_1, \lambda_2, \ldots, \lambda_s$ are the roots of the equation

$$|(1-\lambda)\boldsymbol{I}(\boldsymbol{\theta}_0) - \boldsymbol{J}(\boldsymbol{\theta}_0)| = 0. \tag{10.31}$$

The λ_i, in general, depend on $\boldsymbol{\theta}_0$. Because $\boldsymbol{I}(\boldsymbol{\theta}_0) - \boldsymbol{J}(\boldsymbol{\theta}_0)$ is positive definite, $\boldsymbol{I}(\boldsymbol{\theta}_0) > \boldsymbol{J}(\boldsymbol{\theta}_0)$, and because $(1-\lambda)\boldsymbol{I}(\boldsymbol{\theta}_0) - \boldsymbol{J}(\boldsymbol{\theta}_0)$ is negative definite for $\lambda \geq 1$, we obtain that

$$0 < \lambda_i(\boldsymbol{\theta}_0) < 1, \qquad i = 1, 2, \ldots, s. \tag{10.32}$$

From Eqs. (10.17) and (10.26) to (10.30), it follows that under H_0 as $n \to \infty$ the joint limit distribution of the vector

$$\begin{pmatrix} \boldsymbol{X}_n(\hat{\boldsymbol{\theta}}_n) \\ \sqrt{n}(\hat{\boldsymbol{\theta}}_n - \boldsymbol{\theta}_0) \end{pmatrix} \text{ is } AN\Big(\mathbf{0}_{r+s}, \Gamma(\boldsymbol{\theta}_0)\Big), \tag{10.33}$$

where

$$\Gamma(\boldsymbol{\theta}) = \left[\begin{array}{c|c} \boldsymbol{W}(\boldsymbol{\theta}) & \mathbf{0}_{r\times s} \\ \hline \mathbf{0}_{s\times r} & \boldsymbol{I}^{-1}(\boldsymbol{\theta}) \end{array}\right]. \tag{10.34}$$

From Eq. (10.34), we obtain that under H_0, $\boldsymbol{X}_n(\hat{\boldsymbol{\theta}}_n)$ and $\sqrt{n}(\hat{\boldsymbol{\theta}}_n - \boldsymbol{\theta}_0)$ are asymptotically independent, and hence that the statistic of Pearson

$$X_n^2(\hat{\boldsymbol{\theta}}_n) = \boldsymbol{X}_n^T(\hat{\boldsymbol{\theta}}_n)\boldsymbol{X}_n(\hat{\boldsymbol{\theta}}_n)$$

is also asymptotically independent of $\sqrt{n}(\hat{\boldsymbol{\theta}}_n - \boldsymbol{\theta}_0)$ as $n \to \infty$.

From relations (10.29) to (10.34) and Lemma 2.1 we have the principal statement of this paragraph:

Theorem 10.1 (Chernoff and Lehmann, 1954). *If the conditions 1–7 hold, then*

$$\lim_{n\to\infty} \boldsymbol{P}\{X_n^2(\hat{\boldsymbol{\theta}}_n) < x|H_0\} = \boldsymbol{P}\{\chi_{r-s-1}^2 + \sum_{j=1}^{s} \lambda_j \xi_j^2 < x\} \tag{10.35}$$

where χ_{r-s-1}^2, $\xi_1, \ldots, \xi_s$ are independent, $\xi_i \sim N_1(0,1)$, and $\lambda_i = \lambda_i(\boldsymbol{\theta}_0)$ are roots of Eq. (10.31).

Example. Pearson's Test for Normality. Let $X_1, \ldots, X_n$ be a sample, and suppose that we want to test the hypothesis of normality, H_0:

$$\boldsymbol{P}\{X_i \le x|H_0\} = \Phi(\frac{x-\mu}{\sigma}),$$

where the parameters μ and σ^2 are unknown, $|\mu| < \infty$, $\sigma^2 > 0$.

Under H_0, as in the previous example, the statistic $\theta_n^2 = (\bar{X}_n, s_n^2)^T$ is the maximum likelihood estimator for $\theta = (\mu, \sigma^2)^T$, where

$$\bar{X}_n = \frac{1}{n}\sum_{i=1}^{n} X_i, \qquad s_n^2 = \frac{1}{n}\sum_{i=1}^{n}(X_i - \bar{X}_n)^2.$$

If for testing H_0 we choose Pearson's test based on the statistic (10.1),

$$X_n^2(\tilde{\theta}_n) = \sum_{i=1}^{r} \frac{[\nu_i - np_i(\hat{\theta}_n)]^2}{np_i},$$

then from the Chernoff-Lehmann theorem it follows that

$$\lim_{n\to\infty} \boldsymbol{P}\{X_n^2(\hat{\theta}_n) \le x | H_0\} = \boldsymbol{P}\{\chi_{r-3}^2 + \lambda_1\xi_1^2 + \lambda_2\xi_2 \le x\}.$$

Example. Pearson's Test for the Poisson Distribution. Let $X_1, \ldots, X_n$ be a sample, and suppose that we want to test the hypothesis H_0:

$$\boldsymbol{P}\{X_i = x | H_0\} = \frac{\theta^x}{x!}e^{-\theta},$$

where the parameter θ is unknown, $\theta > 0$. Under H_0 the statistic $\sum_{i=1}^n X_i$ is sufficient for θ, and $\hat{\theta}_n = 1/n \sum_{i=1}^n X_i$ is the maximum likelihood estimator for θ. From the Chernoff-Lehmann Theorem,

$$\lim_{n\to\infty} \boldsymbol{P}\{X_n^2(\hat{\theta}_n) \le x | H_0\} = \boldsymbol{P}\{\chi_{r-2}^2 + \lambda_1\xi_1^2 \le x\}.$$

where $X_n^2(\hat{\theta}_n)$ is the Pearson statistic for testing H_0.

The theorem of Chernoff and Lehmann reveals that the standard Pearson statistic with θ estimated by $\hat{\boldsymbol{\theta}}_n$ is undesirable because its limit law is complicated and depends, in general, on $\boldsymbol{\theta}_0$. Any efficient estimator sequence produces the same problem. For scale and shift families, the limit does not depend on θ_0 (see Sections 11 and 13).

An extension of the Chernoff and Lehmann theorem to a more general setting was found by LeCam, Mahan, and Singh (1983).

Remark 10.1. Consider the Hellinger statistic (see Section 7),

$$F_n^2(\hat{\boldsymbol{\theta}}_n) = 4\sum_{i=1}^{r}\left(\sqrt{\nu_i} - \sqrt{np_i(\hat{\boldsymbol{\theta}}_n)}\right)^2.$$

Because $F_n^2(\hat{\boldsymbol{\theta}}_n)$ and $X_n^2(\hat{\boldsymbol{\theta}}_n)$ have the same asymptotic properties, we obtain

$$\lim_{n\to\infty} \boldsymbol{P}\{F_n^2(\hat{\boldsymbol{\theta}}_n) > x \mid H_0\} = \boldsymbol{P}\{\chi^2_{r-s-1} + \lambda_1\xi_1^2 + \cdots + \lambda_s\xi_s^2 > x\}.$$

Remark 10.2. As remarked by Chibisov (1971), the limiting distribution of $X_n^2(\hat{\boldsymbol{\theta}}_n)$ under contiguous alternatives also involves the λ_j's appearing in Eq. (10.35), which in general depend on the unknown $\boldsymbol{\theta}$. Hence, it is difficult to compute power. The results of Beran (1975) on tail probabilities for positive linear combinations of independent noncentral chi-squared random variables can be helpful here. About BAN estimates for chi-squared tests, see Rao (1973), Bemis and Bhapkar (1983), and Stroud (1971).

Remark 10.3. Let us suppose that conditions 1–7 are satisfied, so that for the minimum chi-squared estimator $\tilde{\boldsymbol{\theta}}_n$ and the maximum likelihood estimator $\hat{\boldsymbol{\theta}}_n$ we have asymptotic normality with limiting covariance matrices $\boldsymbol{J}^{-1}$ and $\boldsymbol{I}^{-1}$ for grouped and ungrouped data, that is, relations (9.24) and (10.16), respectively as $n \to \infty$. Let us suppose now that $r \to \infty$, as $n \to \infty$ so that the conditions of Tumanyan hold:

$$\max_{1\le i\le r} p_i \to 0 \qquad \text{and} \qquad \min_{1\le i\le r} np_i \to \infty \qquad \text{as } n \to \infty.$$

In this case $\boldsymbol{J}(\boldsymbol{\theta}) \to \boldsymbol{I}(\boldsymbol{\theta})$ as $n \to \infty$, the estimators $\tilde{\boldsymbol{\theta}}_n$ and $\hat{\boldsymbol{\theta}}_n$ are asymptotically equivalent, all $\lambda_j \to 0$ in Eq. (10.35), and hence the statistical tests based on $X_n^2(\tilde{\boldsymbol{\theta}}_n)$ and $X_n^2(\hat{\boldsymbol{\theta}}_n)$ are asymptotically equivalent as $r \to \infty$. This was observed by Watson (1958).

11. CHI-SQUARED TEST WITH RANDOM CELL BOUNDARIES: DISTRIBUTION OF TEST STATISTIC AGAIN DEPENDS ON METHOD OF ESTIMATION

Background

The standard statistic of Pearson

$$X_n^2 = \sum_{i=1}^{r} \frac{(\nu_i - np_i)^2}{np_i}$$

is defined in terms of the vector of frequencies $\boldsymbol{\nu} = (\nu_1, \ldots, \nu_r)^T$, where the boundaries, a_i of the grouping intervals, or cells,

$$-\infty = a_0 < a_1 < a_2 < \cdots < a_{r-1} < a_r = +\infty$$

are chosen independently of the observations $X_1, \ldots, X_n$. Theoretically, $a_1, \ldots, a_{r-1}$ are defined in advance of the experiment. In real problems, however, the boundaries are chosen, at least to some extent, in accordance with observed realization of the random variables $X_1, \ldots, X_n$. Some instances and reasons for this were discussed in Chapter 1 and further discussion will appear in this section. The fact that the $a_i = a_i(X_1, \ldots, X_n)$ are random can distort the limit distribution of the statistic X_n^2 or $X_n^2(\boldsymbol{\theta}_n^*)$, since the vector of frequencies $\boldsymbol{\nu}^* = (\nu_1^*, \ldots, \nu_r^*)^T$, based on $a_i(X_1, \ldots, X_n)$ no longer has a multinomial distribution. Early efforts to understand the effect of this phenomenon on the resulting inference were made by Roy (1956), Watson (1957, 1958, 1959), Bolshev (1969a), Dahiya and Gurland (1970a,b, 1972a,b), Chibisov (1971), and Moore (1971). The question was pursued further by Nikulin (1973a,b,c, 1974, 1979), Bofinger (1973), Rao and Robson (1974), Moore and Spruill (1975), Ruymgaart (1974, 1975), and others. In this section we present some of the more important results of Roy, Watson, Dahiya and Gurland, and Chibisov, which display this problem in the context of large-sample theory under a composite null hypothesis.

Let $X_1, \ldots, X_n$ be i.i.d. random variables, with distribution function $F_0(x) = \boldsymbol{P}\{X_1 < x\}$. We want to test the hypothesis H_0 according to which

$$F_0(x) = F(x;\boldsymbol{\theta}_0) \in \{F(x;\boldsymbol{\theta}), \boldsymbol{\theta} = (\theta_1, \ldots, \theta_s)^T \in \boldsymbol{\Theta} \subset \boldsymbol{R}^s\}.$$

Let $\boldsymbol{\theta}_0$ be an interior point of $\boldsymbol{\Theta}$. As in Chapter 1, suppose given $p_1, \ldots, p_r$, $p_i > 0$, $p_1 + \cdots + p_r = 1$, and let u_i denote the partial sums of p_i, $u_i = p_1 + \cdots + p_i$, $i = 1, \ldots, r$. Following the work of Chibisov (1971), we introduce a neighborhood V^δ of the set of quantile points of $F(\cdot;\boldsymbol{\theta})$ corresponding to u_i, $i = 1, \ldots, r$. For $\delta > 0$, let

$$U^\delta = \bigcup_{i=1}^{r} (u_i - \delta, u_i + \delta) = \bigcup_{i=1}^{r} U_i^\delta$$

$$V^\delta = F_0^{-1}(U^\delta) = \bigcup_{i=1}^{r} V_i^\delta.$$

We assume that, for some $\delta > 0$, the function $F_0(x)$ is continuous on V^δ. Let

$$\begin{aligned}
F^{-1}(u;\boldsymbol{\theta}) &= \inf\{x : F(x;\boldsymbol{\theta}) \geq u\}, && 0 < u < 1 \\
a_i(\boldsymbol{\theta}) &= F^{-1}(u_i;\boldsymbol{\theta}), && i = 1, \ldots, r, \\
p_i(\boldsymbol{\theta}', \boldsymbol{\theta}) &= \boldsymbol{P}_{\boldsymbol{\theta}}\{a_{i-1}(\boldsymbol{\theta}') \leq X_1 < a_i(\boldsymbol{\theta}')\} \\
&= \int_{a_{i-1}(\boldsymbol{\theta}')}^{a_i(\boldsymbol{\theta}')-} dF(x;\boldsymbol{\theta}).
\end{aligned} \tag{11.1}$$

As earlier, we take $a_0(\boldsymbol{\theta}) = -\infty$, $a_r(\boldsymbol{\theta}) = +\infty$ for all $\boldsymbol{\theta} \in \boldsymbol{\Theta}$, so that

$$F(a_0(\boldsymbol{\theta}');\boldsymbol{\theta}) = 0, \qquad F(a_r(\boldsymbol{\theta}');\boldsymbol{\theta}) = 1, \qquad \boldsymbol{\theta}', \boldsymbol{\theta} \in \boldsymbol{\Theta}.$$

We shall write

$$a_i = a_i(\boldsymbol{\theta}_0), \quad p_i(\boldsymbol{\theta}) = p_i(\boldsymbol{\theta}_0, \boldsymbol{\theta}),$$

and observe that $p_i(\boldsymbol{\theta}_0) = p_i$, $i = 1, \ldots, r$. Again $F_n(x)$ denotes the empirical distribution function,

$$F_n(x) = \frac{1}{n} \sum_{j=1}^{n} I(X_j < x).$$

Let

$$\nu_i = n \int_{a_{i-1}}^{a_i-} dF_n(x), \qquad \nu_i^* = n \int_{a_{i-1}(\boldsymbol{\theta}_n^*)}^{a_i(\boldsymbol{\theta}_n^*)-} dF_n(x), \qquad i = 1, \ldots, r.$$

The vector $\boldsymbol{\nu} = (\nu_1, \ldots, \nu_r)^T$ is the result of counting the values of $X_1, \ldots, X_n$ in each of the intervals

$$(a_0, a_1), [a_1, a_2), \ldots, [a_{r-1}, a_r),$$

when the class boundaries a_i are independent of the data. The vector $\boldsymbol{\nu}^* = (\nu_1^*, \ldots, \nu_r^*)^T$ is the result of counting the values of $X_1, \ldots, X_n$ in each of the intervals

$$(a_0, a_1(\boldsymbol{\theta}_n^*)), [a_1(\boldsymbol{\theta}_n^*), a_2(\boldsymbol{\theta}_n^*)), \ldots, [a_{r-1}(\boldsymbol{\theta}_n^*), a_r),$$

where $\boldsymbol{\theta}_n^*$ is an estimator of $\boldsymbol{\theta}_0$, computed from $X_1, \ldots, X_n$. The points $a_i(\boldsymbol{\theta}_n^*)$ are approximations for the a_i which would form intervals $[a_{i-1}, a_i)$ with probabilities p_i under the unknown true distribution $F(x; \boldsymbol{\theta}_0)$. The $a_i(\boldsymbol{\theta}_n^*)$ are constructed using $F(x; \boldsymbol{\theta}_n^*)$ in Eq. (11.1). We introduce the order statistics

$$X_{(1)} \leq X_{(2)} \leq \cdots \leq X_{(n)}, \tag{11.2}$$

constructed from the sample $X_1, \ldots, X_n$ and the notation

$$\begin{aligned} &Y_{ni} = X_{([nu_i]+1)}, \qquad u_i = p_1 + \ldots + p_i, \qquad i = 1, \ldots, r-1, \\ &Y_{nr} = -Y_0 = \infty. \end{aligned}$$

In Section 9 we defined the vector

$$\boldsymbol{X}_n(\boldsymbol{\theta}) = \left(\frac{\nu_1 - np_1(\boldsymbol{\theta})}{\sqrt{np_1(\boldsymbol{\theta})}}, \ldots, \frac{\nu_r - np_r(\boldsymbol{\theta})}{\sqrt{np_r(\boldsymbol{\theta})}} \right)^T = \frac{1}{\sqrt{n}} \boldsymbol{P}^{-1/2}(\boldsymbol{\theta})(\boldsymbol{\nu} - n\boldsymbol{p})^T, \tag{11.3}$$

where $\boldsymbol{p} = \boldsymbol{p}(\boldsymbol{\theta}) = (p_1(\boldsymbol{\theta}), \ldots, p_r(\boldsymbol{\theta}))^T$ and $\boldsymbol{P}(\boldsymbol{\theta})$ is the diagonal matrix with elements $p_1(\boldsymbol{\theta}), \ldots, p_r(\boldsymbol{\theta})$ on the main diagonal. The associated quadratic form of Pearson is

$$X_n^2(\boldsymbol{\theta}) = \boldsymbol{X}_n^T(\boldsymbol{\theta}) \boldsymbol{X}_n(\boldsymbol{\theta}) = \sum_{i=1}^{r} \frac{(\nu_i - np_i(\boldsymbol{\theta}))^2}{np_i(\boldsymbol{\theta})}. \tag{11.4}$$

Now, we introduce the analogous vectors

$$\boldsymbol{X}_n(\boldsymbol{\theta}_n^*, \boldsymbol{\theta}) = \left(\frac{(\nu_1^* - np_1(\boldsymbol{\theta}_n^*, \boldsymbol{\theta})}{\sqrt{np_1(\boldsymbol{\theta}_n^*, \boldsymbol{\theta})}}, \ldots, \frac{(\nu_r^* - np_r(\boldsymbol{\theta}_n^*, \boldsymbol{\theta})}{\sqrt{np_r(\boldsymbol{\theta}_n^*, \boldsymbol{\theta})}} \right)^T, \tag{11.5}$$

$$\tilde{\boldsymbol{X}}(\boldsymbol{\theta}_n^*, \boldsymbol{\theta}) = \left(\frac{(\nu_1^* - np_1(\boldsymbol{\theta}_n^*, \boldsymbol{\theta})}{\sqrt{np_1}}, \ldots, \frac{\nu_r^* - np_r(\boldsymbol{\theta}_n^*, \boldsymbol{\theta})}{\sqrt{np_r}} \right)^T, \tag{11.6}$$

and

$$\boldsymbol{Z}_n(\boldsymbol{\theta}) = \left(\sqrt{\frac{n}{p_1}} (p_1 - \Delta F(Y_{n-1}; \boldsymbol{\theta})), \ldots, \sqrt{\frac{n}{p_r}} (p_r - \Delta F(Y_{nr}; \boldsymbol{\theta})) \right)^T, \tag{11.7}$$

where

$$\Delta F(Y_{ni}; \boldsymbol{\theta}) = \int_{Y_{n,i-1}}^{Y_{ni}} dF(x; \boldsymbol{\theta}). \tag{11.8}$$

The quadratic forms associated with Eqs. (11.5) to (11.7), analogous to Eq. (11.4), are

$$X_n^2(\boldsymbol{\theta}_n^*) = \sum_{i=1}^{r} \frac{(\nu_i^* - np_i(\boldsymbol{\theta}_n^*, \boldsymbol{\theta}))^2}{np_i(\boldsymbol{\theta}_n^*, \boldsymbol{\theta})}, \tag{11.9}$$

$$\tilde{X}_n^2(\boldsymbol{\theta}_n^*) = \sum_{i=1}^{r} \frac{(\nu_i^* - np_i(\boldsymbol{\theta}_n^*, \boldsymbol{\theta}))^2}{np_i}, \tag{11.10}$$

$$Z_n^2(\boldsymbol{\theta}) = \sum_{i=1}^{r} \frac{n}{p_i} (p_i - \Delta F(Y_{ni}, \boldsymbol{\theta}))^2. \tag{11.11}$$

But definitions (11.3) through (11.11) depend on the unknown $\boldsymbol{\theta}$. We need to use an additional estimator $\bar{\boldsymbol{\theta}}_n$ for the true value of $\boldsymbol{\theta}$.

The statistics

$$X_n^2(\bar{\boldsymbol{\theta}}_n), \quad X_n^2(\boldsymbol{\theta}_n^*, \bar{\boldsymbol{\theta}}_n), \quad \tilde{X}_n^2(\boldsymbol{\theta}_n^*, \bar{\boldsymbol{\theta}}_n), \quad Z_n^2(\bar{\boldsymbol{\theta}}_n)$$

correspond to distinct statistical procedures. The statistic $X_n^2(\bar{\boldsymbol{\theta}}_n)$ is used when the cell boundaries a_i are fixed in advance and the frequency vector is $\boldsymbol{\nu}$. The probabilities p_i are unknown in this case, since F_0 is unknown. In the construction of $X_n^2(\boldsymbol{\theta}_n^*, \bar{\boldsymbol{\theta}}_n)$ and $\tilde{X}(\boldsymbol{\theta}_n^*, \boldsymbol{\theta}_n)$, the p_i are chosen before the experiment, but the cell boundaries are unknown and approximated by $a_i(\boldsymbol{\theta}_n^*)$.

Next we introduce regularity conditions under which all these statistics are asymptotically the same. Following Birch (1964) and Chibisov (1971), we suppose that there exists $\delta > 0$ such that for $x \in V^\delta$

$$F(x;\boldsymbol{\theta}) = F(x;\boldsymbol{\theta}_0)+(\boldsymbol{\theta}-\boldsymbol{\theta}_0)^T h(x)+r(x;\boldsymbol{\theta}) \tag{11.12}$$

where

$$\sup_{x\in V^\delta} |r(x;\boldsymbol{\theta})| = o(|\boldsymbol{\theta}-\boldsymbol{\theta}_0|) \quad \text{as} \quad |\boldsymbol{\theta}-\boldsymbol{\theta}_0| \to 0, \tag{11.13}$$

and $h(x) = \textbf{grad}\ F(x;\boldsymbol{\theta})|_{\boldsymbol{\theta}=\boldsymbol{\theta}_0}$, that is,

$$\begin{aligned} \boldsymbol{h}(x) &= (h_1(x),\ldots,h_s(x)), \\ h_i(x) &= \left.\frac{\partial F(x;\boldsymbol{\theta})}{\partial \boldsymbol{\theta}_i}\right|_{\boldsymbol{\theta}=\boldsymbol{\theta}_0} = g_i(F_0(x)), \end{aligned} \tag{11.14}$$

and we assume the $g_i(u)$ are bounded and continuous functions on U^δ.

Furthermore, as before, we suppose that the matrix $\boldsymbol{B} = [b_{ij}]$ of order $r \times s$ with elements

$$b_{ij} = \frac{1}{\sqrt{p_i(\boldsymbol{\theta})}} \frac{\partial p_i(\boldsymbol{\theta})}{\partial \theta_j}$$

has rank s.

Finally we assume that for any $\tau > 0$ there exist $\rho > 0$ and $\delta > 0$ such that, for any $x_i \in V_i^\delta$ and $|\boldsymbol{\theta}-\boldsymbol{\theta}_0| > \tau$,

$$\max_{1\le i\le r} |\Delta F(x_i;\boldsymbol{\theta})-p_i| > \rho. \tag{11.15}$$

Remark 11.1. In the case of fixed cell boundaries, only the conditions on $F(a_i;\boldsymbol{\theta})$ are required. In Eq. (11.12), we assume that

$$r(a_i,\boldsymbol{\theta}) = o(|\boldsymbol{\theta}-\boldsymbol{\theta}_0|) \quad \text{as} \quad |\boldsymbol{\theta}-\boldsymbol{\theta}_0| \to o$$

and $h_i(x)$ exist for $i = 1,\ldots,r-1$. Instead of Eq. (11.15), we assume in this case that for any $\tau > 0$ there exists $\rho > 0$ such that from $|\boldsymbol{\theta}-\boldsymbol{\theta}_0| > \tau$ it follows that

$$\max_{1\le i\le r} |\Delta F(a_i;\boldsymbol{\theta})-p_i| > \rho.$$

Now let $\hat{\boldsymbol{\theta}}_n$ be a sequence of $\sqrt{n}$-consistent estimators of $\boldsymbol{\theta}$, and let $\boldsymbol{\theta}_n^*$ be a sequence of consistent estimators of $\boldsymbol{\theta}$. Chibisov (1971) proved the following asymptotic equivalence.

Theorem 11.1. *If H_0 is true, then under the above regularity conditions, as $u \longrightarrow \infty$*

$$\left.\begin{aligned} X_n^2(\boldsymbol{\theta}_n^*, \hat{\boldsymbol{\theta}}_n) - X_n^2(\hat{\boldsymbol{\theta}}_n) &\to 0 \\ \tilde{X}_n^2(\boldsymbol{\theta}_n^*, \hat{\boldsymbol{\theta}}_n) - X_n^2(\hat{\boldsymbol{\theta}}_n) &\to 0 \\ Z_n^2(\hat{\boldsymbol{\theta}}_n) - X_n^2(\hat{\boldsymbol{\theta}}_n) &\to 0 \end{aligned}\right\} \text{ in } \boldsymbol{P}_0 \text{ probability} \tag{11.16}$$

Combining Theorem 11.1 of Chibisov with Theorem 10.1 of Chernoff and Lehmann we see that all the statistics introduced above have the same unsatisfactory limit distribution.

Now we give a result of Dahiya and Gurland (1970a,b) concerning Pearson's chi-squared test with random intervals, which is an extension of an unpublished result of Roy (1956) and Theorem 10.1 of Chernoff and Lehmann.

Following the papers of Dahiya and Gurland (1970a,b, 1972a,b) we suppose that $\boldsymbol{\theta}_n^*$ is an estimator of $\boldsymbol{\theta}_0$ such that

$$\sqrt{n}(\boldsymbol{\theta}_n^* - \boldsymbol{\theta}_0) = \frac{1}{n}\sum_{j=1}^{n} \boldsymbol{v}(X_j) + o_{\boldsymbol{P}_0}(\mathbf{1}_s) \tag{11.17}$$

where $\boldsymbol{v} = (v_1, \ldots, v_s)^T$ and

$$\boldsymbol{E}_{\boldsymbol{\theta}_0} \boldsymbol{v}(X_j) = \mathbf{0}_s, \qquad \mathbf{Var}_{\boldsymbol{\theta}_0} \boldsymbol{v}(X_j) = \boldsymbol{V}$$

is finite. For example, the maximum-likelihood estimator, $\hat{\boldsymbol{\theta}}_n$, and the minimum chi-squared estimator, $\tilde{\boldsymbol{\theta}}_n$, satisfy Eq. (11.17).

Roy's Extension of the Chernoff-Lehmann Theorem

Let us partition the real line into r intervals $J_i(\boldsymbol{\theta}_n^*)$ depending on $\boldsymbol{\theta}_n^*$ such that

$$J_i(\boldsymbol{\theta}_n^*) = \{x : g_{i-1}(\boldsymbol{\theta}_n^*) \le x < g_i(\boldsymbol{\theta}_n^*)\}, \qquad i = 1, \dots, r,$$

where $g_i(\boldsymbol{\theta})$ are continuous functions having partial derivatives. Let $\boldsymbol{\nu}^* = (\nu_1^*, \dots, \nu_r^*)^T$ be the vector of frequencies; that is, ν_i^* is the number of X_j's falling in $J_i(\boldsymbol{\theta}_n^*)$. As in Eq. (11.1), let

$$p_i(\boldsymbol{\theta}_n^*) = p_i(\boldsymbol{\theta}_n^*, \boldsymbol{\theta}_n^*) = F(g_i(\boldsymbol{\theta}_n^*); \boldsymbol{\theta}_n^*) - F(g_{i-1}(\boldsymbol{\theta}_n^*); \boldsymbol{\theta}_n^*). \tag{11.18}$$

The following theorem, analyzed by Roy (1956), gives the asymptotic distribution under H_0 of the statistic

$$X_R^2(\boldsymbol{\theta}_n^*) = X_n^2(\boldsymbol{\theta}_n^*, \boldsymbol{\theta}_n^*) = \sum_{i=1}^{r} \frac{(\nu_i^* - np_i(\boldsymbol{\theta}_n^*))^2}{np_i(\boldsymbol{\theta}_n^*)}. \tag{11.19}$$

Theorem 11.2. *If H_0 is true, then under the above regularity conditions the asymptotic distribution of $X_R^2(\boldsymbol{\theta}_n^*)$ as $n \to \infty$ is that of*

$$\lambda_1 \xi_1^2 + \lambda_2 \xi_2^2 + \cdots + \lambda_r \xi_r^2, \tag{11.20}$$

where $\xi_1, \xi_2, \dots, \xi_r$ are mutually independent standard normal, $N_1(0,1)$, random variables, $\lambda_1, \dots, \lambda_r$ are the characteristic roots of the matrix $\mathbf{P}^{-1}\boldsymbol{\Sigma}$, where $\mathbf{P} = \mathbf{P}(\boldsymbol{\theta}_0)$ is the diagonal matrix with the elements $p_1(\boldsymbol{\theta}_0), \dots, p_r(\boldsymbol{\theta}_0)$ on the main diagonal,

$$\boldsymbol{\Sigma} = \boldsymbol{\Sigma}(\boldsymbol{\theta}_0) = \mathbf{P} - \mathbf{p}\mathbf{p}^T - \mathbf{U}^T\mathbf{W} - \mathbf{W}^T\mathbf{U} + \mathbf{U}^T\mathbf{V}\mathbf{U}, \tag{11.21}$$

where

$$\mathbf{p} = \mathbf{p}(\boldsymbol{\theta}) = (p_1(\boldsymbol{\theta}), \dots, p_r(\boldsymbol{\theta}))^T,$$

$$\mathbf{U}^T = [U_{ij}]_{r\times s}, \quad U_{ij} = U_{ij}(\boldsymbol{\theta}) = \int_{g_{i-1}(\boldsymbol{\theta})}^{g_i(\boldsymbol{\theta})} \frac{\partial f(x;\boldsymbol{\theta})}{\partial \boldsymbol{\theta}}\, dx,$$

$$\mathbf{W}^T = [W_{ij}]_{r\times s}, \quad W_{ij} = W_{ij}(\boldsymbol{\theta}) = \int_{g_{i-1}(\boldsymbol{\theta})}^{g_i(\boldsymbol{\theta})} v_j(x) f(x;\boldsymbol{\theta})\, dx.$$

If $\boldsymbol{\theta}_n^* = \hat{\boldsymbol{\theta}}_n$ is the maximum likelihood estimator, which satisfies Eq. (10.16), then, as was shown by Roy, in Eq. (11.20), $r - s - 1$ of the λ_i are equal to 1, s of the λ_i lie between 0 and 1, and one of the λ_i

is equal to 0. Consequently, under H_0 the asymptotic distribution of $X_R^2(\hat{\boldsymbol{\theta}}_n)$ as $n \to \infty$ is that of

$$\chi^2_{r-s-1} + \lambda_1 \xi_1^2 + \cdots + \lambda_s \xi_s^2, \tag{11.22}$$

where χ^2_{r-s-1}, $\xi_1, \ldots, \xi_s$ are independent, $\xi_i \sim N_1(0,1)$, and, in general, $\lambda_i = \lambda_i(\boldsymbol{\theta}_0)$. In fact this is another extension of the Chernoff-Lehmann Theorem 10.1 to the case of random cell boundaries, similar to Chibisov's Theorem 11.1, and the comment that follows it.

Location and Scale Families

In formulas (11.20) and (11.22) the coefficients $\lambda_i = \lambda_i(\boldsymbol{\theta}_0)$ may depend on $\boldsymbol{\theta}_0$. But, as was shown by Dahiya and Gurland (1970a, 1972b), under certain circumstances when $s = 2$ the asymptotic distribution of X_R^2 can be free of $\boldsymbol{\theta}$. Following Dahiya and Gurland, we suppose that under H_0 the density function $f(x;\boldsymbol{\theta})$ of X_i belongs to a location and scale family

$$\left\{ f(x;\boldsymbol{\theta}) = \frac{1}{\sqrt{\theta_2}} f\left(\frac{x-\theta_1}{\sqrt{\theta_2}}\right) \right\}, \tag{11.23}$$
$$\boldsymbol{\theta} = (\theta_1, \theta_2)^T \in \boldsymbol{\Theta} = \{(\theta_1, \theta_2)^T : |\theta_1| < \infty, \quad \theta_2 > 0\},$$

where θ_1 and $\sqrt{\theta_2}$ are location and scale parameters, respectively, and $f(x;\boldsymbol{\theta})$ is continuous in x and differentiable with respect to the parameters. Moreover, we suppose that the family (11.23) has the property

$$\boldsymbol{E}X = \theta_1, \qquad \textbf{Var } X_1 = \theta_2. \tag{11.24}$$

For example, the normal law $N_1(\theta_1, \theta_2)$ has this property. Let $\boldsymbol{\theta}_n^* = (\theta_{1n}^*, \theta_{2n}^*)^T$, where θ_{1n}^* and θ_{2n}^* are estimators of θ_1 and θ_2 calculated by the *method of moments*:

$$\theta_{1n}^* = \bar{X}_n = \frac{1}{n}\sum_{i=1}^n X_i, \qquad \theta_{2n}^* = \frac{1}{n}\sum_{i=1}^n (X_i - \bar{X}_n)^2.$$

If we choose $g_i(\boldsymbol{\theta}_n^*) = \bar{X}_n + c_i s_n$ $(i = 1, \ldots, r)$, then, as was shown by Dahiya and Gurland (1970a, 1972b), we have

Theorem 11.3. *Under H_0, the asymptotic distribution of $X_R^2(\boldsymbol{\theta}_n^*)$ as*

$n \to \infty$ is that of

$$\lambda_1 \xi_2^2 + \cdots + \lambda_r \xi_r^2$$

as in Theorem 11.2. Furthermore, the distribution of X_R^2 is invariant with respect to location and scale parameters.

From Theorem 11.3 for the family (11.23), under (11.24), λ_i $(i = 1, \ldots, r)$ do not depend on $\boldsymbol{\theta}_0$ if $g_i(\boldsymbol{\theta})$ is chosen to be $g_i(\boldsymbol{\theta}) = \theta_1 + c_i\sqrt{\theta_2}$. Dahiya and Gurland (see also Watson, 1957) used this fact in the normal case, where according to Theorem 11.2 and Theorem 11.3 the statistic X_R^2 is distributed, under H_0, in the limit as $n \to \infty$ as

$$\chi_{r-3}^2 + \lambda_1 \xi_1^2 + \lambda_2 \xi_2^2, \tag{11.25}$$

with χ_{r-3}^2, ξ_1, and ξ_2 independent. Using some results of Watson (1957), Dahiya and Gurland tabulated the distribution of X_R^2 for $r = 3, 4, \ldots, 15$, when the significance level α is equal to 0.1, 0.5, and 0.01, in the case where, in the definition of $g_i(\boldsymbol{\theta}_n^*) = \bar{X}_n + c_i s_n$, the constants c_i $(i = 1, \ldots, r)$ are chosen so that $p_i(\boldsymbol{\theta}_n^*) = 1/r$.

The results of Roy and Watson were generalized also by Moore (1971) and Pollard (1979) to the case of rectangular cells in any number of dimensions. Some problems arising in the proof of the asymptotic normality of the vector $\boldsymbol{\nu}^* = (\nu_1^*, \ldots, \nu_r^*)^T$ in this case were discussed by Ruymgaart (1974) and Pollard (1979).

Remark 11.2. Maximum Likelihood and the Method of Moments. Let us consider the density family

$$f(x; \boldsymbol{\theta}) = h(x) \exp\left\{ \sum_{m=1}^{s} \theta_m x^m + v(\boldsymbol{\theta}) \right\}, \qquad x \in \boldsymbol{X} \subseteq \boldsymbol{R}^1, \tag{11.26}$$

where $\boldsymbol{X} = \{x : f(x; \boldsymbol{\theta}) > 0\}$ is open in $\boldsymbol{R}^1$, $\boldsymbol{\theta} \in \boldsymbol{\Theta}$. The family (11.26) is rich, containing Poisson, normal distributions, etc. It is evident that

$$\boldsymbol{U}_n = \left(\sum_{i=1}^{n} X_i, \sum_{i=1}^{n} X_i^2, \ldots, \sum_{i=1}^{n} X_i^s \right)^T \tag{11.27}$$

is the complete minimal sufficient statistic for $\boldsymbol{\theta}$.

We suppose that

1. the support X does not depend on $\boldsymbol{\theta}$;

2. the matrix

$$\boldsymbol{H}_v(\boldsymbol{\theta}) = - \left\| \frac{\partial^2}{\partial\theta_i \partial\theta_j} v(\boldsymbol{\theta}) \right\|_{s \times s} \tag{11.28}$$

of the function $v(\boldsymbol{\theta})$ is positive definite;

3. the moment $a_s(\boldsymbol{\theta}) = \boldsymbol{E}_{\boldsymbol{\theta}} X_1^s$ exists.

Using the results of Zacks (1971), it is not difficult to show that the maximum-likelihood estimator $\hat{\boldsymbol{\theta}}_n = \hat{\boldsymbol{\theta}}_m(\boldsymbol{U}_n)$ and the method-of-moments estimator $\bar{\boldsymbol{\theta}}_n = \bar{\boldsymbol{\theta}}_n(\boldsymbol{U}_n)$ of $\boldsymbol{\theta}$ coincide; that is, $\bar{\boldsymbol{\theta}}_n = \hat{\boldsymbol{\theta}}_n$. Let

$$a(\boldsymbol{\theta}) = (a_1(\boldsymbol{\theta}), \ldots, a_s(\boldsymbol{\theta}))^T \qquad \text{and} \qquad \boldsymbol{T}_n = \frac{1}{n}\,\boldsymbol{U}_n.$$

Then

$$a(\boldsymbol{\theta}) = -\frac{\partial}{\partial \boldsymbol{\theta}} v(\boldsymbol{\theta}),$$

and hence the likelihood equation is $\boldsymbol{T}_n = \boldsymbol{a}(\boldsymbol{\theta})$, and $\hat{\boldsymbol{\theta}}_n$ is root of this equation. On the other hand, we have $\boldsymbol{E}_{\boldsymbol{\theta}} \boldsymbol{T}_n \equiv \boldsymbol{a}(\boldsymbol{\theta})$. From the properties of the statistic $\boldsymbol{U}_n$ it follows that $\boldsymbol{T}_n$ is the MVUE of $\boldsymbol{a}(\boldsymbol{\theta})$, and $\bar{\boldsymbol{\theta}}_n$ is the root of the same equation, $\boldsymbol{T}_n = \boldsymbol{a}(\boldsymbol{\theta})$, used to find $\hat{\boldsymbol{\theta}}_n$. Hence $\hat{\boldsymbol{\theta}}_n = \bar{\boldsymbol{\theta}}_n$. Some examples:

1. The X_i are Poisson,

$$f(x;\theta) = \frac{\theta^x e^{-\theta}}{x!}, \quad x \in \{0, 1, 2, \ldots\}, \quad \theta > 0.$$

The maximum-likelihood estimator of $\theta = \boldsymbol{E}X_i$ is

$$\hat{\theta}_n = \frac{1}{n} \sum_{i=1}^{n} X_i,$$

which is also the method-of-moments estimator.

2. The X_i are normal,

$$f(x;\boldsymbol{\theta}) = \frac{1}{\sqrt{2\pi}\sigma}\exp\{-\frac{1}{2\sigma^2}(x-\mu)^2\},$$

$$x \in \boldsymbol{R}', \qquad \boldsymbol{\theta} = (\mu\sigma^2), \qquad |\mu| < \infty, \qquad \sigma^2 > 0.$$

The maximum-likelihood estimator of $\boldsymbol{\theta}$ is $\hat{\boldsymbol{\theta}}_n = (\hat{\mu}_n, \hat{\sigma}_n^2)$,

$$\hat{\mu}_n = \bar{X}_n = \frac{1}{n}\sum_{i=1}^{n} X_i, \qquad \hat{\sigma}_n^2 = \frac{1}{n}\sum_{i=1}^{n}(X_i - \bar{X}_n)^2$$

which is also the method-of-moments estimator.

3. The X_i are Bernoulli,

$$f(x;\theta) = \theta^x(1-\theta)^{1-x}, \quad x \in \{0,1\}, \quad 0 < \theta < 1, \quad \theta = \boldsymbol{E}X_i.$$

The maximum-likelihood estimator of θ is

$$\hat{\theta}_n = \bar{X}_n = \frac{1}{n}\sum_{i=1}^{n} X_i,$$

which is also the method-of-moments estimator.

4. The X_i have logistic density

$$f(x;\boldsymbol{\theta}) = \frac{\pi}{\sqrt{3}\sigma}\frac{\exp\left[-\frac{\pi}{\sqrt{3}}(\frac{x-\mu}{\sigma})\right]}{\left[1+\exp[-\frac{\pi}{\sqrt{3}}(\frac{x-\mu}{\sigma})]\right]^2},$$

$$x \in \boldsymbol{R}^d, \qquad \boldsymbol{\theta} = (\mu, \sigma^2), \qquad \mu = \boldsymbol{E}X_i,$$

$$\sigma^2 = \mathbf{Var}X_i, \qquad |\mu| < \infty, \qquad \sigma^2 > 0.$$

The methods-of-moments estimator is

$$\tilde{\boldsymbol{\theta}}_n = (\tilde{\mu}_n, \tilde{\sigma}_n^2), \quad \tilde{\mu}_n = \bar{X}_n, \quad d\tilde{\sigma}_n^2 = \frac{1}{n}\Sigma(X_i - \bar{X}_n)^2.$$

But $\tilde{\boldsymbol{\theta}}_n$ is not efficient nor even asymptotically efficient for $\boldsymbol{\theta}$ and hence is not asymptotically equivalent to the maximum-likelihood estimator $\hat{\boldsymbol{\theta}}_n$. The statistic

$$\boldsymbol{U}_2 = \left(\sum_{i=1}^{n} X_i, \sum_{i=1}^{n} X_i^2\right)$$

is not sufficient for $\boldsymbol{\theta}$, since $f(x;\boldsymbol{\theta})$ does not belong to the family (11.26). However $\tilde{\boldsymbol{\theta}}_n$ is $\sqrt{n}$-consistent for $\boldsymbol{\theta}$. In this example there is no sufficient statistic other than the data itself. More information is at the end of Section 13.

5. The X_i have the Cauchy density

$$f(x;\theta) = \frac{1}{\pi(1+(x-\theta)^2)}, \qquad x \in \boldsymbol{R}', \qquad \theta \in \boldsymbol{R}^1,$$

with shift parameter θ. In this case $\boldsymbol{E}X_i$ does not exist. The method of moments estimator of θ is $\tilde{\theta}_n = \frac{1}{n}\sum_{i=1}^{n} X_i$, which is not even consistent for θ, since $\tilde{\theta}_n$ has distribution $f(x;\theta)$. Of course $f(x;\theta)$ does not belong to (11.26).

Finally we remark that in general an estimator $\tilde{\boldsymbol{\theta}}_n$ based on the method of moments is not asymptotically efficient, and hence does not satisfy the Chernoff-Lehmann theorem. Before applying the Dahiya and Gurland extension of the Chernoff-Lehmann theorem, one needs to study the asymptotic properties of the consistent estimator that one wishes to use.

12. RECAPTURING THE CHI-SQUARED LIMIT: A MODIFICATION OF PEARSON'S STATISTIC

Let us consider again the situation where we have i.i.d. $X_1, \ldots, X_n$ and we want to test the hypothesis H_0 according to which the distribution function of X_1 belongs to a family of continuous distribution functions

$$\{F(x;\boldsymbol{\theta})\}, \quad \boldsymbol{\theta} = (\theta_1, \ldots, \theta_s)^T \in \boldsymbol{\Theta} \subset \boldsymbol{R}^s, \quad x \in \boldsymbol{R}^1. \tag{12.1}$$

We suppose that the regularity conditions of Theorem 10.1 of Chernoff and Lehmann hold. The problem addressed in this section was introduced at the beginning of Section 11. We consider, for instance, the statistic $X_n^2(\boldsymbol{\theta}_n^*, \hat{\boldsymbol{\theta}}_n)$, defined in Section 11, where $\boldsymbol{\theta}_n^*$ is any consistent estimator of $\boldsymbol{\theta}$, and $\hat{\boldsymbol{\theta}}_n$ is the maximum-likelihood estimator of $\boldsymbol{\theta}$. The limit distribution of $X_n^2(\boldsymbol{\theta}_n^*, \hat{\boldsymbol{\theta}}_n)$ is typically not chi-squared, and, what is worse, depends on $\boldsymbol{\theta}$, as shown by Chernoff and Lehmann. The reader interested in scale-and-shift families may be pleased to find that in this case the Chernoff and Lehmann limit does not depend on $\boldsymbol{\theta}$, as was shown by Roy (1956). This case is discussed in some detail in Section 13.

A Modification of Pearson's Statistic

The unfortunate form of the limit distribution of $X_n^2(\boldsymbol{\theta}_n^*, \hat{\boldsymbol{\theta}}_n)$ gave rise to the idea of trying to find some alternate to the standard Pearson statistic for which the limit distribution as $n \to \infty$ is chi-squared. Such an alternate was found by Nikulin (1973a,b,c, 1974). Using the notation of Sections 9, 10, and 11, we consider the vector

$$\mathbf{X}_n(\boldsymbol{\theta}) = \left(\frac{\nu_1 - np_1(\boldsymbol{\theta})}{\sqrt{np_1(\boldsymbol{\theta})}}, \ldots, \frac{\nu_r - np_r(\boldsymbol{\theta})}{\sqrt{np_r(\boldsymbol{\theta})}} \right)^T, \tag{12.2}$$

where $p_i(\theta)$ is defined by Eq. (9.1). Let $\widehat{\boldsymbol{\theta}}_n$ be the maximum-likelihood estimator of $\boldsymbol{\theta}$, which under regularity conditions satisfies the asymptotic relation

$$\sqrt{n}(\hat{\boldsymbol{\theta}}_n - \boldsymbol{\theta}_0) = \frac{1}{\sqrt{n}} \mathbf{I}^{-1}(\boldsymbol{\theta}_0)\Lambda(\boldsymbol{\theta}_0) + o_{\mathbf{P}_0}(1), \tag{12.3}$$

where $\boldsymbol{\theta}_0$ is the true value of the parameter $\boldsymbol{\theta}$, $\mathbf{I}(\boldsymbol{\theta}) = (1/n)\mathbf{E}_{\boldsymbol{\theta}}\Lambda(\boldsymbol{\theta})\Lambda(\boldsymbol{\theta})$, the Fisher information matrix corresponding to one observation X_1 and $\Lambda(\boldsymbol{\theta}) = \mathbf{grad} \log L(\boldsymbol{\theta})$, where $L(\boldsymbol{\theta}) = \prod_{i=1}^n f(X_i; \boldsymbol{\theta})$. As was shown in Section 10 [see Eqs. (10.30), (10.33), and (10.34)] under H_0

$$\mathbf{X}_n(\hat{\boldsymbol{\theta}}_n) \text{ is } AN(\mathbf{0}_r, \mathbf{W}(\boldsymbol{\theta}_0)), \tag{12.4}$$

where

$$\mathbf{W}(\boldsymbol{\theta}_0) = \mathbf{I}_r - \mathbf{q}\mathbf{q}^T - \mathbf{B}\mathbf{I}^{-1}\mathbf{B}^T, \quad \text{rank } \mathbf{W}(\boldsymbol{\theta}_0) = r - 1, \tag{12.5}$$

$$\mathbf{q} = \mathbf{q}(\boldsymbol{\theta}) = (\sqrt{p_1(\boldsymbol{\theta})}, \ldots, \sqrt{p_r(\boldsymbol{\theta})})^T, \quad \mathbf{B} = \mathbf{B}(\boldsymbol{\theta}) = [\mathbf{b}_{ij}]_{r \times s},$$
$$b_{ij} = \frac{1}{\sqrt{p_i(\boldsymbol{\theta})}} \frac{\partial p_i(\boldsymbol{\theta})}{\partial \theta_j}.$$

To test H_0 , we consider the statistic

$$Y_n^2(\hat{\boldsymbol{\theta}}_n) = \boldsymbol{X}_n^T(\hat{\boldsymbol{\theta}}_n)\boldsymbol{W}^-(\hat{\boldsymbol{\theta}}_n)\boldsymbol{X}_n(\hat{\boldsymbol{\theta}}_n), \tag{12.6}$$

where $\boldsymbol{W}^-$, rank $\boldsymbol{W}^-(\boldsymbol{\theta}) = k-1$. The statistic Y_n^2 is invariant under choice of $\boldsymbol{W}^-(\boldsymbol{\theta})$, and one can easily verify that

$$\boldsymbol{W}^-(\boldsymbol{\theta}) = (\boldsymbol{I}_r - \boldsymbol{B}\boldsymbol{I}^{-1}\boldsymbol{B}^T)^{-1}. \tag{12.7}$$

For the definition of $\boldsymbol{W}^-$ see the appendix. For properties in this context see Puri and Russel (1984).

Theorem 12.1 *For any* $x > 0$

$$\lim_{n\to\infty} \boldsymbol{P}\{Y_n^2(\hat{\boldsymbol{\theta}}_n) > x \mid H_0\} = \boldsymbol{P}(\chi_{r-1}^2 > x). \tag{12.8}$$

The proof of this theorem we obtain immediately from Eq. (12.3) and the Lemma of Chernoff and Lehmann in Section 9.

From Eqs. (12.6) and (12.7), it follows (see Nikulin, 1973c) that the statistic $Y_n^2(\hat{\boldsymbol{\theta}})$ can be written in another form more convenient for computation:

$$Y_n^2(\hat{\boldsymbol{\theta}}_n) = X_n^2(\hat{\boldsymbol{\theta}}_n) + \boldsymbol{X}_n^T(\hat{\boldsymbol{\theta}}_n)\boldsymbol{G}(\hat{\boldsymbol{\theta}}_n)\boldsymbol{X}_n(\hat{\boldsymbol{\theta}}_n), \tag{12.9}$$

where

$$\boldsymbol{G}(\boldsymbol{\theta}) = \boldsymbol{B}(\boldsymbol{\theta})(\boldsymbol{I}(\boldsymbol{\theta}) - \boldsymbol{B}^T(\boldsymbol{\theta})\boldsymbol{B}(\boldsymbol{\theta}))^{-1}\boldsymbol{B}^T(\boldsymbol{\theta}),$$

because

$$\left(\boldsymbol{I}_r - \boldsymbol{B}\boldsymbol{I}^{-1}\boldsymbol{B}^T\right)^{-1} = \boldsymbol{I}_r + \boldsymbol{B}(\boldsymbol{I} - \boldsymbol{B}^T\boldsymbol{B})^{-1}\boldsymbol{B}^T = \boldsymbol{I}_r + \boldsymbol{G}. \tag{12.10}$$

As was noted in Section 10, the matrix

$$\boldsymbol{I}(\boldsymbol{\theta}) - \boldsymbol{B}^T(\boldsymbol{\theta})\boldsymbol{B}(\boldsymbol{\theta}) = \boldsymbol{I}(\boldsymbol{\theta}) - \boldsymbol{J}(\boldsymbol{\theta})$$

is positive definite, because $n\boldsymbol{I}(\boldsymbol{\theta})$ is the information matrix of the sufficient (!) statistic $X_1, \ldots, X_n$, and $n\boldsymbol{J}(\boldsymbol{\theta})$ is the information matrix of the vector of frequencies $\boldsymbol{\nu} = (\nu_1, \ldots, \nu_r)^T$, which is not a sufficient statistic for $\boldsymbol{\theta}$ in the continuous case. Hence, there exists $(\boldsymbol{I}(\boldsymbol{\theta}) - \boldsymbol{J}(\boldsymbol{\theta}))^{-1}$. Because

$$\boldsymbol{B}^T(\boldsymbol{\theta})\boldsymbol{X}_n(\boldsymbol{\theta}) = \frac{1}{\sqrt{n}}\boldsymbol{L}(\boldsymbol{\theta}),$$

where

$$\boldsymbol{L}(\boldsymbol{\theta}) = (L_1(\boldsymbol{\theta}), \ldots, L_s(\boldsymbol{\theta}))^T, \qquad L_j(\boldsymbol{\theta}) = \sum_{i=1}^{r} \frac{\nu_i}{p_i} \frac{\partial p_i(\boldsymbol{\theta})}{\partial \theta_j},$$

we obtain from Eqs. (12.9) and (12.10) that

$$Y_n^2(\hat{\boldsymbol{\theta}}_n) = X_n^2(\hat{\boldsymbol{\theta}}_n) + \frac{1}{n} \boldsymbol{L}^T(\hat{\boldsymbol{\theta}}_n)(\boldsymbol{I}(\hat{\boldsymbol{\theta}}_n) - \boldsymbol{J}(\hat{\boldsymbol{\theta}}_n))^{-1} \boldsymbol{L}(\hat{\boldsymbol{\theta}}_n). \tag{12.11}$$

Formula (12.11) shows that the statistic $\boldsymbol{Y}_n^2(\hat{\boldsymbol{\theta}}_n)$ consists of two quadratic forms, one of which is Pearson's statistic $X_n^2(\hat{\boldsymbol{\theta}}_n)$.

Random Cell Boundaries: Wald's Statistic

Instead of defining $Y_n^2(\hat{\boldsymbol{\theta}}_n)$ by Eq. (12.6), one can construct a similar statistic with random cell boundaries. Choose a vector $\boldsymbol{p} = (p_1, \ldots, p_r)^T$ such that $\boldsymbol{p}^T \mathbf{1}_r = 1$ and all $p_i > 0$. If there is an alternative, we would choose $\boldsymbol{p}$ to maximize power (see Section 10). As was noted in Chapter 1, there are considerations in favor of choosing $p_1 = p_2 = \cdots = p_r = 1/r$, if there is no alternative. Set

$$F^{-1}(a, \boldsymbol{\theta}) = \inf\{x : F(x; \boldsymbol{\theta}) \geq a\}, \qquad 0 < a < 1, \quad \boldsymbol{\theta} \in \boldsymbol{\Theta},$$

and let

$$a_i(\boldsymbol{\theta}) = F^{-1}(p_1 + \cdots + p_i; \boldsymbol{\theta}), \qquad i = 1, \ldots, r-1, \quad a_r = -a_0 = +\infty.$$

For each $\boldsymbol{\theta} \in \boldsymbol{\Theta}$ a partition of the real line into r intervals is formed, with boundary points $a_0, a_1(\boldsymbol{\theta}), \ldots, a_{r-1}(\boldsymbol{\theta}), a_r$. We assume that the matrix $\boldsymbol{\Omega} = [\omega_{ij}]_{s \times r}$ with elements

$$\omega_{ij} = \int_{a_{j-1}(\boldsymbol{\theta})}^{a_j(\boldsymbol{\theta})} \frac{\partial f(x; \boldsymbol{\theta})}{\partial \theta_i} \, dx$$

is of rank s, and that the derivatives

$$\frac{\partial a_i(\boldsymbol{\theta})}{\partial \theta_j}, \qquad i = 1, \ldots, r; \quad j = 1, \ldots, s,$$

exist and are continuous for all $\boldsymbol{\theta} \in \boldsymbol{\Theta}$. Let $\boldsymbol{\nu}^* = (\nu_1^*, \ldots, \nu_r^*)^T$ be a vector of frequencies obtained as a result of grouping the random variables $X_1, \ldots, X_n$ using the random partition obtained from the one above by replacing the unknown $\boldsymbol{\theta}$ in $a_i(\boldsymbol{\theta})$ by the maximum-likelihood estimator $\hat{\boldsymbol{\theta}}_n$. We have here the same regularity conditions as in the theorem of Roy (Section 11), and

$$\omega_{ij} = -f(a_j(\boldsymbol{\theta}); \boldsymbol{\theta}) \frac{\partial a_j(\boldsymbol{\theta})}{\partial \theta_i} + f(a_{j-1}(\boldsymbol{\theta}); \boldsymbol{\theta}) \frac{\partial a_{j-1}(\boldsymbol{\theta})}{\partial \theta_i} \tag{12.12}$$

holds for elements of the matrix $\boldsymbol{\Omega}$. Further, let

$$\boldsymbol{X}_n = \left(\frac{\nu_1^* - np_1}{\sqrt{np_1}}, \ldots, \frac{\nu_r^* - np_r}{\sqrt{np_r}} \right)^T, \tag{12.13}$$

$$\boldsymbol{P} = \begin{bmatrix} p_1 & 0 & \ldots & 0 \\ 0 & p_2 & \ldots & 0 \\ \vdots & \vdots & \vdots & \vdots \\ 0 & 0 & \ldots & p_r \end{bmatrix}.$$

Then from the results of Section 10 and Section 11, it follows that

$$\boldsymbol{X}_n(\hat{\boldsymbol{\theta}}_n) \quad \text{is} \quad AN(\mathbf{0}_r, \boldsymbol{W}(\boldsymbol{\theta}_0)), \tag{12.14}$$

where

$$\boldsymbol{W}(\boldsymbol{\theta}) = \boldsymbol{I}_r - \boldsymbol{q}\boldsymbol{q}^T - \boldsymbol{P}^{-1/2}\boldsymbol{\Omega}^T(\boldsymbol{\theta})\boldsymbol{I}^{-1}(\boldsymbol{\theta})\boldsymbol{\Omega}(\boldsymbol{\theta})\boldsymbol{P}^{-1/2}, \tag{12.15}$$

$\boldsymbol{q} = (\sqrt{p_1}, \sqrt{p_2}, \ldots, \sqrt{p_r})^T$ does not depend on $\boldsymbol{\theta}$, rank $\boldsymbol{\Omega} = r - 1$. From expressions (12.13) and (12.14), it follows that under H_0 the statistic, known also as the Wald statistic (see Wald, 1943),

$$\boldsymbol{Y}_n^2(\hat{\boldsymbol{\theta}}_n) = \boldsymbol{X}_n^T \boldsymbol{W}^-(\hat{\boldsymbol{\theta}}_n)\boldsymbol{X}_n$$

has in the limit as $n \to \infty$ a chi-squared distribution with $r-1$ degrees of freedom, as does the statistic (12.6) where $\boldsymbol{W}$ is defined by Eq. (12.5). Using Eqs. (12.7 and (12.12), we can write the statistic $Y_n^2(\hat{\boldsymbol{\theta}}_n)$ in the form

$$Y_n^2(\hat{\boldsymbol{\theta}}_n) = X_n^2 + \frac{1}{n}\boldsymbol{\alpha}^T \boldsymbol{G}^{-1}(\hat{\boldsymbol{\theta}}_n)\boldsymbol{\alpha}, \tag{12.16}$$

where

$$X_n^2 = \sum_{i=1}^{r} \frac{(\nu_i^* - np_j)^2}{np_j}, \tag{12.17}$$

$$\boldsymbol{\alpha} = (\alpha_1, \ldots, \alpha_s)^T, \alpha_j = \frac{\omega_{ji}\nu_1^*}{p_1} + \ldots + \frac{\omega_{jr}\nu_r^*}{p_r},$$

$$\boldsymbol{G}(\boldsymbol{\theta}) = \boldsymbol{I}(\boldsymbol{\theta}) - \boldsymbol{\Omega}(\boldsymbol{\theta})\boldsymbol{P}^{-1}\boldsymbol{\Omega}^T(\boldsymbol{\theta}). \tag{12.18}$$

The elements g_{uv} of $\boldsymbol{G}$ are

$$g_{uv} = i_{uv} - \sum_{j=1}^{r} \frac{\omega_{uj}\omega_{vj}}{p_j},$$

where $i_{uv} = i_{uv}(\boldsymbol{\theta})$ is an element of the matrix $\boldsymbol{I}(\boldsymbol{\theta})$. In formulas (12.13) and (12.17), we have no parameter $\boldsymbol{\theta}$, but the frequencies ν_i^* were constructed using the random cell boundaries $a_j(\hat{\boldsymbol{\theta}}_n)$. With a fortunate choice of the vector $\boldsymbol{p}$, which depends on the user, sometimes the statistic Y_n^2 can be written in a form more convenient for computation. The form also depends on the density $f(x;\boldsymbol{\theta})$. The Wald statistic was studied by Nikulin (1973a,b,c), Rao and Robson (1974), Moore (1977), and Drost (1988), among others.

Remarks About Power

Remark 12.1. If we wish to test the hypothesis H_0, according to which the density function of X_1 is $f(x;\boldsymbol{\theta})$, $\boldsymbol{\theta} \in \boldsymbol{\Theta}$, against H_n, according to which the density of X_1 is $g_n(x;\boldsymbol{\theta}) = (1-\eta_n)f(x;\boldsymbol{\theta}) + \eta_n g(x;\boldsymbol{\theta})$, $\boldsymbol{\theta} \in \boldsymbol{\Theta}$, $\eta_n \to 0$, as $n \to \infty$, where

$$\int g(x;\boldsymbol{\theta})\, dx = 1, \qquad g(x;\boldsymbol{\theta}) \geq 0,$$

then we can construct a chi-squared test with two grouping classes, as was noted in Remark 4.1. To do this we define

$$I_1 = \{x : f(x;\boldsymbol{\theta}) \geq g_n(x;\boldsymbol{\theta})\} \quad \text{and} \quad I_2 = \{x : f(x;\boldsymbol{\theta}) < g_n(x;\boldsymbol{\theta})\},$$

for which we need to solve the equation

$$f(x;\boldsymbol{\theta}) = g_n(x;\boldsymbol{\theta}),$$

which is equivalent to

$$f(x;\boldsymbol{\theta}) = g(x;\boldsymbol{\theta}).$$

Of course the roots, $a_i(\boldsymbol{\theta})$, of this equation depend on $\boldsymbol{\theta}$ which we estimate by $\hat{\boldsymbol{\theta}}_n$. To construct I_1 and I_2 we unify the intervals $[a_{i-1}(\hat{\boldsymbol{\theta}}_n); a_i(\hat{\boldsymbol{\theta}}_n))$ where $f(x;\hat{\boldsymbol{\theta}}_n) \geq g_n(x;\hat{\boldsymbol{\theta}}_n)$ and those where $f(x;\hat{\boldsymbol{\theta}}_n) < g_n(x;\hat{\boldsymbol{\theta}}_n)$. Then we construct the statistic $Y_n^2(\hat{\boldsymbol{\theta}}_n)$, based on two classes with random cell boundaries $a_i(\hat{\boldsymbol{\theta}}_n)$. Under H_0 , in the limit as $n \to \infty$, this statistic has the chi-squared distribution with 1 degree of freedom. This procedure is particularly useful when n is not large because the numbers of observations in just two grouping intervals is larger and the relevant central limit theorem is the de Moivre-Laplace version, e.g. for testing for normality against the family of logistic distributions.

Remark 12.2. Distribution of $Y_n^2(\hat{\boldsymbol{\theta}}_n)$ under Alternative Hypotheses. We wish to test $H_0 : X_i \sim f(x;\boldsymbol{\theta})$, $\boldsymbol{\theta} = (\theta_1, \ldots, \theta_s)^T \in \Theta \subset \boldsymbol{R}^s$ against a contiguous sequence of alternatives of the form

$$H_n : X_i \sim g(x;\boldsymbol{\theta}) = f(x,\boldsymbol{\theta}) + \frac{1}{\sqrt{n}}\delta(x;\boldsymbol{\theta}),$$

where $\int \delta(x;\boldsymbol{\theta})dx = 0$.

Assume that the partial derivatives

$$\frac{\partial\delta(x_i;\boldsymbol{\theta})}{\partial\theta_i\partial\theta_j}, \qquad i,j = 1,\ldots,s,$$

exist. For $j = 1, 2, \ldots, r$,

$$\int_{a_{j-1}(\boldsymbol{\theta})}^{a_j(\boldsymbol{\theta})} g(x;\boldsymbol{\theta})\,dx = p_j + \frac{1}{\sqrt{n}}c_j(\boldsymbol{\theta}), \qquad c_j(\boldsymbol{\theta}) = \int_{a_{j-1}(\boldsymbol{\theta})}^{a_j(\boldsymbol{\theta})} \delta(x;\boldsymbol{\theta})\,dx.$$

Then

$$\lim_{n\to\infty} \boldsymbol{P}\{Y_n^2(\hat{\boldsymbol{\theta}}_n) \geq x | H_n\} = \boldsymbol{P}\{\chi_{r-1}^2(\lambda) \geq x\} \tag{12.19}$$

where

$$\lambda = \lambda(\boldsymbol{\theta}) = \sum_{j=1}^{r} \frac{c_j^2(\boldsymbol{\theta})}{p_j} + \boldsymbol{d}^T(\boldsymbol{\theta})\boldsymbol{G}^{-1}(\boldsymbol{\theta})\boldsymbol{d}(\boldsymbol{\theta}), \qquad (12.20)$$

$$\boldsymbol{d}(\boldsymbol{\theta}) = (d_1(\boldsymbol{\theta}), \ldots, d_s(\boldsymbol{\theta}))^T, \qquad d_j(\boldsymbol{\theta}) = \sum_{i=1}^{r} \frac{w_{ji}(\boldsymbol{\theta})c_i(\boldsymbol{\theta})}{p_i},$$

$\boldsymbol{G}(\boldsymbol{\theta})$ is given by Eq. (12.18).

The noncentrality parameter λ can be estimated by $\lambda(\hat{\boldsymbol{\theta}}_n)$. Power is maximized by maximizing the parameter of noncentrality. How this is done depends on what constraints and other information we have. For instance if r is fixed, we choose grouping sets in such a way that λ is maximized.

Remark 12.3. Asymptotic Power when r Increases with n. Drost (1988, Paragraph II.2) studies asymptotic power of chi-squared type tests against an alternative in a "contamination" neighborhood where H_0 is a location and scale family, $F(\frac{\cdot-\mu}{\sigma})$, that is, the alternative is

$$H_{1n} : G_{\eta_n} = (1-\eta_n)F\left(\frac{\cdot-\mu}{\sigma}\right) + \eta_n G\left(\frac{\cdot-\mu}{\sigma}\right)$$

where $\eta_n \to 0$, at a rate, for example, of $n^{-1/2}$. He finds that for a class of statistics including Eq. (12.6), under regularity conditions, asymptotic power is determined by the asymptotic ratio of the noncentrality parameter and $r^{1/2}$, both growing with n.

Remark 12.4. Power against Heavy-tailed Alternatives. Drost, Kallenberg, and Oosterhoff (1990) include the generalized chi-squared statistic $Y_n^2(\hat{\theta}_n^2)$ in their study of EDF tests for shift and scale families. They are interested in alternative distributions with heavy tails. They note that little attention has been paid to the effect of the method of estimation on the power of EDF tests. In particular, they show that an estimator that is good under H_0 and not so good under the alternative may result in good power against heavy-tailed alternatives. This topic is also studied by Drost (1988).

Sutrick (1986) studies the asymptotic power of chi-squared compared to the likelihood ratio test; see also Bolshev (1987).

Quantile Chi-Squared Tests

Again we wish to test the hypothesis that the distribution of the X_i belongs to a family of continuous functions

$$\{F(x;\boldsymbol{\theta}),\ \boldsymbol{\theta}=(\theta_1,\dots,\theta_s)^T\in\Theta\subset \boldsymbol{R}^s\}. \tag{12.21}$$

A chi-squared test can be based on sample quantiles instead of on grouping intervals. Let $(X_{n1},\dots,X_{nn})^T$ denote the vector of order statistics obtained from the sample $X_1,X_2,\dots,X_n$:

$$-\infty < X_{n1}\le X_{n2}\le\cdots\le X_{n,n-1}\le X_{nn}<\infty. \tag{12.22}$$

Fix a vector $\boldsymbol{p}=(p_1,\dots,p_r)^T$ of positive probabilities p_i such that $\boldsymbol{p}^T\mathbf{1}_r=1$, $r>s+1$. Next consider the sample quantiles

$$\begin{aligned} &Y_{n0}=-\infty, \qquad Y_{ni}=X_{n,[nu_i]+1}, \qquad i=1,\dots,r-1,\\ &Y_{nr}=+\infty, \end{aligned} \tag{12.23}$$

where $u_i=p_1+\cdots+p_i$, $[a]$ is the integral part of a. It is well known that $\{Y_{ni}\}$ is a consistent sequence of estimators for the quantiles of $F(x,\boldsymbol{\theta})$,

$$\begin{aligned} x_i(\boldsymbol{\theta}) &= F^{-1}(p_1+\cdots+p_i;\boldsymbol{\theta})\\ &= F^{-1}(u_i;\boldsymbol{\theta}), \qquad i=1,\dots,r-1. \end{aligned} \tag{12.24}$$

We suppose that the function $F(x;\boldsymbol{\theta})$ satisfies conditions (10.5) to (10.7) and in addition that the matrix $\Omega=\|\omega_{ij}\|$ with elements

$$\begin{aligned} \omega_{ij}=\omega_{ij}(\boldsymbol{\theta}) &= \int_{x_{j-1}(\boldsymbol{\theta})}^{x_j(\boldsymbol{\theta})}\frac{\partial}{\partial\theta_i}f(x;\boldsymbol{\theta})\,dx,\\ &\qquad i=1,2,\dots,s, \quad j=1,2,\dots,r, \end{aligned} \tag{12.25}$$

has rank s, $\boldsymbol{\theta}\in\Theta$.

Now let us consider the vector $\boldsymbol{Z}_n=(Z_1,Z_2,\dots,Z_{r-1})^T$, where

$$\begin{aligned} &Z_i=\sqrt{n}[F(Y_{ni};\boldsymbol{\theta})-u_i],\\ &u_i=p_1+\cdots+p_i, \qquad i=1,2,\dots,r-1. \end{aligned} \tag{12.26}$$

Under H_0 the vector $\boldsymbol{Z}_n=\boldsymbol{Z}_n(\boldsymbol{\theta})$ is asymptotically normally distributed as $n\to\infty$ with zero mean vector and covariance matrix

$$\boldsymbol{\Sigma} = \left\| \begin{matrix} u_1 & u_1 & u_1 & \dots & u_1 \\ u_1 & u_2 & u_2 & \dots & u_2 \\ u_1 & u_2 & u_3 & \dots & u_3 \\ \vdots & \vdots & \vdots & & \\ u_1 & u_2 & u_r & & \end{matrix} \right\| - \left\| \begin{matrix} u_1 \\ u_2 \\ u_3 \\ \vdots \\ u_r \end{matrix} \right\| \|u_1, u_2, u_3, \dots, u_r\|.$$

The covariance matrix $\boldsymbol{\Sigma}$ is nonsingular, its inverse $\boldsymbol{\Sigma}^{-1}$ is given by Eq. (7.67). From the theorem of Mosteller (1946) (see also Wilks, 1962, Bolshev, 1969a), it follows that the quadratic form

$$X_n^2(\boldsymbol{\theta}) = \boldsymbol{Z}_n^2 \boldsymbol{\Sigma}^{-1} \boldsymbol{Z}_n = n \sum_{i=1}^{r} \frac{[\Delta F(Y_{ni}, \boldsymbol{\theta}) - p_i]^2}{p_i}$$
$$= \sum_{i=1}^{r} \frac{1}{p_i} (Z_i - Z_{i-1})^2 \tag{12.27}$$

has in the limit as $n \to \infty$ a chi-squared distribution with $r-1$ degrees of freedom, where

$$\Delta F(Y_{ni}; \boldsymbol{\theta}) = F(Y_{ni}; \boldsymbol{\theta}) - F(Y_{n,i-1}; \boldsymbol{\theta}). \tag{12.28}$$

The parameter $\boldsymbol{\theta}$ must be estimated. The maximum-likelihood estimator $\hat{\boldsymbol{\theta}}_n$ of $\boldsymbol{\theta}$ is asymptotically efficient; that is, $\{\hat{\boldsymbol{\theta}}_n\}$ satisfies condition (10.16). Using the same technique as before one sees that the asymptotic distribution of $\boldsymbol{Z}_n(\hat{\boldsymbol{\theta}}_n)$, is normal, $N(\boldsymbol{0}_{r-1} \boldsymbol{W})$, with covariance matrix $\boldsymbol{W} = \boldsymbol{W}(\boldsymbol{\theta})$. Our quantile-based chi-squared statistic is

$$Y_n^2(\hat{\boldsymbol{\theta}}_n) = \boldsymbol{Z}_n^T(\hat{\boldsymbol{\theta}}_n) \boldsymbol{\Sigma}^{-1}(\hat{\boldsymbol{\theta}}_n) \boldsymbol{Z}_n(\hat{\boldsymbol{\theta}}_n)$$
$$= X_n^2(\hat{\boldsymbol{\theta}}_n) + n \sum_{i,j=1}^{s} \omega_{ij}(\hat{\boldsymbol{\theta}}_n) \gamma_i(\hat{\boldsymbol{\theta}}_n) \gamma_j(\hat{\boldsymbol{\theta}}_n), \tag{12.29}$$

where $X_n^2(\boldsymbol{\theta})$ is given by Eq. (12.27), $\omega_{ij}(\boldsymbol{\theta})$ are the elements of the matrix

$$\Omega = \|\omega_{ij}\| = [\boldsymbol{I} - G]^{-1}, \tag{12.30}$$

$\boldsymbol{I} = \boldsymbol{I}(\boldsymbol{\theta})$ is the Fisher information matrix of X_1, $G = \|g_{ij}(\boldsymbol{\theta})\|$,

$$g_{ij} = \sum_{k=1}^{r} w_{ik}(\boldsymbol{\theta}) w_{jk}(\boldsymbol{\theta}) / p_k,$$

w_{ij} being given by Eq. (12.25), and

$$\gamma_i(\theta) = \sum_{k=1}^{r} \frac{w_{ik}(\boldsymbol{\theta})\Delta F(X_{nk};\boldsymbol{\theta})}{p_k},$$

$\Delta F(Y_{n,i};\boldsymbol{\theta})$ being given by Eq. (12.28).

Theorem.

$$\lim_{n\to\infty} \boldsymbol{P}\{Y_n^2(\hat{\boldsymbol{\theta}}_n) \geq x|H_0\} = \boldsymbol{P}\{\chi_{r-1}^2 \geq x\}. \tag{12.31}$$

The power function of a quantile-based chi-squared test against a wide class of close alternatives can be expressed in terms of a non-central chi-squared distribution with $(r-1)$ degrees of freedom. Quantile tests have been studied by Weiss (1957), Witting (1959), Bolshev (1969a), Gebert and Kale (1969), Chibisov (1971), Hartley and Pfaffenberger (1972), Nikulin (1974), Bofinger (1973), Miyamoto (1976), see also Read and Cressie (1988). Using the results of Ogawa (1976), Miyamoto (1976) studied the problem of optimum spacing for chi-squared tests based on sample quantiles. The asymptotic behavior of empirical quantile processes is studied by Csörgo (1983).

13. THE CHI-SQUARED TEST FOR CONTINUOUS DISTRIBUTIONS WITH LOCATION AND SCALE PARAMETERS

In this section, we consider the problem of testing the hypothesis H_0 that the distribution function of the i.i.d. $X_1, \ldots, X_n$ belongs to a family

$$\{F(x;\boldsymbol{\theta})\}, \quad \boldsymbol{\theta} = (\theta_1, \theta_2)^T \in \boldsymbol{\Theta} \subset \boldsymbol{R}^2$$

of continuous functions, depending on the location parameter, θ_1, and scale parameter θ_2,

$$F(x;\boldsymbol{\theta}) = F\left(\frac{x-\theta_1}{\theta_2}\right), \quad \boldsymbol{\Theta} = \{\boldsymbol{\theta} = (\theta_1, \theta_2) : |\theta_1| < \infty, \quad \theta_2 > 0\},$$

where $F(x)$ is a given function. The important case of the normal family is treated in Section 14. The chi-squared test for location and scale families was developed by Nikulin (1973b,c) and further studied by Rao and Robson (1974), Moore and Spruill (1975), and Drost (1988).

Construction of the Test Statistic

Let

$$f(x;\boldsymbol{\theta}) = \frac{1}{\theta_2} f\left(\frac{x-\theta_1}{\theta_2}\right)$$

be the density function of X_i. Let us assume that $F(x,\boldsymbol{\theta})$ satisfies the regularity conditions of Sections 2 and 3 and define the Fisher information matrix $\mathbf{I}(\boldsymbol{\theta})$, corresponding to one observation X_j.

$$\mathbf{I}(\boldsymbol{\theta}) = \mathbf{E}_{\boldsymbol{\theta}} \boldsymbol{\Lambda}_j \boldsymbol{\Lambda}_j^T, \tag{13.1}$$

where

$$\boldsymbol{\Lambda}_j = \textbf{grad}\ \log\left[\frac{1}{\theta_2} f\left(\frac{X_j-\theta_1}{\theta_2}\right)\right]$$

$$= \left(\frac{\partial}{\partial\theta_1}\log\frac{1}{\theta_2} f\left(\frac{X_j-\theta_1}{\theta_2}\right), \frac{\partial}{\partial\theta_2}\log\frac{1}{\theta_2} f\left(\frac{X_j-\theta_1}{\theta_2}\right)\right)^T. \tag{13.2}$$

We compute $\mathbf{I}(\boldsymbol{\theta})$:

$$\mathbf{E}_\theta\left\{\frac{\partial}{\partial\theta_1}\log\frac{1}{\theta_2} f\left(\frac{X_j-\theta_1}{\theta_2}\right)\right\}^2$$

$$= \int_{-\infty}^{\infty}\left[\frac{f'\left(\frac{x-\theta_1}{\theta_2}\right)}{\theta_2 f\left(\frac{x-\theta_1}{\theta_2}\right)}\right]^2 \frac{1}{\theta_2} f\left(\frac{x-\theta_1}{\theta_2}\right) dx$$

$$= \frac{1}{\theta_2^2}\int_{-\infty}^{\infty}\left[\frac{f'(x)}{f(x)}\right]^2 f(x)\,dx = \frac{1}{\theta_2^2} I_{11}, \tag{13.3}$$

$$\boldsymbol{E}_{\boldsymbol{\theta}}\left\{\frac{\partial}{\partial\theta_1}\log\frac{1}{\theta_2}f\left(\frac{X_j-\theta_1}{\theta_2}\right)\frac{\partial}{\partial\theta_2}\log\frac{1}{\theta_2}f\left(\frac{X_j-\theta_1}{\theta_2}\right)\right\}$$

$$=\frac{1}{\theta_2^2}\int_{-\infty}^{\infty}\left[\frac{f'(x)}{f(x)}+x\left(\frac{f'(x)}{f(x)}\right)^2\right]f(x)\,dx$$

$$=\frac{1}{\theta_2^2}\int_{-\infty}^{\infty}x\left[\frac{f'(x)}{f(x)}\right]^2 f(x)\,dx=\frac{1}{\theta_2^2}I_{21}=\frac{1}{\theta_2^2}I_{12}, \tag{13.4}$$

$$\boldsymbol{E}_{\boldsymbol{\theta}}\left\{\frac{\partial}{\partial\theta_2}\log\frac{1}{\theta_2}f\left(\frac{X_j-\theta_1}{\theta_2}\right)\right\}^2$$

$$=\int_{-\infty}^{\infty}\left[\frac{1}{\theta_2}+\frac{x-\theta_1}{\theta_2}\cdot\frac{f'\left(\frac{x-\theta_1}{\theta_2}\right)}{f\left(\frac{x-\theta_1}{\theta_2}\right)}\right]^2\frac{1}{\theta_2}f\left(\frac{x-\theta_1}{\theta_2}\right)dx$$

$$=\frac{1}{\theta_2^2}\int_{-\infty}^{\infty}\left[1+x\frac{f'(x)}{f(x)}\right]^2 f(x)\,dx$$

$$=\frac{1}{\theta_2^2}\left[\int_{-\infty}^{\infty}x^2\left(\frac{f'(x)}{f(x)}\right)^2 f(x)\,dx-1\right]=\frac{1}{\theta_2^2}I_{22}, \tag{13.5}$$

and hence

$$\boldsymbol{i}(\boldsymbol{\theta})=\boldsymbol{E}_{\boldsymbol{\theta}}\boldsymbol{\Lambda}_j\boldsymbol{\Lambda}_j^T=\frac{1}{\theta_2^2}\begin{bmatrix}I_{11} & I_{12}\\ I_{21} & I_{22}\end{bmatrix}, \tag{13.6}$$

where the elements I_{ij} are defined by formulas (13.3), (13.4), and (13.5).

The maximum-likelihood estimator $\hat{\boldsymbol{\theta}}_n=(\hat{\theta}_{1n},\hat{\theta}_{2n})^T$ of the vector $\boldsymbol{\theta}=(\theta_1,\theta_2)^T$ is a solution of the equation

$$\Lambda(\boldsymbol{\theta})=\mathbf{0}_2,$$

where

$$\boldsymbol{\Lambda}(\theta)=\mathbf{grad}\,\log L(\theta),\quad L(\theta)=\frac{1}{\theta_2^n}\prod_{i=1}^{n}f\left(\frac{X_i-\theta_1}{\theta_2}\right),$$

and hence

$$\boldsymbol{\Lambda}(\theta) = \sum_{j=1}^{n} \boldsymbol{\Lambda}_j = \sum_{j=1}^{n} \textbf{grad} \log\left[\frac{1}{\theta_2} f\left(\frac{X_j - \theta_1}{\theta_2}\right)\right].$$

As $n \to \infty$ (see Section 2),

$$\sqrt{n}(\hat{\boldsymbol{\theta}}_n - \boldsymbol{\theta}_0) \xrightarrow{\boldsymbol{P}_0} N_2(\mathbf{0}_2, \boldsymbol{I}^{-1}(\boldsymbol{\theta}_0)).$$

Following Sections 11 and 12, we choose a vector $\boldsymbol{p} = (p_1, \ldots, p_r)^T$ of positive probabilities, $p_i > 0$, such that $p_1 + p_2 + \cdots + p_r = 1$, and define the quantiles

$$x_0 < x_1 < \cdots < x_{r-1} < x_r$$

by $x_0 = -\infty$, $x_r = \infty$,

$$x_i = F^{-1}(p_1 + \cdots + p_i) = \inf\{x : F(x) \geq p_1 + \cdots + p_i\}, \quad i = 1, \ldots, r-1.$$

Furthermore, let $\boldsymbol{\nu}^* = (\nu_1^*, \ldots, \nu_r^*)^T$ be the frequency vector arising from grouping the random variables $X_1, \ldots, X_n$ over the intervals

$$(-\infty; x_1\hat{\theta}_{2n} + \hat{\theta}_{1n}], (x_1\hat{\theta}_{2n} + \hat{\theta}_{1n}; x_2\hat{\theta}_{2n} + \hat{\theta}_{1n}], \ldots, (x_{r-1}\hat{\theta}_{2n} + \hat{\theta}_{1n}; +\infty).$$

From Section 12, we obtain, as $n \to \infty$,

$$\frac{1}{\sqrt{n}}(\boldsymbol{\nu}^* - n\boldsymbol{p}) \xrightarrow{\boldsymbol{P}_0} N_r(\mathbf{0}_r, \boldsymbol{W}), \tag{13.7}$$

where

$$\boldsymbol{W} = \boldsymbol{P} - \boldsymbol{p}\boldsymbol{p}^T - \boldsymbol{C}^T(\boldsymbol{\theta}_0)\boldsymbol{I}^{-1}(\boldsymbol{\theta}_0)\boldsymbol{C}(\boldsymbol{\theta}_0), \tag{13.8}$$

$$\boldsymbol{p} = (p_1, \ldots, p_r)^T, \text{ rank } \boldsymbol{W} = r - 1,$$

$$\boldsymbol{P} = \begin{bmatrix} p_1 & 0 & \ldots & 0 \\ 0 & p_2 & \ldots & 0 \\ \vdots & \vdots & \ddots & \vdots \\ 0 & 0 & \ldots & p_r \end{bmatrix},$$

$$\boldsymbol{I}(\boldsymbol{\theta}) = \frac{1}{\theta_2^2}\begin{bmatrix} I_{11} & I_{12} \\ I_{21} & I_{22} \end{bmatrix}_{2\times 2}, \qquad \boldsymbol{C}(\boldsymbol{\theta}) = -\frac{1}{\theta_2}\begin{bmatrix} \boldsymbol{c}_1 \vdots \boldsymbol{c}_2 \end{bmatrix}_{r\times 2},$$

$$\boldsymbol{c}_1 = (c_{11}, \ldots, c_{1r})^T, \qquad \boldsymbol{c}_2 = (c_{21}, \ldots, c_{2r})^T, \qquad c_{1j} = f(x_j) - f(x_{j-1}),$$
$$c_{2j} = x_j f(x_j) - x_{j-1} f(x_{j-1}), \quad j = 1, \ldots, \tag{13.9}$$

Let us prove Eqs. (13.9). Using Sections 11 and 12, because

$$c_{ij} = \int_{a_{j-1}(\boldsymbol{\theta})}^{a_j(\boldsymbol{\theta})} \frac{\partial}{\partial \theta_i} f(x_j, \boldsymbol{\theta}) = \int_{x_{j-1}\theta_2+\theta_1}^{x_j\theta_2+\theta_1} \frac{\partial}{\partial \theta_i} \left[\frac{1}{\theta_2} f\left(\frac{x-\theta_1}{\theta_2}\right)\right] dx$$

$$= -\left\{\frac{1}{\theta_2} f\left(\frac{(x_j\theta_2+\theta_1)-\theta_1}{\theta_2}\right) \frac{\partial(x_j\theta_2+\theta_1)}{\partial\theta_i}\right.$$

$$\left. -\frac{1}{\theta_2} f\left(\frac{(x_{j-1}\theta_2+\theta_1)-\theta_1}{\theta_2}\right) \frac{\partial(x_{j-1}\theta_2+\theta_1)}{\partial\theta_i}\right\},$$

$$i = 1, 2; \quad j = 1, \ldots, r,$$

we obtain

$$c_{ij} = \begin{cases} -\frac{1}{\theta_2}[f(x_j) - f(x_{j-1})], & \text{if } i = 1;\ j = 1, \ldots, r, \\ -\frac{1}{\theta_2}[x_j f(x_j) - x_{j-1} f(x_{j-1})], & \text{if } i = 2;\ j = 1, \ldots, r, \end{cases}$$

and hence

$$\boldsymbol{C}(\boldsymbol{\theta}) = -\frac{1}{\theta_2}[\boldsymbol{c}_1 \vdots \boldsymbol{c}_2]_{r\times 2}.$$

It is evident that

$$\boldsymbol{I}^{-1}(\boldsymbol{\theta}) = \frac{\theta_2^2}{I_{11}\cdot I_{22} - I_{12}^2} \begin{bmatrix} I_{22} & -I_{12} \\ -I_{12} & I_{11} \end{bmatrix},$$

and hence the product in Eq. (13.8),

$$\boldsymbol{C}^T(\boldsymbol{\theta})\boldsymbol{I}^{-1}(\boldsymbol{\theta})\boldsymbol{C}(\boldsymbol{\theta}),$$

does not depend on $\boldsymbol{\theta}$, so that the matrix $\boldsymbol{W}$ does not depend on $\boldsymbol{\theta}_0$. We conclude that the limit distribution of the vector $1/\sqrt{n}(\boldsymbol{\nu}^* - n\boldsymbol{p})$ does not depend on the unknown parameter $\boldsymbol{\theta}_0$!

Our new test statistic is

$$Y_n^2 = \frac{1}{n}(\boldsymbol{\nu}^* - n\boldsymbol{p})^T \boldsymbol{W}^-(\boldsymbol{\nu}^* - n\boldsymbol{p}), \tag{13.10}$$

which does not depend on $\boldsymbol{\theta}_0$.

A Useful Formula for the Test Statistic

From the results of Section 12, we know that under H_0 the limit distribution of the statistic Y_n^2 as $n \to \infty$ is the chi-squared distribution with $(r-1)$ degrees of freedom:

$$\lim_{n\to\infty} \boldsymbol{P}\{Y_n^2 > x | H_0\} = \boldsymbol{P}\{\chi_{r-1}^2 > x\}. \tag{13.11}$$

Let us give another representation of Y_n^2. We denote

$$\begin{aligned}
X^2 &= \sum_{i=1}^{r} \frac{(\nu_i^* - np_i)^2}{np_i}, \\
\lambda_1 &= I_{11} - \boldsymbol{c}_1^T \boldsymbol{P}^{-1} \boldsymbol{c}_1 = I_{11} - \sum_{i=1}^{r} \frac{c_{1i}^2}{p_i}, \\
\lambda_2 &= I_{22} - \boldsymbol{c}_2^T \boldsymbol{P}^{-1} \boldsymbol{c}_2 = I_{22} - \sum_{i=1}^{r} \frac{c_{2i}^2}{p_i}, \\
\lambda_3 &= I_{12} - \boldsymbol{c}_1^T \boldsymbol{P}^{-1} \boldsymbol{c}_2 = I_{12} - \sum_{i=1}^{r} \frac{c_{1i}c_{2i}}{p_i}, \\
\alpha(\boldsymbol{\nu}^*) &= \boldsymbol{c}_1^T \boldsymbol{P}^{-1} \boldsymbol{\nu}^* = \sum_{i=1}^{r} \frac{c_{1i}\nu_i^*}{p_i}, \\
\beta(\boldsymbol{\nu}^*) &= \boldsymbol{c}_2^T \boldsymbol{P}^{-1} \boldsymbol{\nu}^* = \sum_{i=1}^{r} \frac{c_{2i}\nu_i^*}{p_i}.
\end{aligned} \tag{13.12}$$

By simple calculation one can see that

$$Y_n^2 = X^2 + \frac{1}{n(\lambda_1\lambda_2 - \lambda_3^2)}[\lambda_1\beta^2(\boldsymbol{\nu}^*) - 2\lambda_3\alpha(\boldsymbol{\nu}^*)\beta(\boldsymbol{\nu}^*) + \lambda_2\alpha^2(\boldsymbol{\nu}^*)]. \tag{13.13}$$

Remark. Neyman's (1937) smooth test is a modification of Pearson's test to take alternatives into account. It was developed further, particularly for location and scale families, by Barton (1956), Watson (1959), Thomas and Pierce (1979), Best and Rayner (1985), Rayner and Best (1986), Koziol (1987), Jacque and Bera (1987). The book by Rayner and Best (1989) is devoted to the circle of problems connected with smooth tests.

Symmetric Distributions

If the density $f(x)$ is an even function, that is, $f(x) = f(-x)$ for all admissible x, and $\boldsymbol{p}=(p_1,\ldots,p_r)^T$ is such that $p_1=p_r$, $p_2=p_{r-1}$, $p_3=p_{r-2}$, and so on, then $\lambda_3 = 0$. In particular, $\lambda_3 = 0$ if $p_1 = p_2 = \cdots = p_r = 1/r$. To see this, note that

$$I_{12} = \frac{1}{\theta_2}\int_{-\infty}^{\infty} x\left[\frac{f'(x)}{f(x)}\right]^2 f(x)\,dx = 0,$$

and hence, from Eq. (13.12), we have

$$\lambda_3 = -\boldsymbol{c}_1^T\boldsymbol{P}^{-1}\boldsymbol{c}_2.$$

But from Eq. (13.9)

$$\begin{aligned}\boldsymbol{c}_1^T\boldsymbol{P}^{-1}\boldsymbol{c}_2 &= \sum_{i=1}^{r}\frac{c_{1i}c_{2i}}{p_i} = \sum_{i=1}^{r}\frac{1}{p_i}\,[f(x_i)-f(x_{i-1}][x_if(x_i)-x_{i-1}f(x_{i-1})] \\ &= \sum_{i=1}^{r-1}\frac{x_if^2(x_i)}{p_i} \\ &\quad -\sum_{i=2}^{r-1}\frac{1}{p_i}[x_if(x_{i-1})f(x_i)+x_{i-1}f(x_{i-1})f(x_i)-x_{i-1}f^2(x_{i-1})] \\ &= 2\sum_{i=1}^{r-1}\frac{x_if^2(x_i)}{p_i}-\sum_{i=2}^{r-1}\frac{1}{p_i}(x_i+x_{i+1})f(x_i)f(x_{i-1}),\end{aligned}$$

where $x_i = F^{-1}(p_1+\cdots+p_i)$, $i = 1,\ldots,r-1$; $x_0 = -\infty$, $x_r = \infty$. If $r = 2m$, then $x_i = -x_{r-i}$ $(i = 1,\ldots,r-1)$, and hence

$$\sum_{i=1}^{r-1} \frac{x_i f^2(x_i)}{p_i} = 0, \qquad \sum_{i=2}^{r-2} \frac{(x_i + x_{i-1})}{p_i} f(x_i) f(x_{i-1}) = 0. \tag{13.14}$$

In the same way, one can prove that Eqs. (13.14) are true if $r = 2m+1$.

Hence, if $f(x)$ is even and $p_1 = p_r$, $p_2 = p_{r-1}, \ldots$, then $\lambda_3 = 0$ and

$$Y_n^2 = X^2 + \frac{1}{n\lambda_1} \left(\boldsymbol{c}_1^T \boldsymbol{P}^{-1} \boldsymbol{\nu}^* \right)^2 + \frac{1}{n\lambda_2} \left(\boldsymbol{c}_2^T \boldsymbol{P}^{-1} \boldsymbol{\nu}^* \right)^2;$$

that is,

$$Y_n^2 = X^2 + \frac{1}{n\lambda_1} \left(\sum_{i=1}^{r} \frac{c_{1i} \nu_i^*}{p_i} \right)^2 + \frac{1}{n\lambda_2} \left(\sum_{i=1}^{r} \frac{c_{2i} \nu_i^*}{p_i} \right)^2, \tag{13.15}$$

where

$$\lambda_1 = I_{11} - \sum_{i=1}^{r} \frac{c_{1i}^2}{p_i}, \qquad \lambda_2 = I_{22} - \sum_{i=1}^{r} \frac{c_{2i}^2}{p_i}.$$

In particular, if $p_1 = p_2 = \cdots = p_r = 1/r$, then

$$Y_n^2 = X^2 + \frac{r^2}{n} \left[\frac{1}{\lambda_1} \left(\sum_{i=1}^{r} c_{1i} \nu_i^* \right)^2 + \frac{1}{\lambda_2} \left(\sum_{i=1}^{r} c_{2i} \nu_i^* \right)^2 \right], \tag{13.16}$$

where

$$X^2 = \frac{r}{n} \sum_{i=1}^{r} (\nu_i^*)^2 - n, \quad \lambda_1 = I_{11} - r \sum_{i=1}^{r} c_{1i}^2, \quad \lambda_2 = I_{22} - r \sum_{i=1}^{r} c_{2i}^2.$$

For example, if $r = 2$, $p_1 = p_2 = 0.5$, $\nu_1^* = n - \nu_2^*$,

$$\boldsymbol{c}_1 = (f(0), -f(0))^T, \qquad \boldsymbol{c}_2 = (0, 0)^T, \qquad \lambda_1 = I_{11} - 4f^2(0),$$

and hence in this case $[f(x) = f(-x),\ r = 2]$

$$\begin{aligned} Y_n^2 &= \frac{2}{n} [(\nu_1^*)^2 + (\nu_2^*)^2] - n + \frac{4}{n\lambda_1} (c_{11} \nu_1^* + c_{12} \nu_2^*)^2 \\ &= \frac{(\nu_1^* - \nu_2^*)^2 I_{11}}{n[I_{11} - 4f(0)]}. \end{aligned} \tag{13.17}$$

Formula (13.17) gives a form of the statistic Y_n^2 that one can use for testing the symmetry of the distribution of X_j. Of course, if the hypothesis H_0 is true, then

$$\lim_{n\to\infty} \boldsymbol{P}\{Y_n^2 > x\} = \boldsymbol{P}\{\chi_1^2 > x\}.$$

Shift Families

Let us suppose that θ_2 is known, for example, that $\theta_2 = 1$. In this case we have

$$F\left(\frac{x-\theta_1}{\theta_2}\right) = F(x-\theta_1)$$

and the vector $\boldsymbol{\nu}^* = (\nu_1^*, \dots, \nu_r^*)^T$ is constructed by grouping the observations $X_1, \dots, X_n$ using the intervals

$$(-\infty, x_1+\hat{\theta}_{1n}], (x_1+\hat{\theta}_{1n}, x_2+\hat{\theta}_{1n}], \dots, (x_{r-1}+\hat{\theta}_{1n}; +\infty),$$

where $\hat{\theta}_{1n}$ is the maximum-likelihood estimator for θ_1, $x_j = F^{-1}(p_1 + \cdots + p_j)$. In this case

$$\frac{1}{\sqrt{n}}(\boldsymbol{\nu}^* - n\boldsymbol{p}) \overset{\boldsymbol{P}_0}{\Longrightarrow} N(\mathbf{0}_r, \boldsymbol{W}),$$

where [compare with Eq. (13.8)]

$$\boldsymbol{W} = \boldsymbol{P} - \boldsymbol{p}\boldsymbol{p}^T - \frac{1}{I_{11}}\boldsymbol{c}_1\boldsymbol{c}_1^T.$$

The statistic Y_n^2 is

$$Y_n^2 = X^2 + \frac{1}{n\lambda_1}\left(\sum_{i=1}^{r} \frac{c_{1i}\nu_i^*}{p_i}\right)^2, \tag{13.18}$$

where $\lambda_1 = I_{11} - \boldsymbol{c}_1^T \boldsymbol{P}^{-1} \boldsymbol{c}_1$.

Scale Families

If the parameter θ_2 is unknown, and θ_1 is known, for example, $\theta_1 = 0$, the vector $\boldsymbol{\nu}^* = (\nu_1^*, \dots, \nu_r^*)^T$ is constructed by grouping the observations

$X_1, \dots, X_n$ using intervals

$$(-\infty; x_1\hat{\theta}_{2n}], (x_1\hat{\theta}_{2n}; x_2\hat{\theta}_{2n}], \dots, (x_{r-1}\hat{\theta}_{2n}; +\infty),$$

where $\hat{\theta}_{2n}$ is the maximum-likelihood estimator of θ_2. In this case

$$\frac{1}{\sqrt{n}}(\boldsymbol{\nu}^* - n\boldsymbol{p}) \xrightarrow{\boldsymbol{P}_0} N(\boldsymbol{0}_r, \boldsymbol{W}),$$

where

$$\boldsymbol{W} = \boldsymbol{P} - \boldsymbol{p}\boldsymbol{p}^T - \frac{1}{I_{22}}\boldsymbol{c}_2^T\boldsymbol{c}_2,$$

and hence

$$Y_n^2 = X^2 + \frac{1}{n\lambda_2}\left(\sum_{i=1}^{r}\frac{c_{2i}\nu_i^*}{p_i}\right)^T, \tag{13.19}$$

where $\lambda_2 = I_{22} - \boldsymbol{c}_2^T\boldsymbol{P}^{-1}\boldsymbol{c}_2$.

Example. Gamma Scale Family. Let $X_1, X_2, \dots, X_n$ be i.i.d. with density $f(x, \theta)$, $\theta > 0$ where, under H_0,

$$f(x,\theta) = \begin{cases} 0, & x < 0, \\ \frac{1}{\Gamma(m)\theta^m}\, x^{m-1}e^{-x/\theta}, & x \geq 0, \end{cases}$$

$$\text{some fixed } m \in \boldsymbol{X}_1 = \{1, 2, \dots, \}, \qquad \Gamma(m) = \int_0^\infty x^{m-1}e^{-x}\,dx.$$

We construct statistic (13.19). First, we calculate

$$\begin{aligned} I_{22} &= \int_0^\infty \left[1 + x\frac{f'(x;1)}{f(x;1)}\right]^2 f(x;1)\,dx \\ &= 1 + 2\int_0^\infty (m-1-x)\frac{x^{m-1}e^{-x}}{\Gamma(m)}\,dx \\ &\quad + \int_0^\infty \frac{x^{m-1}e^{-x}}{\Gamma(m)}(m-1-x)^2\,dx = m. \end{aligned}$$

The likelihood function $L(\theta)$ is

$$L(\theta) = \prod_{i=1}^{n} f(X_i; \theta) = \frac{1}{\Gamma^n(m)\theta^{mn}} \left(\prod_{i=1}^{n} X_i^{m-1} \right) \exp\left\{ -\frac{1}{\theta} \sum_{i=1}^{n} X_i \right\}.$$

Since

$$\frac{\partial}{\partial\theta} L(\theta) = -\frac{mn}{\theta} + \frac{1}{\theta^2} \sum_{i=1}^{n} X_i,$$

the maximum-likelihood estimator $\hat{\theta}_n$ for θ is

$$\hat{\theta}_n = \frac{1}{m} \bar{X}_n, \qquad \text{where} \qquad \bar{X}_n = \frac{1}{n} \sum_{i=1}^{n} X_i.$$

Now choose the numbers $x_1, \ldots, x_{r-1}$ to satisfy the equations

$$\frac{i}{r} = \int_0^{x_i} f(x, 1)\, dx = \int_0^{x_i} \frac{x^{m-1}}{\Gamma(m)} e^{-x}\, dx, \qquad i = 1, \ldots, r-1,$$

and let $x_0 = 0$, $x_r = +\infty$. This choice of x_i gives us an equiprobable partition,

$$p_1 = p_i = \cdots = p_r = 1/r.$$

Let $\boldsymbol{\nu}^* = (\nu_1, \ldots, \nu_r)^T$ be the vector of frequencies obtained by grouping the random variables $X_1, \ldots, X_n$ using

$$(0; x_1\hat{\theta}_n], (x_1\hat{\theta}_n; x_2\hat{\theta}_n], \ldots, (x_{r-1}\hat{\theta}_n; +\infty).$$

To form Y_n^2 by Eq. (13.19) we calculate

$$\lambda_2 = I_{22} - \sum_{i=1}^{r} \frac{c_{2i}^2}{p_i} = m - r \sum_{i=1}^{r} c_{2i}^2,$$

because $p_i = 1/r$, where

$$c_{2i} = x_i f(x_i; 1) - x_{i-1} f(x_{i-1}; 1),$$

and hence

$$Y_n^2 = \frac{r}{n} \sum_{i=1}^{n} \nu_i^{*2} - n + \frac{r^2}{n\lambda_2} \left(\sum_{i=1}^{r} c_{2i} \nu_i^* \right)^2.$$

If H_0 is true then

$$\lim_{n\to\infty} \boldsymbol{P}\{Y_n^2 > x\} = \boldsymbol{P}\{\chi_{r-1}^2 > x\}.$$

Example. Exponential Families with Scale and/or Shift. Consider the problem of testing the hypothesis H_0 according to which the distribution function of the independent identically distributed random variables $X_1, \ldots, X_n$ belongs to the family $F(x;\boldsymbol{\theta})$, $\boldsymbol{\theta} = (\theta_1, \theta_2)^T \in \Theta \subset \boldsymbol{R}^2$ where

$$F(x;\boldsymbol{\theta}) = F\left(\frac{x-\theta_1}{\theta_2}\right) = \begin{cases} 1 - e^{-\frac{x-\theta_1}{\theta_2}}, & x \geq \theta_1, \\ 0 & x < \theta_1. \end{cases}$$

$$\boldsymbol{\Theta} = \{\boldsymbol{\theta} = (\theta_1, \theta_2)^T : \theta_1 \geq 0, \theta_2 > 0\}.$$

This family is important for applications (see e.g., Lawless, 1982). Other goodness-of-fit tests for exponential distributions are found in Stephens (1986c) and Rayner and Best (1982, 1989). In other words we want to test the hypothesis that the X_i have an exponential law. There are four cases:

i. θ_1 and θ_2 are known,

ii. θ_1 is unknown, θ_2 is known,

iii. θ_1 is known, θ_2 is unknown,

iv. θ_1 and θ_2 are unknown,

If H_0 is true then the density function of X_i is

$$f(x;\boldsymbol{\theta}) = \frac{1}{\theta_2} f\left(\frac{x-\theta_1}{\theta_2}\right) = \begin{cases} 0 & \text{if } x < \theta_1, \\ \frac{1}{\theta_2}\exp\left\{-\frac{x-\theta_1}{\theta_2}\right\}, & \text{if } x \geq \theta_1, \end{cases}$$

Denote $f(x;(0,1)^T)$ by $f(x)$ and consider the equiprobable case, $p = (1/r, \ldots, 1/r)^T$. The definition of x_i by $x_0 = 0$, $x_r = \infty$ and

$$\frac{j}{r} = \int_0^{x_j} f(x)\, dx, \qquad j = 1, \ldots, r-1.$$

gives

$$x_j = -\ln(1-\frac{j}{r}), \qquad j=1,\ldots,r-1.$$

To test H_0 in case i we group $X_1,\ldots,X_n$ using the intervals

$$(0;x_1\theta_2+\theta_1],(x_1\theta_2+\theta_1;x_2\theta_2+\theta_1],\ldots,(x_{r-1}\theta_2+\theta_1;+\infty).$$

Let $\boldsymbol{\nu}^* = (\nu_1^*,\ldots,\nu_r^*)^T$ be the vector of frequencies. To test H_0 we use the standard statistic of Pearson,

$$Y_n^2 = X^2 = \sum_{i=1}^{r} \frac{(\nu_i^* - np_i)^2}{np_i} = \frac{r}{n}\sum_{i=1}^{r}\nu_i^{*2} - n.$$

Now consider case iii, where θ_1 is known (see also Rao and Robson, 1974). The maximum-likelihood estimator $\hat{\theta}_{2n}$ for θ_2 is

$$\hat{\theta}_{2n} = \bar{X}_n - \theta_1 = \frac{1}{n}\sum_{i=1}^{r} X_i - \theta_1.$$

We construct the vector of frequencies $\boldsymbol{\nu}^* = (\nu_1^*,\ldots,\nu_r^*)^T$ by grouping the random variables $X_1,\ldots,X_n$ using the intervals

$$(0;x_1\hat{\theta}_{2n}+\theta_1],(x_1\hat{\theta}_{2n}+\theta_1;x_2\hat{\theta}_{2n}+\theta_1],\ldots,(x_{r-1}\hat{\theta}_{2n}+\theta_1;+\infty).$$

We calculate λ_2. From Eq. (13.12), we have

$$I_{22} = \int_0^\infty x^2\left[\frac{f'(x)}{f(x)}\right]^2 f(x)\,dx = \int_0^\infty x^2 e^{-x}\,dx = 1,$$

and hence

$$\begin{aligned}
\lambda_2 &= I_{22} - \sum_{i=1}^{r}\frac{c_{2i}^2}{p_i} \\
&= 1 - r\sum_{i=1}^{r}[x_i f(x_i) - x_{i-1}f(x_{i-1})]^2 \\
&= 1 - \frac{1}{r}\left[\ln^2(1-\frac{1}{r}) + \ln^2\frac{1}{r}\right] \\
&\quad - \frac{1}{r}\sum_{i=2}^{r-1}\left[(1-\frac{i}{r})\ln\frac{r-i+1}{r-i} + \frac{1}{r}\ln(1-\frac{i-1}{r})\right]^2.
\end{aligned}$$

Now we can construct the test statistic

$$Y_n^2 = \frac{r}{n}\sum_{i=1}^{r}\nu_i^{*2} - n + \frac{r^2}{n\lambda_2}\left(\sum_{i=1}^{r}\nu_i^* c_{2i}\right)^2.$$

Consider case ii. We have θ_1 unknown, θ_2 known (e.g., $\theta_2 = 1$). This case is not regular; the support of the distribution of X_i depends on θ. Consider the order statistics

$$X_{(1)} \leq X_{(2)} \leq \cdots \leq X_{(n)},$$

constructed from our sample $X_1, \ldots, X_n$. The density function of the vector $(X_{(1)} \ldots, X_{(n)})^T$ is

$$\begin{aligned} p(z_1, \ldots, z_n) &= n! \exp\{-\Sigma(z_i - \theta_1)\}, \\ &= ne^{-n(z_1-\theta_1)} \cdot (n-1)! \exp\left\{-\sum_{i=2}^{n}(z_i - z_1)\right\}, \quad (13.20) \end{aligned}$$

$$\theta_1 \leq z_1 \leq \cdots \leq z_n.$$

Hence, $X_{(1)}$ is a sufficient statistic for θ_1. In fact $X_{(1)}$ is a superefficient estimator for θ_1 and

$$\boldsymbol{E}X_{(1)} = \theta_1 + \frac{1}{n}, \quad \textbf{Var}\ X_{(1)} = \frac{1}{n^2}.$$

From Eq. (13.20) we have that if H_0 is true then, given $X_{(1)}$, the random variables

$$U_2 = X_{(2)} - X_{(1)}, U_3 = X_{(3)} - X_{(1)}, \ldots, U_n = X_{(n)} - X_{(1)}$$

are independent with law

$$U_i \sim e^{-(x - X_{(1)})}, \qquad i = 2, \ldots, n.$$

Construct the vector $\boldsymbol{\nu}^* = (\nu_1^*, \ldots, \nu_r^*)^T$ by grouping $U_2, \ldots, U_n$ using the intervals

$$(X_{(1)}, x_1 + X_{(1)}]; (x_1 + X_{(1)}; x_2 + X_{(1)}], \ldots, (x_{r-1} + X_{(1)}; +\infty),$$

Pearson's statistic is

$$X^2 = \frac{r}{n-1}\sum_{i=1}^{r} \nu_i^{*2} - (n-1).$$

If H_0 is true then

$$\lim_{n\to\infty} \mathbf{P}\{X^2 > x\} = \mathbf{P}\{\chi^2_{r-1} > x\}.$$

Now we consider the last case (iv). As in case ii, let

$$U_2 = X_{(2)} - X_{(1)}, \ldots, U_n = X_{(n)} - X_{(1)}.$$

If H_0 is true, then, given $X_{(1)}$, the random variables $U_2, \ldots, U_n$ form an independent sample from the law

$$F(\frac{x}{\theta_2}) = 1 - \exp\{-\frac{x}{\theta_2}\}, \quad x \geq 0.$$

The maximum-likelihood estimator $\hat{\theta}_{2n}$ calculated from $U_2, \ldots, U_n$ is

$$\hat{\theta}_{2n} = \frac{1}{n-1}\sum_{i=2}^{n} U_i = \frac{n}{n-1}[\bar{X}_n - X_{(1)}].$$

Construct the frequency vector $\boldsymbol{\nu}^* = (\nu_1^*, \ldots, \nu_r^*)^T$ by grouping $U_2, \ldots, U_n$ using the intervals

$$(X_{(1)}, x_1\hat{\theta}_{2n} + X_{(1)}], (x_1\hat{\theta}_{2n} + X_{(1)}, x_2\hat{\theta}_{2n} + X_{(1)}], \ldots, (x_{r-1}\hat{\theta}_{2n} + X_{(1)}, +\infty),$$

and let

$$Y_n^2 = \frac{r}{(n-1)}\sum_{i=1}^{r} \nu_i^{*2} - (n-1) + \frac{r^2}{(n-1)\lambda_2}\left(\sum_{i=1}^{r} \nu_i^* c_{2i}\right)^2,$$

where

$$c_{2i} = x_i f(x_i) - x_{i-1} f(x_{i-1}), \qquad \lambda_2 = 1 - r\sum_{i=1}^{r} c_2^{2i}.$$

If H_0 is true, then

$$\lim_{n\to\infty} \boldsymbol{P}\{Y_n^2 > x\} = \boldsymbol{P}\{\chi_{r-1}^2 > x\}.$$

Example. Cauchy Shift Family. Let $X_1, \ldots, X_n$ be i.i.d., and let the hypothesis H_0 be that X_i has density $f(x;\theta)$, where

$$f(x;\theta) = \frac{1}{\pi[1+(x-\theta)^2]}, \qquad |x| < \infty, \qquad |\theta| < \infty.$$

In this case the Fisher information

$$\boldsymbol{I}(\theta) = I_{11} = \boldsymbol{E}\left\{\frac{\partial \log f(X_j;\theta)}{\partial \theta}\right\}^2 = \frac{4}{\pi}\int_{-\infty}^{\infty} \frac{(x-\theta)^2}{[1+(x-\theta)^2]^3}\,dx = \frac{1}{2}.$$

Let $\boldsymbol{p} = (1/r, \ldots, 1/r)^T$ and define the x_i by

$$x_0 = -\infty, \quad x_r = +\infty, \quad \frac{i}{r} = \frac{1}{\pi}\int_{-\infty}^{x_i} \frac{dx}{1+x^2}, \quad i = 1, \ldots, r-1.$$

The maximum-likelihood estimator $\hat{\theta}_n$ of θ is the solution of

$$\theta = \frac{\sum_{i=1}^{n} X_i/[1+(X_i-\theta)^2]}{\sum_{i=1}^{n} 1/[1+(X_i-\theta)^2]}.$$

If $\boldsymbol{\nu}^* = (\nu_1^*, \ldots, \nu_r^*)^T$ is the frequency vector obtained using the intervals

$$(-\infty; x_1+\hat{\theta}_n], (x_1+\hat{\theta}_n; x_2+\hat{\theta}_n], \ldots, (x_{r-1}+\hat{\theta}_n; +\infty),$$

then

$$Y_n^2 = \frac{r}{n}\sum_{i=1}^{r} \nu_i^{*2} - n + \frac{r^2}{n\lambda_1}\left(\sum_{i=1}^{r} \nu_i^* c_{1i}\right)^2$$

has the chi-squared distribution, in the limit as $n \to \infty$, with $(r-1)$ degrees of freedom, where

$$c_{1i} = f(x_i) - f(x_{i-1}) = \frac{1}{\pi}\left[\frac{1}{1+x_i^2} - \frac{1}{1+x_{i-1}^2}\right],$$

$$\lambda_1 = I_{11} - r\sum_{i=1}^{r} c_{1i}^2 = \frac{1}{2} - r\sum_{i=1}^{r} c_{1i}^2.$$

In particular, if $r = 4$, $p_1 = p_2 = p_3 = p_4 = 1/4$, then

$$x_0 = -\infty \qquad c_{11} = c_{12} = \frac{1}{2\pi},$$
$$x_1 = -1,$$
$$x_2 = 0, \qquad c_{13} = c_{14} = -\frac{1}{2\pi},$$
$$x_3 = 1,$$
$$x_4 = +\infty, \qquad \lambda_1 = \frac{1}{2} - \frac{4}{\pi^2},$$

and

$$Y_n^2 = \frac{4}{n}\sum_{i=1}^{4} \nu_i^{*2} - n + \frac{8}{n(\pi^2 - 8)}(\nu_1^* + \nu_2^* - \nu_3^* - \nu_4^*)^2.$$

If H_0 is true

$$\lim_{n\to\infty} \boldsymbol{P}\{Y_n^2 > x\} = \boldsymbol{P}\{\chi_3^2 > x\}.$$

Example. Logistic Distribution. Suppose $X_1, \ldots, X_n$ are i.i.d. with logistic density as in Remark 11.2. Balakrishnan and Cohen (1990) proposed an approximate solution $\hat{\theta}_n$ of the maximum-likelihood equation and showed it to be asymptotically efficient. Aguirre and Nikulin (1994) base grouping intervals on this $\hat{\theta}_n$ and show that the associated Pearson's statistic has limiting chi-squared distribution with $(r-1)$ degrees of freedom. A comprehensive study of applications and methods for the logistic distribution can be found in the *Handbook of the Logistic Distribution* edited by Balakrishnan (1992).

14. THE CHI-SQUARED TEST OF NORMALITY

Construction of Test Statistic – Univariate Case

Let $X_1, \ldots, X_n$ be i.i.d. We wish to test the hypothesis, H_0 , that

$$X_i \sim \Phi(\frac{x-\mu}{\sigma}), \qquad |\mu| < \infty, \quad \sigma > 0,$$

where $\Phi(x)$ is the distribution function of the standard normal law. The Fisher information matrix is

$$\boldsymbol{I}(\boldsymbol{\theta}) = \boldsymbol{I}(\mu, \sigma) = \frac{1}{\sigma^2}\begin{bmatrix} I_{11} & I_{12} \\ I_{21} & I_{22} \end{bmatrix} = \frac{1}{\sigma^2}\begin{bmatrix} 1 & 0 \\ 0 & 2 \end{bmatrix}, \quad \boldsymbol{\theta} = (\mu, \sigma)^T,$$

because

$$I_{11} = \int_{-\infty}^{\infty} \left[\frac{\varphi'(x)}{\varphi(x)}\right]^2 \varphi(x)\,dx = 1$$

$$I_{12} = \int_{-\infty}^{\infty} x\left[\frac{\varphi'(x)}{\varphi(x)}\right]^2 \varphi(x) = 0,$$

$$I_{22} = \int_{-\infty}^{\infty} \left[1 + x\frac{\varphi'(x)}{\varphi(x)}\right]^2 \varphi(x)\,dx = 2.$$

Set

$$\eta_i = \begin{cases} \frac{1}{\sigma}(X_i - \mu), & \text{if } \mu \text{ and } \sigma \text{ are known,} \\ \frac{1}{\sigma}\sqrt{\frac{n}{n-1}}(X_i - \bar{X}_n), & \text{if } \mu \text{ is unknown, } \sigma \text{ is known,} \\ \frac{1}{\tilde{s}}(X_i - \mu), & \text{if } \mu \text{ is known, } \sigma \text{ is unknown,} \\ \frac{1}{s}(X_i - \bar{X}_n), & \text{if } \mu \text{ and } \sigma \text{ are unknown,} \end{cases}$$

where $i = 1, 2, \ldots, n$ and

$$\bar{X}_n = \frac{1}{n}\sum_{i=1}^{n} X_i, \qquad \tilde{s}^2 = \frac{1}{n}\sum_{i=1}^{n}(X_i - \mu)^2, \qquad s^2 = \frac{1}{n}\sum_{i=1}^{n}(X_i - \bar{X}_n)^2.$$

The statistics $\bar{X}_n$, $\tilde{s}^2$, and s^2 are the maximum-likelihood estimators for the parameters μ and σ^2. Let $\boldsymbol{p} = (p_1, \ldots, p_r)^T$ be the vector of positive probabilities such that

$$p_1 + \cdots + p_r = 1, \qquad p_i > 0, \qquad i = 1, \ldots, r,$$

and let

$$x_j = \Phi^{-1}(p_1 + \cdots + p_j), \qquad j = 1, \ldots, r-1.$$

Setting $x_0 = -\infty$, $x_r = +\infty$, consider $\boldsymbol{\nu}^* = (\nu_1^*, \ldots, \nu_r)^T$, the frequency vector obtained by grouping $\eta_1, \eta_2, \ldots, \eta_n$ using the intervals

$(x_0, x_1], (x_1, x_2], \ldots, (x_{r-1}, x_r)$. It is evident that the same vector ν^* is obtained by grouping $X_1, \ldots, X_n$ using the intervals ($i = 1, \ldots, r$):

$$J_i = \begin{cases} (x_{i-1}\sigma + \mu, x_i\sigma + \mu], & \text{if } \mu \text{ and } \sigma \text{ are known,} \\ (x_{i-1}\sigma + \bar{X}_n, x_i\sigma + \bar{X}_n], & \text{if } \mu \text{ is unknown, } \sigma \text{ is known,} \\ (x_{i-1}\tilde{s} + \mu, x_i\tilde{s} + \mu], & \text{if } \mu \text{ is known, } \sigma \text{ is unknown,} \\ (x_{i-1}s + \bar{X}_n, x_i s + \bar{X}_n], & \text{if } \mu \text{ and } \sigma \text{ are unknown.} \end{cases}$$

If H_0 is true then

$$\boldsymbol{P}\{\eta_j \in (x_{i-1}, x_i]\} = \boldsymbol{P}\{X_j \in J_i\} = p_i, \qquad i = 1, \ldots, r.$$

As in Section 6 we consider the statistic

$$Y_n^2 = \begin{cases} X^2, & \text{if } \mu \text{ and } \sigma^2 \text{ are known,} \\ X^2 + \frac{1}{n\lambda_1}\alpha^2(\boldsymbol{\nu}^*), & \text{if } \mu \text{ is unknown, } \sigma \text{ is known,} \\ X^2 + \frac{1}{n\lambda_2}\beta^2(\boldsymbol{\nu}^*), & \text{if } \mu \text{ is known, } \sigma \text{ is unknown,} \\ X^2 + \frac{\lambda_1\beta^2(\boldsymbol{\nu}^*) - 2\lambda_3\alpha(\boldsymbol{\nu}^*)\beta(\boldsymbol{\nu}^*) + \lambda_2\alpha^2(\boldsymbol{\nu}^*)}{n[\lambda_1\lambda_2 - \lambda_3^2]}, & \\ & \text{if } \mu \text{ and } \sigma \text{ are unknown,} \end{cases} \tag{14.1}$$

where, as in Eq. (13.12), we have

$$X^2 = \sum_{i=1}^{r} \frac{(\nu_i^* - np_i)^2}{np_i},$$

$$\alpha(\nu^*) = \boldsymbol{c}_1^T \boldsymbol{P}^{-1} \boldsymbol{\nu}^* = \sum_{j=1}^{r} \frac{c_{1j}\nu_j^*}{p_j},$$

$$\beta(\nu^*) = \boldsymbol{c}_2^T \boldsymbol{P}^{-1} \boldsymbol{\nu}^* = \sum_{j=1}^{r} \frac{c_{2j}\nu_j^*}{p_j},$$

$$\lambda_1 = I_{11} - \boldsymbol{c}_1^T \boldsymbol{P}^{-1} \boldsymbol{c}_1 = 1 - \sum_{j=1}^{r} \frac{c_{1j}^2}{p_j},$$

$$\lambda_2 = I_{22} - \boldsymbol{c}_2^T \boldsymbol{P}^{-1} \boldsymbol{c}_2 = 2 - \sum_{j=1}^{r} \frac{1}{p_j} c_{2j}^2,$$

$$\lambda_3 = I_{12} - \boldsymbol{c}_1^T \boldsymbol{P}^{-1} \boldsymbol{c}_2 = - \sum_{j=1}^{r} \frac{1}{p_j} c_{1j} c_{2j},$$

$$\begin{aligned}
\boldsymbol{c}_1 &= (c_{11}, \dots, c_{1r})^T, \\
c_{1j} &= \varphi(x_j) - \varphi(x_{j-1}), \\
\boldsymbol{c}_2 &= (c_{21}, \dots, c_{2r})^T, \\
c_{2j} &= -x_j \varphi(x_j) + x_{j-1} \varphi(x_{j-1}) = \varphi'(x_j) - \varphi'(x_{j-1}), \\
\varphi(x) &= \Phi'(x).
\end{aligned}$$

From Section 12 we know that under H_0 as $n \to \infty$ the statistic Y_n^2 has an asymptotically chi-squared distribution with $(r-1)$ degrees of freedom.

Some Particular Cases

1. If the vector $\boldsymbol{p} = (p_1, \dots, p_r)^T$ is chosen symmetric, so that $p_1 = p_r$, $p_2 = p_{r-1}$, $p_3 = p_{r-2}$ and so on, then $\lambda_3 = 0$. Ogawa (1976) found that this symmetric choice is optimal for testing normality.

2. Let $p_1 = p_2 = \cdots = p_r = 1/r$. In this case $\lambda_3 = 0$, and instead of Eq. (14.1) we have

$$Y^2 = \begin{cases} X^2, & \text{if } \mu \text{ and } \sigma^2 \text{ are known,} \\ X^2 + \frac{1}{n}(\boldsymbol{\varepsilon}^T \boldsymbol{\nu}^*)^2, & \text{if } \mu \text{ is unknown, } \sigma \text{ is known,} \\ X^2 + \frac{1}{n}(\boldsymbol{w}^T \boldsymbol{\nu}^*)^2, & \text{if } \mu \text{ is known, } \sigma \text{ is unknown,} \\ X^2 + \frac{1}{n}(\boldsymbol{\varepsilon}^T \boldsymbol{\nu}^*)^2 + \frac{1}{n}(\boldsymbol{w}^T \boldsymbol{\nu}^*)^2, & \text{if } \mu \text{ and } \sigma \text{ are unknown,} \end{cases}$$

where

$$X^2 = \frac{r}{n} \sum_{i=1}^{r} (\nu_i^* - \frac{n}{r})^2 = \frac{r}{n} \sum_{i=1}^{r} \nu_i^{*2} - n,$$

$$\boldsymbol{\varepsilon} = (\varepsilon_1, \dots, \varepsilon_r)^T = \frac{r}{\sqrt{\lambda_1}} \boldsymbol{c}_1,$$

$$\boldsymbol{w} = (w_1, \dots, w_r)^T = \frac{r}{\sqrt{\lambda_2}} \boldsymbol{c}_2,$$

$$\lambda_1 = 1 - r\boldsymbol{c}_1^T\boldsymbol{c}_1 = 1 - r\sum_{j=1}^{r} c_{1j}^2 = 1 - r\sum_{j=1}^{r}[\varphi(x_j) - \varphi(x_{j-1})]^2,$$

$$\lambda_2 = 2 - r\boldsymbol{c}_2^T\boldsymbol{c}_2 = 2 - r\sum_{j=1}^{r} c_{2j}^2 = 2 - r\sum_{j=1}^{r}[\varphi'(x_j) - \varphi'(x_{j-1})]^2,$$

$$x_j = \Phi^{-1}(\frac{j}{k}), \quad j = 1, \ldots, r-1, \quad x_0 = -\infty, \quad x_r = +\infty.$$

If $r = 2$, we have

$$p_1 = p_2 = 0.5, \qquad \nu_1^* + \nu_2^* = n, \qquad c_1 = \left(\frac{1}{\sqrt{2\pi}}, -\frac{1}{\sqrt{2\pi}}\right)^T$$

and

$$Y^2 = \begin{cases} \dfrac{(\nu_1^* - \nu_2^*)^2}{n}, & \text{if } \mu \text{ and } \sigma \text{ are known,} \\ \dfrac{(\nu_1^* - \nu_2^*)^2}{n(1-\frac{2}{\pi})}, & \text{otherwise.} \end{cases}$$

3. We compute $\alpha(\boldsymbol{\nu}^*)$ appearing in Eq. (14.1) when

$$p_1 = p_2 = \cdots = p_r = 1/r.$$

Then

$$\begin{aligned} \alpha(\boldsymbol{\nu}^*) &= \sum_{i=1}^{r} \frac{c_{1i}\nu_i^*}{p_i} \\ &= \sum_{i=1}^{r} \frac{c_{1i}(\nu_i^* - np_i)}{p_i} \\ &= r\sum_{i=1}^{r} c_{1i}(\nu_i^* - \frac{n}{r}) = \boldsymbol{c}_1^T(\boldsymbol{\nu}^* - \frac{n}{r}\mathbf{1}_r) \end{aligned}$$

Since

$$\boldsymbol{E}\boldsymbol{\nu}^* = n\boldsymbol{p} = \frac{n}{r}\mathbf{1}_r$$

we have

$$\boldsymbol{E}\alpha(\boldsymbol{\nu}^*) = r\boldsymbol{E}\boldsymbol{c}_1^T(\boldsymbol{\nu}^* - \frac{n}{r}\mathbf{1}_r) = 0$$

and

$$\mathbf{Var}\ \alpha(\boldsymbol{\nu}^*) = r^2\ \mathbf{Var}\ \boldsymbol{c}_1^T(\boldsymbol{\nu}^* - \frac{n}{r}\mathbf{1}_r) = r^2\boldsymbol{c}_1^T\boldsymbol{E}(\boldsymbol{\nu}^* - \frac{n}{r}\mathbf{1}_r)(\boldsymbol{\nu}^* - \frac{n}{r}\mathbf{1}_r)^T\boldsymbol{c}_1$$

$$= r^2 n\boldsymbol{c}_1^T[\frac{1}{r}\boldsymbol{I}_r - \frac{1}{r^2}\mathbf{1}_r\mathbf{1}_r^T - \boldsymbol{c}_1\boldsymbol{c}_1^T]\boldsymbol{c}$$

$$= r^2 n[\frac{1}{r}\boldsymbol{c}_1^T\boldsymbol{c}_1 - (\boldsymbol{c}_1^T\boldsymbol{c}_1)^2] = \frac{n}{r}\boldsymbol{c}_1^T\boldsymbol{c}_1(1 - r\boldsymbol{c}_1^T\boldsymbol{c}_1)$$

$$= rn\lambda_1\left(\sum_{i=1}^{r} c_{1i}^2\right).$$

By the central limit theorem, $\alpha(\boldsymbol{\nu}^*)$ is asymptotically normal and hence the statistic

$$Z^2 = \frac{r}{n}\frac{(\boldsymbol{c}_1^T\boldsymbol{\nu}^*)^2}{(\boldsymbol{c}_1^T\boldsymbol{c}_1)[1 - r\boldsymbol{c}_1^T\boldsymbol{c}_1]} = \frac{r(\boldsymbol{c}_1^T\boldsymbol{\nu}^*)^2}{n\lambda_1(\boldsymbol{c}_1^T\boldsymbol{c})}$$

has in the limit as $n \to \infty$ a chi-squared distribution with 1 degree of freedom,

$$\lim_{n\to\infty} \boldsymbol{P}\{Z^2 \le x | H_0\} = \boldsymbol{P}\{\chi_1^2 \le x\}.$$

This statistic is useful if the frequencies $\boldsymbol{\nu}^*$ are too small for direct application of the central limit theorem.

Table 14.1 gives values of the vectors

$$\boldsymbol{x} = (x_1, \ldots, x_{r-1})^T, \quad \boldsymbol{\varepsilon} = (\varepsilon_1, \ldots, \varepsilon_r)^T = \frac{r}{\sqrt{\lambda_1}}\boldsymbol{c}_1,$$
$$\boldsymbol{w} = (w_1, \ldots, w_r)^T = \frac{r}{\sqrt{\lambda_2}}\boldsymbol{c}_2$$

for $r = 3(1)25$. Because we have

$$x_i = -x_{r-i}, \qquad \varepsilon_i = -\varepsilon_{r-i+1}, \qquad w_i = w_{r-i+1},$$

for even $r = 2\ell$, the values of x_i, ε_i, and w_i are given in the table only for $i = 1, 2, \ldots, \ell$. For odd $r = 2\ell+1$, the values of x_i are given only for $i = 1, \ldots, \ell$ and the values of ϵ_i and w_i for $i = 1, 2, \ldots, \ell+1$.

This section follows papers of Nikulin (1973a,b). The table is from Nikulin (1973b).

Other tests for normality are studied in the work of Shapiro and Wilk (1965), Sharkadi (1966), Jaque and Bera (1987), and others. In Shapiro, Wilk, and Chen (1968) various tests are compared. Watson (1957) studied chi-squared tests for normality. O'Reilly and Quesenberry (1973) and Dzhaparidze and Nikulin (1982a) used the Rosenblatt transformation (1952) to construct tests. More information is in the work of Stephens (1986d).

Multivariate case: Best Unbiased Estimator for the Density Test Statistic

Let $\boldsymbol{X}_1, \boldsymbol{X}_2, \ldots, \boldsymbol{X}_n$ be a sample, $\boldsymbol{X}_i = (X_{i1}, \ldots, X_{irn})^T$. We wish to test the hypothesis H_0 according to which $\boldsymbol{X}_i$ has an m-dimensional normal distribution $N_m(\boldsymbol{a}, \boldsymbol{A})$, with density

$$f(\boldsymbol{x}|\boldsymbol{a}, \boldsymbol{A}) = \frac{1}{(2\pi)^{m/2}\sqrt{\det \boldsymbol{A}}} \exp\{-\frac{1}{2}\{(\boldsymbol{x}-\boldsymbol{a})^T \boldsymbol{A}^{-1}(\boldsymbol{x}-\boldsymbol{a})\},$$

$\boldsymbol{a} = \boldsymbol{E}\boldsymbol{X}_i$, $\boldsymbol{A} = \boldsymbol{E}(\boldsymbol{X}_i - \boldsymbol{a})(\boldsymbol{X}_i - \boldsymbol{a})^T$ is the covariance matrix of $\boldsymbol{X}_i$. We suppose that $\boldsymbol{A}$ is nonsingular. In this model, under H_0 there exists a sufficient statistic $\boldsymbol{U} = (\overline{\boldsymbol{X}}_n, \boldsymbol{S})^T$ for $\boldsymbol{\theta} = (\boldsymbol{a}, \boldsymbol{A})^T$, where

$$\overline{\boldsymbol{X}}_n = \frac{1}{n}\sum_{i=1}^{n} \boldsymbol{X}_i, \qquad \boldsymbol{S} = \frac{1}{n-1}\sum_{i=1}^{n}(\boldsymbol{X}_i - \overline{\boldsymbol{X}}_n)(\boldsymbol{X}_i - \overline{\boldsymbol{X}}_n)^T.$$

It is known that $\boldsymbol{U}$ is the best unbiased estimator for $\boldsymbol{\theta}$. Let us denote by $\hat{f}(x|\boldsymbol{U})$ the best unbiased estimator for $f(\boldsymbol{x}|\boldsymbol{a}, \boldsymbol{A})$:

$$\hat{f}(\boldsymbol{x}|\boldsymbol{U}) = \begin{cases} 0, & \text{if } \boldsymbol{x} \notin \boldsymbol{J}, \\ \dfrac{\Gamma(\frac{n-1}{2})[\pi(n-1)]^{-m/2}}{\Gamma(\frac{n-m-1}{2})|\boldsymbol{S}|^{1/2}} \left[1 - \dfrac{(\boldsymbol{x}-\overline{\boldsymbol{X}}_n)^T \boldsymbol{S}^{-1}(\boldsymbol{x}-\overline{\boldsymbol{X}}_n)}{n-1}\right]^{\frac{(n-m-3)}{2}} & \text{, if } \boldsymbol{x} \in \boldsymbol{J} \end{cases}$$

where $\boldsymbol{J}$ represents the sample space of the random vector $\boldsymbol{X}_i$ given $\boldsymbol{U}$,

$$\boldsymbol{J} = \{\boldsymbol{x} : (\boldsymbol{x} - \overline{\boldsymbol{X}}_n)^T \boldsymbol{S}^{-1}(\boldsymbol{x} - \overline{\boldsymbol{X}}_n) < n-1, \boldsymbol{x} \in \boldsymbol{R}^m\}.$$

As in Section 12, we choose a vector $\boldsymbol{p} = (p_1, \ldots, p_r)^T$ such that $\boldsymbol{p}^T \boldsymbol{1}_r = 1$, $p_i > 0$, and find a partition of $\boldsymbol{R}^m$ into $r \geq 2$ mutually disjoint classes $J_1, J_2, \ldots, J_r$ such that

Table 14.1. Boundaries of class-intervals of x_i and coefficients of ε_i and w_i.

r \ i	1	2	3	4	5	6	7	8	9	10	11	12	13
	−0.4307												
3	2.3988	0											
	0.3764	−0.7527											
	−0.6745	0											
4	3.4040	0.8694											
	0.7623	−0.7623											
	−0.8416	−0.2533											
5	4.3607	1.6570	0										
	1.1424	−0.6678	−0.9491										
	−0.9674	−0.4307	0										
6	5.2793	2.4035	0.7468										
	1.5139	−0.5330	−0.9809										
	−1.0676	−0.5659	−0.1800										
7	6.1636	3.1211	1.4365	0									
	1.8766	−0.3780	−0.9481	−1.1009									
	−1.1503	−0.6745	−0.3186	0									
8	7.0243	3.8191	2.0958	0.6738									
	2.2307	−0.2116	−0.8809	−1.1382									
	−1.2206	−0.7647	−0.4307	−0.1397									
9	7.8590	4.4983	2.7303	1.3058	0								
	2.5769	−0.0385	−0.7927	−1.1305	−1.2377								
	−1.2816	−0.8416	−0.5244	−0.2523	0								
10	8.6722	5.1620	3.3469	1.9099	0.6226								
	2.9158	0.1389	−0.6909	−1.0949	−1.2689								
	−1.3352	−0.9085	−0.6046	−0.3488	−0.1142								
11	9.4669	5.8118	3.9487	2.4936	1.2119	0							
	3.2479	0.3188	−0.5795	−1.0405	−1.2737	−1.3458							
	−1.3830	−0.9674	−0.6745	−0.4307	−0.2104	0							
12	10.2424	6.4494	4.5379	3.0613	1.7775	0.5836							
	3.5737	0.5002	−0.4614	−0.9729	−1.2557	−1.3839							
	−1.4261	−1.0201	−0.7363	−0.5024	−0.2934	−0.0966							
13	11.0026	7.0755	5.1160	3.6159	2.3251	1.1398	0						
	3.8936	0.6825	−0.3382	−0.8955	−1.2213	−1.3957	−1.4508						

For each r the values of x_i are given in the first row, the values of ε_i and w_i' in the second and third rows.

r \ i	1	2	3	4	5	6	7	8	9	10	11	12	13
	−1.4652	−1.0676	−0.7916	−0.5659	−0.3661	−0.1800	0						
14	11.7841	7.6911	5.6842	4.1593	2.8582	1.6754	0.5524						
	4.2080	0.8651	−0.2112	−0.8107	−1.1747	−1.3884	−1.4881						
	1.5011	−1.1108	−0.8416	−0.6229	−0.4307	−0.2533	−0.0837						
15	12.4800	8.2971	6.2434	4.6929	3.3793	2.1950	1.0816	0					
	4.5173	1.0477	−0.0814	−0.7200	−1.1188	−1.3669	−1.5040	−1.5479					
	−1.5341	−1.1503	−0.8871	−0.6745	−0.4888	−0.3186	−0.1573						
16	13.1992	8.8940	6.7942	5.2179	3.8903	2.7015	1.5928	0.5265					
	4.8218	1.2301	−0.0506	−0.6248	−1.0554	−1.3344	−1.5038	−1.5842					
	−1.5647	−1.1868	−0.9289	−0.7215	−0.5414	−0.3774	−0.2220	−0.0738					
17	13.9066	9.4823	7.3374	5.7349	4.3923	3.1969	2.0896	1.0331	0				
	5.1218	1.4120	0.1843	−0.5259	−0.9862	−1.2931	−1.4910	−1.6026	−1.6387				
	1.5932	−1.2206	−0.9674	−0.7647	−0.5985	−0.4307	−0.2822	−0.1397	0				
18	14.6028	10.0627	7.8734	6.2447	4.8863	3.6829	2.5744	1.5238	0.5046				
	5.4176	1.5933	0.3193	−0.4239	−0.9120	−1.2447	−1.4684	−1.6072	−1.6738				
	−1.6199	−1.2521	−1.0031	−0.8046	−0.6336	−0.4795	−0.3360	−0.1992	−0.0660				
19	15.2887	10.6356	8.4026	6.7479	5.3732	4.1605	3.0490	2.0014	0.9918	0			
	5.7093	1.7739	0.4552	−0.3196	−0.8339	−1.1905	−1.4376	1.6008	−1.6939	−1.7242			
	−1.6449	−1.2816	−1.0364	−0.8416	−0.6745	−0.5244	−0.3853	−0.2533	−0.1257	0			
20	15.9647	11.2012	8.9254	7.2448	5.8534	4.6308	3.5148	2.4679	1.4647	0.4856			
	5.9972	1.9538	0.5919	−0.2132	−0.7525	−1.1315	−1.4002	−1.5853	−1.7019	−1.7853			
	−1.6684	−1.3092	−1.0676	−0.8761	−0.7124	−0.5659	−0.4307	−0.3030	−0.1800	−0.0597			
21	16.6314	11.7602	9.4423	7.7359	6.3277	5.0944	3.9728	2.9251	1.9257	0.9559	0		
	6.2816	2.1329	0.7291	−0.1052	−0.6682	−1.0684	−1.3572	−1.5624	−1.7001	−1.7794	−1.8053		
	−1.6906	−1.3352	−1.0968	−0.9085	−0.7479	−0.6046	−0.4728	−0.3488	−0.2299	−0.1142	0		
22	17.2893	12.3126	9.9534	8.2216	6.7964	5.5519	4.4239	3.3740	2.3766	1.4133	0.4690		
	6.5624	2.3112	0.8667	−0.0042	−0.5815	−1.0018	−1.3095	−1.5333	−1.6901	−1.7899	−1.8384		
	−1.7117	−1.3597	−1.1243	−0.9388	−0.7810	−0.6407	−0.5519	−0.3912	−0.2759	−0.1642	−0.0545		
23	17.9389	12.8590	10.4591	8.7020	7.2598	6.0038	4.8687	3.8156	2.8187	1.8598	0.9243	0	
	6.7689	2.4627	0.9942	−0.1136	−0.4877	−0.9226	−1.2447	−1.4833	−1.6558	−1.7729	−1.8408	−1.8631	
	−1.7317	−1.3830	−1.1503	−0.9674	−0.8122	−0.6745	−0.5485	−0.4307	−0.3186	−0.2104	−0.1046	0	
24	18.5805	13.3994	10.9596	9.1775	7.7182	6.4505	5.3078	4.2506	3.2531	2.2969	1.3679	0.4543	
	7.1145	2.6651	1.1426	0.2264	−0.4024	−0.8602	−1.2025	−1.4599	−1.6506	−1.7857	−1.8724	−1.9148	
	−1.7507	−1.4051	−1.1750	−0.9945	−0.8416	−0.7063	−0.5828	−0.4677	−0.3585	−0.2533	−0.1510	−0.0502	
25	19.2146	13.9342	11.4552	9.6484	8.1721	6.8924	5.7416	4.6797	3.6806	2.7258	1.8015	0.8961	0
	7.3860	2.8407	1.2807	−0.3387	−0.3105	−0.7858	−1.1443	−1.4170	−1.6229	−1.7736	−1.8767	−1.9369	−1.9567

For each r the values of x_i are given in the first row, the values of ε_i and w_i' in the second and third rows.

$$\boldsymbol{J} = J_1 \cup J_2 \cup \cdots \cup J_r.$$

and

$$p_i = \int_{J_i} \hat{f}(\boldsymbol{x}|\boldsymbol{U})\,dx, \qquad i = 1, 2, \ldots, r.$$

Now we consider the vector of frequencies $\boldsymbol{\mu} = \mu_1, \ldots, \mu_r)^T$, obtained by grouping the data $\boldsymbol{X}_1, \boldsymbol{X}_2, \ldots, \boldsymbol{X}_n$ over the classes $J_1, J_2, \ldots, J_r$. As before, one can verify that

$$\lim_{n\to\infty} \boldsymbol{P}\{\{Y_n^2 \geq x|H_0\} = \boldsymbol{P}\{\chi_{r-1}^2 \geq x\},$$

where

$$Y_n^2 = \frac{1}{n}(\boldsymbol{\mu} - n\boldsymbol{p})^T \boldsymbol{B}^-(\boldsymbol{U})(\boldsymbol{\mu} - n\boldsymbol{p}),$$

$\boldsymbol{B}(\boldsymbol{U})$ estimates $\boldsymbol{B}(\boldsymbol{\theta})$,

$$\boldsymbol{B}(\boldsymbol{\theta}) = \lim_{n\to\infty} \boldsymbol{E}_\theta(\boldsymbol{\mu} - n\boldsymbol{p})(\boldsymbol{\mu} - n\boldsymbol{p})^T.$$

As was shown by Tseranbat (1990)

$$\boldsymbol{P}\{Y_n^2 \leq x|H_0\} = \boldsymbol{P}\{\chi_{r-1}^2 \leq x\} + \begin{cases} O(n^{-3/2}), & \text{if } m = 1, \\ O(n^{-1+\varepsilon}), & \text{if } m \geq 2, \end{cases}$$

uniformly with respect to x as $n \to \infty$ for any $\varepsilon \in (0, 1/4)$.

This technique was used by Nikulin and Seddick-Amer (1991) for the statistical analysis of the data from a two-dimensional normal distrubtion. See also Nikulin (1984b).

The same technique can be used for any family $f(\boldsymbol{x}; \boldsymbol{\theta})$ of multivariate distributions having a sufficient statistic $\boldsymbol{U}$. If there is no sufficient statistic one can use $f(\boldsymbol{x}, \hat{\boldsymbol{\theta}}_n)$ instead of $\hat{f}(\boldsymbol{x}|\boldsymbol{U})$, where $\hat{\boldsymbol{\theta}}_n$ is the maximum-likelihood estimator.

15. OTHER $\sqrt{n}$-CONSISTENT ESTIMATORS: MODIFY THE TEST STATISTIC OR IMPROVE THE ESTIMATOR

We have seen a variety of ways to test composite hypotheses H_0 (Sections 9 to 12). If we have the minimum chi-squared estimator $\tilde{\boldsymbol{\theta}}_n$ for $\boldsymbol{\theta} = (\theta_1, \ldots, \theta_s)^T$, then we can use the standard statistic of Pearson

$$X_n^2(\tilde{\boldsymbol{\theta}}_n) = \sum_{i=1}^{r} \frac{(\nu_i - np_i(\tilde{\boldsymbol{\theta}}_n))^T}{np_i(\tilde{\boldsymbol{\theta}}_n)}, \qquad r \geq s+2, \tag{15.1}$$

and apply the theorem of Fisher, according to which

$$\lim_{n\to\infty} \boldsymbol{P}\{X_n^2(\tilde{\boldsymbol{\theta}}_n) > x \mid H_0\} = \boldsymbol{P}\{\chi_{r-s-1}^2 > x\}. \tag{15.2}$$

If we have the maximum-likelihood estimator $\hat{\boldsymbol{\theta}}_n$, then we can use the statistic Y_n^2, possibly with random cell boundaries, and apply Theorem 12.1, according to which

$$\lim_{n\to\infty} \boldsymbol{P}\{Y_n^2(\hat{\boldsymbol{\theta}}_n) > x \mid H_0\} = \boldsymbol{P}\{\chi_{r-1}^2 > x\}. \tag{15.3}$$

But what if we have an $\sqrt{n}$-consistent estimator $\boldsymbol{\theta}_n^*$, which does not possess the properties (9.24) and (10.16) of $\tilde{\boldsymbol{\theta}}_n$ or $\hat{\boldsymbol{\theta}}_n$ (i.e., θ_n^* is not efficient relative to grouped or ungrouped data, respectively)? This section is devoted to the problem of testing the hypothesis H_0 when we use any $\sqrt{n}$-consistent sequence $\boldsymbol{\theta}_n^*$ of estimators of the parameter $\boldsymbol{\theta}$. By *$\sqrt{n}$-consistent* we mean the somewhat stronger condition that the random vector $\sqrt{n}(\boldsymbol{\theta}_n^* - \boldsymbol{\theta})$ has in the limit, as $n \to \infty$, zero mean and bounded covariance matrix. In the next section we follow the paper of Dzhaparidze and Nikulin (1974), where this problem is considered (see also Dudley, 1976, 1979 and LeCam et al., 1983). Then we look at a feasible method of improving an $\sqrt{n}$-consistent estimator to an efficient one.

Another Modification of Pearson's Statistic

We expect that the statistic $X_n^2(\boldsymbol{\theta}_n^*)$ may be stochastically larger than $X_n^2(\tilde{\boldsymbol{\theta}}_n)$. The question arises: is it possible to change the statistic $X_n^2(\boldsymbol{\theta}_n^*)$ so that the modified statistic, which we denote by $W^2(\boldsymbol{\theta}_n^*)$, has the same chi-squared asymptotic distribution as that of $X_n^2(\tilde{\boldsymbol{\theta}}_n)$ appearing in Fisher's theorem? The following is such a modification.

Theorem 15.1. From Dzhaparidze and Nikulin (1974). *Let $\boldsymbol{\theta}_n^*$ be $\sqrt{n}$-consistent as defined above. Under the regularity conditions of Cramer (9.13),*

$$\lim_{n\to\infty} \boldsymbol{P}\{W_n^2(\boldsymbol{\theta}_n^*) > x \mid H_0\} = \boldsymbol{P}\{\chi_{r-s-1}^2 > x\}, \tag{15.4}$$

where

$$\begin{aligned} W_n^2(\boldsymbol{\theta}) &= \boldsymbol{X}_n^T(\boldsymbol{\theta})(\boldsymbol{I}_r - \boldsymbol{B}(\boldsymbol{\theta})\boldsymbol{J}^{-1}(\boldsymbol{\theta})\boldsymbol{B}^T(\boldsymbol{\theta}))\boldsymbol{X}_n(\boldsymbol{\theta}) \\ &= X_n^2(\boldsymbol{\theta}) - \boldsymbol{X}_n^T(\boldsymbol{\theta})\boldsymbol{B}(\boldsymbol{\theta})\boldsymbol{J}^{-1}(\boldsymbol{\theta})\boldsymbol{B}^T(\boldsymbol{\theta})\boldsymbol{X}_n(\boldsymbol{\theta}), \end{aligned} \tag{15.5}$$

$$\boldsymbol{B}(\boldsymbol{\theta}) = [b_{ij}]_{r\times s}, \qquad b_{ij} = \frac{1}{\sqrt{p_i(\boldsymbol{\theta})}}\frac{\partial p_i(\boldsymbol{\theta})}{\partial \theta_j}, \quad i = 1,\ldots,r; \quad j = 1,\ldots,s;$$

$$\boldsymbol{J}(\boldsymbol{\theta}) = \boldsymbol{B}^T(\boldsymbol{\theta})\boldsymbol{B}(\boldsymbol{\theta}).$$

Proof. From Section 9 we know [see Eq. (9.19)] that

$$\sqrt{n}(\tilde{\boldsymbol{\theta}}_n - \boldsymbol{\theta}_0) = \boldsymbol{J}^{-1}(\boldsymbol{\theta}_0)\boldsymbol{B}^T(\boldsymbol{\theta}_0)\boldsymbol{X}_n(\boldsymbol{\theta}_0) + o_{\boldsymbol{P}_0}(\mathbf{1}_s) \tag{15.6}$$

But, as was shown by LeCam (1956), the estimator

$$\breve{\boldsymbol{\theta}}_n = \boldsymbol{\theta}_n^* + \boldsymbol{J}^{-1}(\boldsymbol{\theta}_n^*)\boldsymbol{B}^T(\boldsymbol{\theta}_n^*)\boldsymbol{X}_n(\boldsymbol{\theta}_n^*) \tag{15.7}$$

satisfies the asymptotic relation (15.6).

Comparison of Eqs. (15.4) and (15.2) shows that we need to prove that

$$X_n^2(\breve{\boldsymbol{\theta}}_n) = W_n^2(\boldsymbol{\theta}_n^*) + o_{\boldsymbol{P}_0}(1). \tag{15.8}$$

The asymptotic relation (15.8) follows from the straightforward (see, e.g., Cramer, 1946) expansion

$$\begin{aligned} X_n^2(\breve{\boldsymbol{\theta}}_n) = X^2(\boldsymbol{\theta}_n^*) &+ \frac{1}{\sqrt{n}}\boldsymbol{U}^T(\boldsymbol{\theta}_n^*)\boldsymbol{J}^{-1}(\boldsymbol{\theta}_n^*)\boldsymbol{B}^T(\boldsymbol{\theta}_n^*)\boldsymbol{X}_n(\boldsymbol{\theta}_n^*) \\ &+ \frac{1}{2n}\boldsymbol{X}^T(\boldsymbol{\theta}_n^*)\boldsymbol{B}(\boldsymbol{\theta}_n^*)\boldsymbol{J}^{-1}(\boldsymbol{\theta}_n^*)\boldsymbol{V}(\boldsymbol{\theta}_n^*)\boldsymbol{J}^{-1}(\boldsymbol{\theta}_n^*)\boldsymbol{B}^T(\boldsymbol{\theta}_n^*)\boldsymbol{X}_n(\boldsymbol{\theta}_n^*) \\ &+ o_{\boldsymbol{P}_0(1)}, \end{aligned} \tag{15.9}$$

$$\frac{1}{n}\boldsymbol{V}(\boldsymbol{\theta}_n^*) = 2\boldsymbol{J}(\boldsymbol{\theta}_n^*) + o_{\boldsymbol{P}_0}(\mathbf{1}_{s\times s}), \tag{15.10}$$

$$\boldsymbol{U}(\boldsymbol{\theta}) = \left(\frac{\partial}{\partial\theta_1} X_n^2(\boldsymbol{\theta}), \ldots, \frac{\partial}{\partial\theta_s} X_n^2(\boldsymbol{\theta})\right)^T, \tag{15.11}$$

$$\boldsymbol{V}(\boldsymbol{\theta}) = \left[\frac{\partial^2 X_n^2(\boldsymbol{\theta})}{\partial\theta_i \partial\theta_j}\right]_{s\times s}, \tag{15.12}$$

$$\frac{1}{\sqrt{n}} \boldsymbol{U}(\boldsymbol{\theta}_n^*) = -2\boldsymbol{B}^T(\boldsymbol{\theta}_n^*)\boldsymbol{X}_n(\boldsymbol{\theta}_n^*) + o_{\boldsymbol{P}_0}(\mathbf{1}_s). \tag{15.13}$$

Remark 15.1. Of course, the statement of Theorem 15.1 is valid for the minimum chi-squared estimator $\tilde{\boldsymbol{\theta}}_n$ and the maximum-likelihood estimator $\hat{\boldsymbol{\theta}}_n$, because they are $\sqrt{n}$-consistent estimators of the parameter $\boldsymbol{\theta}$.

Remark 15.2. From Eq. (12.9), we have

$$\begin{aligned}(\boldsymbol{I}_r - \boldsymbol{B}\boldsymbol{I}^{-1}\boldsymbol{B}^T)^{-1} &= \boldsymbol{I}_r + \boldsymbol{B}(\boldsymbol{I} - \boldsymbol{J})^{-1}\boldsymbol{B}^T \\ &= (\boldsymbol{I}_r - \boldsymbol{B}\boldsymbol{J}^{-1}\boldsymbol{B}^T) + \boldsymbol{B}[(\boldsymbol{I} - \boldsymbol{J})^{-1} + \boldsymbol{J}^{-1}]\boldsymbol{B}^T,\end{aligned} \tag{15.14}$$

and hence, from Eqs. (12.5), (12.6), (12.8), (15.5), and (15.14), we obtain yet another decomposition of the statistic $Y_n^2(\hat{\boldsymbol{\theta}}_n)$:

$$Y_n^2 = W_n^2(\hat{\boldsymbol{\theta}}_n) + \boldsymbol{L}^T(\hat{\boldsymbol{\theta}}_n)[(\boldsymbol{I}(\hat{\boldsymbol{\theta}}_n) - \boldsymbol{J}(\hat{\boldsymbol{\theta}}_n))^{-1} + \boldsymbol{J}^{-1}(\hat{\boldsymbol{\theta}}_n)]\boldsymbol{L}(\hat{\boldsymbol{\theta}}_n), \tag{15.15}$$

where $\boldsymbol{L}(\boldsymbol{\theta}) = \boldsymbol{B}^T(\boldsymbol{\theta})\boldsymbol{X}_n(\boldsymbol{\theta})$. Because for any $x > 0$

$$\lim_{n\to\infty} \boldsymbol{P}\{W_n^2(\hat{\boldsymbol{\theta}}_n) > x \mid H_0\} = \boldsymbol{P}\{\chi^2_{r-s-1} > x\}$$

$$\lim_{n\to\infty} \boldsymbol{P}\{Y_n^2(\hat{\boldsymbol{\theta}}_n) > x \mid H_0\} = \boldsymbol{P}\{\chi^2_{r-1} > x\}$$

from the theorem of Cochran (see, for example, Lancaster, 1969), it follows that under H_0 the statistic

$$M_n^2(\hat{\boldsymbol{\theta}}_n) = \boldsymbol{L}^T(\hat{\boldsymbol{\theta}}_n)[(\boldsymbol{I}(\hat{\boldsymbol{\theta}}_n) - \boldsymbol{J}(\hat{\boldsymbol{\theta}}_n))^{-1} - \boldsymbol{J}^{-1}(\hat{\boldsymbol{\theta}})]\boldsymbol{L}(\hat{\boldsymbol{\theta}}_n) \tag{15.16}$$

is asymptotically independent of the statistic $W_n^2(\hat{\boldsymbol{\theta}}_n)$ as $n \to \infty$ and

$$\lim_{n\to\infty} \boldsymbol{P}\{M_n^2(\hat{\boldsymbol{\theta}}_n) > x \mid H_0\} = \boldsymbol{P}\{\chi^2_s > x\} \tag{15.17}$$

C. McCulloch (1985) gave an interesting interpretation of the statistic (15.16) in terms of the estimators $\tilde{\boldsymbol{\theta}}_n$ and $\hat{\boldsymbol{\theta}}_n$. Namely, McCulloch showed that as $n \to \infty$

$$\sqrt{n}(\hat{\boldsymbol{\theta}}_n - \tilde{\boldsymbol{\theta}}_n) \quad \text{is} \quad AN(\mathbf{0}_s, \boldsymbol{J}^{-1}(\boldsymbol{\theta}_0) - \boldsymbol{I}^{-1}(\boldsymbol{\theta}_0)), \tag{15.18}$$

and hence from Eqs. (15.16) and (15.18) we have

$$M_n^2(\hat{\theta}_n) = n(\hat{\boldsymbol{\theta}}_n - \tilde{\boldsymbol{\theta}}_n)^T(\boldsymbol{J}^{-1}(\boldsymbol{\theta}_0) - \boldsymbol{I}^{-1}(\boldsymbol{\theta}_0))^{-1}(\hat{\boldsymbol{\theta}}_n - \tilde{\boldsymbol{\theta}}_n) + o_{\boldsymbol{P}_0}(1), \tag{15.19}$$

as $n \to \infty$. Moreover, we have, as $n \to \infty$,

$$\boldsymbol{V} = \begin{pmatrix} \sqrt{n}(\hat{\boldsymbol{\theta}}_n - \boldsymbol{\theta}_0) \\ \sqrt{n}(\tilde{\boldsymbol{\theta}}_n - \boldsymbol{\theta}_0) \end{pmatrix} \quad \text{is} \quad AN\left(\mathbf{0}_{2s}, \boldsymbol{\Sigma}(\boldsymbol{\theta}_0)\right), \tag{15.20}$$

where

$$\boldsymbol{\Sigma}(\boldsymbol{\theta}_0) = \begin{bmatrix} \boldsymbol{I}^{-1}(\boldsymbol{\theta}) & \boldsymbol{I}^{-1}(\boldsymbol{\theta}) \\ \boldsymbol{I}^{-1}(\boldsymbol{\theta}) & \boldsymbol{J}^{-1}(\boldsymbol{\theta}) \end{bmatrix}_{2s \times 2s}. \tag{15.21}$$

From Eq. (15.20), we obtain that, as $n \to \infty$, $\boldsymbol{V}^T \boldsymbol{\Sigma}^{-1} \boldsymbol{V}$ is distributed in the limit as χ^2_{2s}, where

$$\boldsymbol{\Sigma}^{-1} = \boldsymbol{\Sigma}^{-1}(\boldsymbol{\theta}_0) = \begin{bmatrix} \boldsymbol{I}(\boldsymbol{\theta}_0 + (\boldsymbol{J}(\boldsymbol{\theta}_0) - \boldsymbol{I}(\boldsymbol{\theta}_0))^{-1} & -(\boldsymbol{J}(\boldsymbol{\theta}_0) - \boldsymbol{I}(\boldsymbol{\theta}_0))^{-1} \\ -(\boldsymbol{J}(\boldsymbol{\theta}_0) - \boldsymbol{I}(\boldsymbol{\theta}_0))^{-1} & (\boldsymbol{J}(\boldsymbol{\theta}_0) - \boldsymbol{I}(\boldsymbol{\theta}_0))^{-1} \end{bmatrix}.$$

The Davidon-Fletcher-Powell Numerical Method to Construct Efficient Estimator Sequence

If s, the dimension of $\boldsymbol{\theta} = (\theta_1, \ldots, \theta_s)^T$, is large, it may be computationally easier to use the Davidon-Fletcher-Powell method of improving an $\sqrt{n}$-consistent estimator of $\boldsymbol{\theta}$ to an efficient estimator, rather than the Newton-Raphson method (see, e.g., Ortega and Rheinboldt, 1979). The Davidon-Fletcher-Powell method is a particular version of the quasi-Newton method, which avoids not only the solution of the likelihood equation but also the inversion of the estimated Fisher information matrix, which appears in the statistics Y_n^2 and W_n^2.

Suppose $\boldsymbol{\theta}_n^*$ is an $\sqrt{n}$-consistent estimator for θ and $\boldsymbol{H}_0$ is a symmetic positive definite $s \times s$ matrix, for example, the identity matrix $\boldsymbol{I}_r$. Define iteratively

$$\boldsymbol{\theta}_n^{j+1} = \boldsymbol{\theta}_n^{(j)} - \frac{\alpha_j}{\sqrt{n}} \boldsymbol{H}_{(j)} \boldsymbol{\Lambda}(\boldsymbol{\theta}_n^{(j)}), \qquad j = 0, 1, 2, \ldots,$$

where $\boldsymbol{\theta}_n^{(o)} = \boldsymbol{\theta}_n^*$, $\boldsymbol{\Lambda}(\boldsymbol{\theta}) = \mathbf{grad}\ \ln L(\boldsymbol{\theta})$ and $\boldsymbol{H}_j$ is defined iteratively by

$$\begin{aligned}
\boldsymbol{H}_{j+1} &= \boldsymbol{H}_j + \frac{\boldsymbol{u}_j \boldsymbol{u}_j^T}{\boldsymbol{u}_j^T \boldsymbol{v}_j} - \frac{(\boldsymbol{H}_j \boldsymbol{v}_j)(\boldsymbol{H}_j \boldsymbol{v}_j)^T}{\boldsymbol{v}_j^T \boldsymbol{H}_j \boldsymbol{v}_j}, \\
\boldsymbol{u}_j &= \sqrt{n}(\boldsymbol{\theta}_n^{(j+1)} - \boldsymbol{\theta}_n^{(j)}), \qquad \boldsymbol{v}_j = \boldsymbol{\Lambda}(\boldsymbol{\theta}_n^{(j+1)}) - \boldsymbol{\Lambda}(\boldsymbol{\theta}_n^{(j)}), \\
\alpha_j &= \frac{\boldsymbol{\Lambda}^T(\boldsymbol{\theta}_n^{(j)}) \boldsymbol{H}_j \boldsymbol{\Lambda}(\boldsymbol{\theta}_n^{(j)})}{\boldsymbol{\Lambda}^T(\boldsymbol{\theta}_n^{(j)}) \boldsymbol{H}_j^T \boldsymbol{I}(\boldsymbol{\theta}_n^{(j)}) \boldsymbol{H}_j \boldsymbol{\Lambda}(\boldsymbol{\theta}_n^{(j)})}.
\end{aligned}$$

Then $\{-\alpha_{s-1} \boldsymbol{H}_{s-1}\}$ is a consistent sequence of estimators of $\boldsymbol{I}^{-1}(\boldsymbol{\theta})$ and $\sqrt{n}(\boldsymbol{\theta}_n^{(s)} - \boldsymbol{\theta})$ is asymptotically normal $N(\mathbf{0}_s, \boldsymbol{I}^{-1}(\boldsymbol{\theta}))$. If $\Lambda(\boldsymbol{\theta}) =$ **grad** $\ln \ell(\boldsymbol{\theta})$, then $\{-\alpha_{s-1} \boldsymbol{H}_{s-1}\}$ a consistent sequence of estimators of $\boldsymbol{J}^{-1}(\boldsymbol{\theta})$ and $\sqrt{n}(\boldsymbol{\theta}^{(s)} - \boldsymbol{\theta})$ is asymptotically normal $N(\mathbf{0}_s, \boldsymbol{J}^{-1}(\boldsymbol{\theta}))$, see Dzhaparidze (1983). From these results follows an analogue of Theorem 15.1: The statistic

$$W_n^2 = \boldsymbol{X}_n^T(\boldsymbol{\theta}_n^*)[\boldsymbol{I}_r + \alpha_{s-1} \boldsymbol{B}(\boldsymbol{\theta}_n^*) \boldsymbol{H}_s \boldsymbol{B}^T(\boldsymbol{\theta}_n^*)] \boldsymbol{X}_n(\boldsymbol{\theta}_n^*)$$

has limit law χ^2_{r-s-1}, under H_0 as $n \to \infty$.

More information can be found in Dzhaparidze (1983) and Dzhaparidze and Nikulin (1982b, 1992).

CHAPTER 3

The Chi-Squared Test for an Exponential Family of Distributions

16. A MODIFIED POWER SERIES DISTRIBUTION

We continue with the problem of first estimating θ and then applying a χ^2 test. In this chapter we consider estimation by the Rao-Blackwell approach, based on sufficient statistics. For a large exponential family of distributions, called the modifed power series distributions, Theorem 17.1 allows us to estimate probabilities by conditional expectations given a minimal sufficient statistic.

Definition of the Class

We consider the problem of testing the hypothesis H_0 according to which the independent identically distributed random variables $X_1, \ldots, X_n$ follow a so-called *modified power-series distribution*:

$$\boldsymbol{P}_\theta\{X_i = x\} = p_\theta(x) = \frac{a(x)\psi^x(\theta)}{f(\theta)}, \quad x \in \boldsymbol{X} \subseteq \boldsymbol{X}_0, \ \theta \in \Theta \subset \boldsymbol{R}^1, \quad (16.1)$$

where $\boldsymbol{X}_0 = \{0, 1, 2, \ldots\}$, $\Theta = \{\theta : 0 \leq |\theta| \leq \rho\}$, ρ is the radius of convergence of the functional series

$$f(\theta) = \sum_{x \in \boldsymbol{X}} a(x)\psi^x(\theta),$$

and $a(x)$ is a positive function on $\boldsymbol{X}$; $a(x) = 0$ if $x \in \boldsymbol{X}_0 - \boldsymbol{X}$, $\psi(\theta)$ is a positive, bounded, and differentiable function on $\boldsymbol{\Theta}$. For each $\alpha \in \boldsymbol{X}_0$, $\beta \in \boldsymbol{X}_1 = \{1, 2, \ldots\}$, let $\boldsymbol{X}_\alpha = \{\alpha, \alpha+1, \ldots\}$, $\boldsymbol{X}^\beta = \boldsymbol{X}_0^\beta = \{0, 1, \ldots, \beta\}$, $\boldsymbol{X}_\alpha^\beta = \{\alpha, \alpha+1, \ldots, \beta\}$, $\alpha < \beta$. The distribution (16.1) is called *non-truncated*, *left-* or *right-truncated*, or *double-truncated* according to the form of $\boldsymbol{X}$. The class of modified power series distributions (16.1), introduced by Gupta (1974, 1977), is a natural generalization of the *power series distribution*, of Noack (1950), according to which

$$\boldsymbol{P}_\theta\{X_i = x\} = \frac{a(x)\theta^x}{f(\theta)}, \; x \in \boldsymbol{X} \subseteq \boldsymbol{X}_0, \; \theta \in \Theta = \{\theta : 0 \leq |\theta| \leq \rho\}. \quad (16.1')$$

In this case, $f(\theta) = \sum_{x \in \boldsymbol{X}} a(x)\theta^x$. If ψ is invertible then (16.1) and (16.1′) define the same family of distributions. If the distribution is truncated, it is evident that the function $f(\theta)$ depends on the truncation parameters α and β.

In the case of power series distributions (16.1′) it is easy to verify (see Noack, 1950) that

$$\boldsymbol{E}_\theta X_i = \frac{\theta f'(\theta)}{f(\theta)}, \; \boldsymbol{E}_\theta X_i^2 = \boldsymbol{E}_\theta X_i + \frac{\theta^2 f''(\theta)}{f(\theta)}, \quad (16.2)$$

and hence

$$\mathbf{Var}_\theta X_i = \frac{\theta f'(\theta)}{f(\theta)} + \frac{\theta^2 f''(\theta)}{f(\theta)} - \left[\frac{\theta f'(\theta)}{f(\theta)}\right]^2.$$

Example 16.1. In Eq. (16.1), let $a(x) = 1/x!$, $\psi(\theta) = \theta$, $f(\theta) = e^\theta$. In this case, $\Theta = (0, \infty)$, and we obtain

$$\boldsymbol{P}_\theta\{X_i = x\} = \frac{\theta^x e^{-\theta}}{x!}, \; x \in \boldsymbol{X}_0, \quad (16.3)$$

that is, X_i has a *Poisson distribution* with parameter θ, $\theta > 0$.

Example 16.2. In Eq. (16.1), if we choose $a(x) = \binom{m}{x} = m!/(x!(m-x)!)$, $m \in \boldsymbol{X}_1 = \{1, 2, \ldots\}$, $x \in \boldsymbol{X}_0^m = \{0, 1, \ldots, m\}$, $\psi(\theta) = \theta/(1-\theta)$, $f(\theta) = (1-\theta)^{-m}$, $\theta \in \Theta = (0, 1)$, then

$$\boldsymbol{P}_\theta\{X_i = x\} = \binom{m}{x}\theta^x(1-\theta)^{m-x}, \; x \in \boldsymbol{X}_0^m; \quad (16.4)$$

that is, we have a *binomial distribution* with parameters m and θ, a right-truncated distribution.

Example 16.3. If we choose

$$a(x) = \frac{\Gamma(k+x)}{\Gamma(x+1)\Gamma(x)}, \qquad x \in \boldsymbol{X}_0, k \in \boldsymbol{X}_1,$$

$$f(\theta) = (1-\theta)^{-k}, \qquad \psi(\theta) = \theta, \qquad \theta \in \Theta = (0,1),$$

then Eq. (16.1) is

$$\boldsymbol{P}_\theta\{X_i = x\} = \binom{k+x-1}{k-1}\theta^x(1-\theta)^k, \tag{16.5}$$

the *negative binomial* distribution with parameters k and θ. If $k = 1$, we have the *geometric* distribution with parameter θ.

Example 16.4. If we choose

$$a(x) = \frac{n\Gamma(n+\beta x)}{x!\Gamma(n+\beta x - x + 1)}, \qquad \psi(\theta) = \theta(1-\theta)^{\beta-1},$$

$$f(x) = (1-\theta)^{-n}, \qquad \theta \in \Theta = (0,1), \qquad x \in \boldsymbol{X}_0, \quad |\theta\beta| < 1,$$

then Eq. (16.1) is

$$\boldsymbol{P}_\theta\{X_1 = x\} = \frac{n\Gamma(n+\beta x)[\theta(1-\theta)^{\beta-1}]^x}{x!\Gamma(n+\beta x - x+1)(1-\theta)^{-n}}; \tag{16.6}$$

that is, we have the *generalized negative binomial distribution* as in Jain and Consul (1971).

Example 16.5. If

$$a(x) = \frac{\binom{\gamma x}{x}}{\gamma x} = \frac{\Gamma(\gamma x+1)}{\Gamma(x+1)\Gamma(\gamma x - x+1)},$$

$$f(\theta) = \ln\frac{1}{1-\theta} - \sum_{k=1}^{\alpha-1}\frac{1}{\gamma k}\binom{\gamma k}{k}[\theta(1-\theta)^{\gamma-1}]^k, \ x \in \boldsymbol{X}_\alpha, \ \alpha \geq 1,$$

$$\psi(\theta) = \theta(1-\theta)^{\gamma-1}, \quad \theta \in \Theta = (0,1), \quad 0 < |\gamma\theta| < 1,$$

then Eq. (16.1) gives us the so-called *left-truncated generalized logarithmic series distribution* with parameters θ and α:

$$\boldsymbol{P}_\theta\{X_i = x\} = \frac{\binom{\gamma x}{x}[\theta(1-\theta)^{\gamma-1}]^x}{\gamma x f(\theta)}, \quad x \in \boldsymbol{X}_\alpha \tag{16.7}$$

In case $\gamma=1$, formula (16.7) gives the so-called *left-truncated logarithmic series distribution*:

$$\boldsymbol{P}_\theta\{X_i = x\} = \frac{\theta^x}{x f(\theta)}, \qquad x \in \boldsymbol{X}_\alpha,$$

$$f(\theta) = \ln\frac{1}{1-\theta} - \sum_{x=1}^{\alpha-1} \frac{\theta^x}{x}.$$

The class (16.1) is rich. It includes, in addition to the examples above, the generalized Poisson distribution, introduced by Consul and Jain (1973), the Borel-Tanner distribution (see Haight and Breuer, 1960; Jain, 1974), other distributions, and their truncated forms. For additional examples of discrete probability distributions (truncated or not) of the form (16.1) or (16.1′) see, for example, Johnson and Kotz (1969, 1982) or Voinov and Nikulin (1989, 1993).

A Sufficient Statistic: The Sum S and its Distribution

Because in Eq. (16.1) the factor $\psi^x(\theta)$ can be written as

$$\psi^x(\theta) = \exp\{x \ln \psi(\theta)\}, \tag{16.8}$$

formula (16.1) defines a probability distribution belonging to the family called the *discrete exponential distributions of rank one.* Using Eq. (16.8), we can write the *likelihood function* $L(\theta)$, constructed from the sample $X_1, X_2, \ldots, X_n$, as

$$L(\theta) = \prod_{i=1}^{n} p_\theta(X_i) = \frac{1}{f^n(\theta)} \exp\{S \ln \psi(\theta)\} \prod_{i=1}^{n} a(X_i), \tag{16.9}$$

where $S=X_1+\cdots+X_n$. From the Neyman-Sheffe factorization theorem, it follows that the family of distributions (16.1) has the *sufficient statistic* S. If hypothesis H_0 is true then the distribution of S is given by

$$\boldsymbol{P}_\theta\{S=s\} = \frac{b(s,n)}{f^n(\theta)}\psi^s(\theta), \; s \in n[\boldsymbol{X}] \in \begin{cases} \boldsymbol{X}_0, & \text{if } \boldsymbol{X} = \boldsymbol{X}_0, \\ \boldsymbol{X}_{n\alpha}, & \text{if } \boldsymbol{X} = \boldsymbol{X}_\alpha, \\ \boldsymbol{X}_{n\alpha}^{n\beta}, & \text{if } \boldsymbol{X} = \boldsymbol{X}_\alpha^\beta, \end{cases} \tag{16.10}$$

where the coefficients $b(s,n)$ are defined by the condition

$$\sum_{s\in n[\boldsymbol{X}]} b(s,n)\psi^s(\theta) = f^n(\theta) = \Big(\sum_{x\in\boldsymbol{X}} a(x)\psi^x(\theta)\Big)^n. \tag{16.11}$$

We use here the notation $n[\boldsymbol{X}]$ and others introduced by Patil (1963). In the notation of Patil, let

$$A^{(i)} = \{a_1^{(i)}, a_2^{(i)}, \ldots\} \; (i = 1, 2, \ldots, n)$$

be an arbitrary subset of $\boldsymbol{X}_0$. The "sum" $A_n = \sum_{i=1}^n A^{(i)}$ of the given subsets $A^{(1)}, \ldots, A^{(n)}$ is defined as the set of all integers of the form $\sum_{i=1}^n a^{(i)}$, where $a^{(i)} \in A^{(i)}$. If all $A^{(i)}$ are the same, and equal to A, then their sum is

$$A_n = \sum_{i=1}^n A^{(i)} = \sum_{i=1}^n A = n[A].$$

If some set contains only one element, a, we denote the set by $\{a\}$.

Example 16.6. Let $X_1, \ldots, X_n$ be a sample from a negative binomial distribution with parameters k and θ (see Example 16.3). In this case the distribution of the sufficient statistic S is given by the formula

$$\boldsymbol{P}_\theta\{S=s\} = \binom{nk+s-1}{nk-1}\theta^s(1-\theta)^{nk}, \; s \in \boldsymbol{X}_0. \tag{16.12}$$

In the case $\psi(\theta) \equiv \theta$, we have from Eq. (16.11)

$$b(s,n) = \frac{1}{s!}\frac{d^s}{d\theta^s} f^n(\theta)\Big|_{\theta=0}. \tag{16.13}$$

If the distribution (16.1) is not truncated, that is, $x \in \boldsymbol{X}_0$, then from Eq. (16.11) we find that

$$\begin{aligned} b(s,n) &= \sum_{x_1+\cdots+x_n=s} \prod_{i=1}^{n} a(x_i) \\ &= \sum_{x_1=0}^{s} \sum_{x_2=0}^{s-x_1} \cdots \sum_{x_{n-1}=0}^{s-x_1-\cdots-x_{n-1}} \prod_{i=1}^{n-1} a(x_i) a(s-x_1-\cdots-x_{n-1}), \end{aligned} \tag{16.14}$$

$$x_i \in \boldsymbol{X}_0,$$

because in this case

$$\begin{aligned} f^n(\theta) &= \left(\sum_{x=0}^{\infty} a(x)\psi^x(\theta) \right)^n \\ &= \sum_{s=0}^{\infty} \left(\sum_{x_1+\cdots+x_n=s} a(x_1)a(x_2)\cdots a(x_n) \right) \psi^s(\theta). \end{aligned}$$

Formula (16.14) is effective for calculating the coefficients $b(s,n)$ when n is not large and $f(\theta)$ is not very complicated. There are other formulae for $b(s,n)$, see for example, Voinov and Nikulin (1993). One of them is

$$\begin{aligned} b(s,n) &= \sum_{x_1=0}^{n} a(x_1) \sum_{x_2+\cdots+x_n=s-x_1} \prod_{i=1}^{n} a(x_i) \\ &= \sum_{x=0}^{s} a(x) b(s-x, n-1), \quad n \geq 2, \quad s \in \boldsymbol{X}_0 \end{aligned} \tag{16.15}$$

according to which of the numbers $b(s,n)$ satisfy a simple recurrence relation. When n is large or the expression for the function $f(\theta)$ is complicated, it is useful to define the numbers $b(s,n)$ by the alternate relations:

$$b(m,n) = \frac{1}{ma(0)} \sum_{k=1}^{m} (k-m+k) a(k) b(m-k, n), \; m \geq 1, \; m \in \boldsymbol{X}_0,$$

$$b(0,n) = a^n(0).$$

Example 16.7. Let $X_1, X_2, \ldots, X_n$ be a sample from the Poisson distribution with parameter θ (see Example 16.1). The sufficient statistic $S = X_1 + \cdots + X_n$ also has the Poisson distribution but with parameter $n\theta$, that is,

$$\mathbf{P}_\theta\{S = s\} = \frac{(n\theta)^s}{s!} e^{-n\theta}, \quad s \in \mathbf{X}_0. \tag{16.16}$$

On the other hand, from Eq. (16.10), we have

$$\mathbf{P}_\theta\{S = s\} = \frac{b(s,n)}{f^n(\theta)} \psi^s(\theta) = \frac{b(s,n)\theta^s}{e^{n\theta}},$$

and hence, from Eq. (16.16), we obtain

$$b(s,n) = \frac{n^s}{s!}. \tag{16.17}$$

If the distribution (16.1) is left-truncated, that is, if $x \in \mathbf{X}_\alpha$, $\alpha \geq 1$, then, from Eq. (16.11) we find that

$$\begin{aligned} b(s,n) = & \sum_{x_1=\alpha}^{s-(n-1)\alpha} \sum_{x_2=\alpha}^{s-x_1-(n-2)\alpha} \\ & \cdots \sum_{x_{n-1}=\alpha}^{s-x_1-\cdots-x_{n-2}-\alpha} a(x_1)\ldots a(x_{n-1}) a(s - \sum_{i=1}^{n-1} x_i), \end{aligned} \tag{16.18}$$

because

$$\begin{aligned} f^n(\theta) &= \sum_{s=n\alpha}^{\infty} b(s,n)\psi^s(\theta) = \left(\sum_{x=\alpha}^{\infty} a(x)\psi^x(\theta)\right)^n \\ &= \sum_{s=n\alpha}^{\infty} \left(\sum_{x_1+\cdots+x_n=s} a(x_1)a(x_2)\ldots a(x_n) \right) \psi^s(\theta). \end{aligned}$$

From Eq. (16.18), it follows that for the left-truncated distribution (16.1), we have

$$b(s,n) = \sum_{x=2}^{s-(n-1)\alpha} a(x)b(s-x, n-1) \tag{16.19}$$

which is similar to Eq. (16.15). If $\alpha=0$, formula (16.19) gives Eq. (16.15). Park (1973b) and Roy and Mitra (1957) have given other recurrence formulae for calculating $b(s,n)$.

Finally, if the distribution (16.1) is double truncated with truncation parameters α and β, $\beta > \alpha$, then as before we have

$$b(s,n) = \sum_{x_1+\cdots+x_n=s} \prod_{i=1}^{n} a(x_i),\ x_i \in \boldsymbol{X}_{\alpha}^{\beta},\ s \in n[\boldsymbol{X}_{\alpha}^{\beta}] = \boldsymbol{X}_{n\alpha}^{n\beta}, \tag{16.20}$$

and one can see that instead of Eq. (16.15) we have

$$b(s,n) = \sum_{x=\max\{\alpha, s-(n-1)\beta\}}^{\min\{(\beta, s-(n-1)\alpha\}} a(x)b(s-x,n-1) \tag{16.21}$$

Example 16.8. Let $X_1,\ldots,X_n$ be a sample from a Bernoulli law with parameter θ, $\theta \in \Theta = (0,1)$, that is,

$$\boldsymbol{P}_{\theta}(X=x) = p_{\theta}(x) = \binom{1}{x}\theta^x(1-\theta)^{1-x},\ x \in \{0,1\} = \boldsymbol{X}_0^1 \tag{16.22}$$

This distribution belongs to the family (16.1$'$) of right-truncated distributions. To see this, let

$$\vartheta = \frac{\theta}{1-\theta},\ \theta = \frac{\vartheta}{1+\vartheta}. \tag{16.23}$$

We have from Eq. (16.22) that

$$p_{\theta}(x) = \frac{\binom{1}{x}\vartheta^x}{1+\vartheta},\ x \in X_0^1, \tag{16.24}$$

which belongs to the family (16.1$'$) with $f(\vartheta) = 1+\vartheta$, $a(x) = 1$. The sufficient statistic $S = X_1+\cdots+X_n$ has a binomial distribution with parameters n and θ, that is,

$$\boldsymbol{P}_{\theta}\{S=s\} = \frac{\binom{n}{s}}{(1+\vartheta)^n}\vartheta^s = \binom{n}{s}\theta^s(1-\theta)^{n-s},\ s \in n[\boldsymbol{X}_0^1] = \{0,1,\ldots,n\} \tag{16.25}$$

Conditional Distribution of Frequencies Given S

Let $X_1, \ldots, X_n$ be a sample from Eq. (16.1) and let $S = X_1 + \cdots + X_n$, a minimal sufficient statistic for θ. Furthermore,

$$\mu_{ix} = \begin{cases} 1, & \text{if } X_i = x, \\ 0, & \text{if } X_i \neq x, \end{cases} \qquad i = 1, 2, \ldots, n \tag{16.26}$$

for any $x \in \boldsymbol{X}_0$, and let

$$\mu_x = \sum_{i=1}^{n} \mu_{ix}, \qquad \boldsymbol{\mu} = (\mu_0, \mu_1, \mu_2, \ldots); \tag{16.27}$$

that is, $\boldsymbol{\mu}$ is the vector of frequencies,

$$\mu_x = \#\{i : X_i = x,\ i = 1, \ldots, n\}, \quad \text{for any } x \in \boldsymbol{X}_0.$$

Then the likelihood function $L(\theta)$ has the form

$$\begin{aligned} L(\theta) &= \prod_{i=1}^{n} \boldsymbol{P}_\theta\{X_i = x\} = [\psi(\theta)]^S \frac{1}{f^n(\theta)} \prod_{i=1}^{n} a(X_i) \\ &= \prod_{x=0}^{\infty} [p_x(\theta)]^{\mu_x} = \exp\left\{ \sum_{x=0}^{\infty} \mu_x \ln p_x(\theta) \right\}. \end{aligned} \tag{16.28}$$

If the observed value s of S is fixed, then we have the relations among s, μ, and the sample size n:

$$\begin{aligned} s &= X_1 + \cdots + X_n = \mu_1 + 2\mu_2 + \cdots + s\mu_s, \\ \mu_0 + \mu_1 + \cdots + \mu_s &= n, \ \mu_x = 0 \text{ if } x > s. \end{aligned} \tag{16.29}$$

The joint conditional distribution of the frequencies $\mu_0, \mu_1, \ldots, \mu_n$ is given by

$$\boldsymbol{P}\left\{ \bigcap_{x=0}^{s} (\mu_x = n_x) \mid S = s \right\} = \frac{n!}{b(s,n)} \prod_{x=0}^{s} \frac{1}{n_x!} [a(x)]^{n_x} \tag{16.30}$$

In fact, the joint law of $(\mu_0, \mu_1, \ldots, \mu_S, S)$ is the multinomial

$$\begin{aligned} \boldsymbol{P}_\theta(\mu_0 &= n_0, \mu_1 = n_1, \ldots, \mu_s = n_s,\ S = s\} \\ &= \frac{n!}{n_0! \ldots n_s!}[p_0(\theta)]^{n_0} \ldots [p_s(\theta)]^{n_s} \end{aligned} \tag{16.31}$$

where $s, n_x \in \boldsymbol{X}_0$, $n_0 + \cdots + n_s = n$, and the law of S is given by Eq. (16.10). From Eqs. (16.1), (16.28), and (16.31) we have

$$\begin{aligned} &\boldsymbol{P}\Big\{ \bigcap_{x=0}^{s} (\mu_x = n_x) \mid S = s \Big\} \\ &\quad = \frac{\boldsymbol{P}_\theta\left\{\cap_{x=0}^{s}(\mu_x = n_x),\ S = s\right\}}{\boldsymbol{P}_\theta\{S = s\}} \\ &\quad = \frac{f^n(\theta)}{b(s,n)\psi^s(\theta)} \frac{n!}{n_0! \ldots n_s! f^n(\theta)} \prod_{x=0}^{s} [a(x)]^{n_x} [\psi(\theta)]^{x n_x}. \end{aligned} \tag{16.32}$$

Using Eq. (16.29) we obtain Eq. (16.30).

Example 16.9. In the Poisson case

$$\boldsymbol{P}_\theta\{X_i = x_i\} = \frac{\theta^x e^{-\theta}}{x!}, \quad i = 1, \ldots, n,$$

using Eqs. (16.16) and (16.30), and Example 16.1 we obtain

$$\boldsymbol{P}\left\{ \bigcap_{x=0}^{s} (\mu_x = n_x) \mid S = s \right\} = \frac{n! s!}{n^s} \prod_{x=0}^{s} \frac{1}{n_x!} \left(\frac{1}{x!}\right)^{n_x}, \tag{16.33}$$

where n_x satisfies Eq. (16.29) and $s \in \boldsymbol{X}_0$.

Example 16.10. If we have

$$\boldsymbol{P}_\theta\{X_i = x\} = \binom{k+x-1}{t-1} \theta^x (1-\theta^k), \qquad i = 1, \ldots, n, \qquad x \in \boldsymbol{X}_0,$$

as in Eq. (16.5), that is, if we have a sample from the negative binomial distribution with the parameters k and θ, then in place of Eq. (16.30) we have

$$\boldsymbol{P}\left\{\bigcap_{x=0}^{s}(\mu_x = n_x) \mid S = s\right\} = \frac{n!s!}{\Gamma(kn+s)} \prod_{x=1}^{s} \left[\frac{1}{n_x!}\binom{k+x-1}{k-1}^{n_x}\right], \tag{16.34}$$

where n_x satisfies Eq. (16.29), $s \in \boldsymbol{X}_0$.

Example 16.11. If

$$\boldsymbol{P}_\theta\{X_i = x\} = \binom{m}{x}\theta^x(1-\theta)^{m-x}, \; x \in \boldsymbol{X}_0^m, \; m \in \boldsymbol{X}_1,$$

that is, we have a sample from a binomial distribution with parmeters m and θ, then from Eq. (16.30) we have

$$\boldsymbol{P}\left\{\bigcap_{x=0}^{s}(\mu_x = n_x) \mid S = s\right\} = \frac{n!s!(nm-s)!}{(nm)!} \prod_{x=0}^{s} \left[\frac{1}{n_x!}\binom{m}{x}^{n_x}\right], \tag{16.35}$$

where n_x satisfies Eq. (16.29), $s \in \boldsymbol{X}_0^{nm}$.

Example 16.12. Consider the Borel-Tanner distribution with parameter k, $k > 0$,

$$\boldsymbol{P}_\theta\{X_i = x\} = k(k+x)^{x-1}\theta^x \frac{1}{x!}\, e^{-\theta(k+x)}, \; x \in \boldsymbol{X}_0, \; \theta \in \Theta = (0,1). \tag{16.36}$$

In this case, the analogue of Eq. (16.30) is

$$\begin{aligned} &\boldsymbol{P}\left\{\bigcap_{x=0}^{s}(\mu_x = n_x) \mid S = s\right\} \\ &\qquad = \frac{(n-1)!s!k^{n-1}}{(nk+s)^{s-1}} \prod_{x=0}^{s} \left\{\frac{1}{n_x!}\left[\frac{(k+x)^{x-1}}{x!}\right]^{n_x}\right\}, \end{aligned} \tag{16.37}$$

where n_x satisfies Eq. (16.29), $s \in \boldsymbol{X}_0$.

17. MINIMUM VARIANCE UNBIASED ESTIMATION FOR A PARAMETRIC FAMILY OF MODIFIED POWER SERIES DISTRIBUTIONS

Sufficiency and Minimum Variance Unbiased Estimation

In the next section (Section 18) we describe a chi-squared goodness-of-fit test of the hypothesis H_0 using the best (in the sense of the minimum quadratic risk) unbiased estimators W_x of the probabilities

$$p_x(\theta) = \boldsymbol{P}_\theta\{X_i = x\} = \frac{a(x)\psi^x(\theta)}{f(\theta)}, \quad x \in \boldsymbol{X}_0. \tag{17.1}$$

The traditional procedure would be, as discussed earlier, to find a point estimator θ_n^* of the unknown parameter θ and then use $p_x(\theta_n^*)$ as estimator of $p_x(\theta)$. Instead, in this section, we formulate an easy generalization of well-known results of Patil (1963) and Joshi and Park (1974) for the modified power series distribution. This theorem tells us how to use the Rao-Blackwell approach for constructing the best unbiased estimator $T = T(S)$, in terms of the sufficient statistic $S = X_1 + \cdots + X_n$, for the value $g(\theta)$ of a given function $g(\cdot)$, $\theta \in \Theta$.

Theorem 17.1. *A necessary and sufficient condition for the existence of the minimum variance unbiased estimator* (MVUE) *of the value $g(\theta)$ of $g(\cdot)$ is*

$$\boldsymbol{W}[g(\theta)f^n(\theta)] \subseteq \boldsymbol{W}[f^n(\theta)], \tag{17.2}$$

where, for any argument $G(\theta)$ with series expansion $G(\theta) = \sum \gamma_r \psi^r(\theta)$,

$$\boldsymbol{W}[G(\theta)] = \{r : r \in \boldsymbol{X}_0,\ \gamma_r > 0\}.$$

If the MVUE, T, *of $g(\theta)$ exists, then the statistic has the form:*

$$T = T(S) = \begin{cases} \frac{c(S,n)}{b(S,n)}, & S \in \boldsymbol{W}[g(\theta)f^n(\theta)], \\ 0, & \textit{otherwise}, \end{cases} \tag{17.3}$$

where the numbers $c(s,t)$ are the coefficients of $\psi^s(\theta)$ in the series expansion of the function $G(\theta) = g(\theta)f^n(\theta)$ in powers of $\psi(\theta)$, and the $b(s,n)$ are defined by Eq. (16.11).

Example 17.1. Bernoulli Distribution. Let us consider again the independent Bernoulli random variables with parameter θ, $\theta \in \Theta = (0,1)$, as in Example 16.8; that is,

$$\boldsymbol{P}_\theta\{X_i = x\} = \theta^x(1-\theta)^{1-x},\ x \in \{0,1\} = \boldsymbol{X}_0^1.$$

Suppose that we wish to estimate θ^k, $k \in \boldsymbol{X}_1$. If we change variables according to Eq. (16.23), then our problem is to estimate

$$g(\vartheta) = \frac{\vartheta^k}{(1+\vartheta)^k} = \theta^k.$$

Here $f(\vartheta) = 1+\vartheta$, and hence $g(\vartheta)f^n(\vartheta) = \vartheta^k(1+\vartheta)^{n-k}$. If $n \geq k$, we have

$$\begin{aligned} g(\vartheta)f^n(\vartheta) &= \vartheta^k \sum_{m=0}^{n-k} \binom{n-k}{m} \vartheta^m \\ &= \sum_{m=k}^{n} \binom{n-k}{m-k} \vartheta^m, \end{aligned} \tag{17.4}$$

and

$$\begin{aligned} \boldsymbol{W}[g(\vartheta)f^n(\vartheta)] = \boldsymbol{X}_k^n &= \{k, k+1, \ldots, n\} \subseteq \boldsymbol{W}[f^n(\vartheta)] \\ &= \boldsymbol{X}_0^n = \{0, 1, \ldots, n\}. \end{aligned}$$

Hence, from Eq. (17.2) it follows that the MVUE exists for $g(\vartheta) = \theta^k$ if $n \geq k$, and it is easy to verify that the MVUE is

$$T(S) = \begin{cases} \binom{n-k}{S-k} / \binom{n}{S}, & \text{if } k \leq S \leq n, \\ 0, & \text{otherwise.} \end{cases} \tag{17.5}$$

On the other hand, if $k > n$, we have

$$\begin{aligned} g(\vartheta)f^n(\vartheta) &= \vartheta^k/(1+\vartheta)^{n-k} \\ &= \sum_{m=0}^{\infty} (-1)^m \binom{k-n+m-1}{m} \theta^{k+m} \sum_{x=k}^{\infty} (-1)^{x-k} \binom{x-n-1}{x-k} \theta^x, \end{aligned}$$

and we obtain

$$\boldsymbol{W}[g(\vartheta)f^n(\vartheta)] = \boldsymbol{X}_k = \{k, k+1, \ldots\} \not\subseteq \boldsymbol{W}[f^n(\vartheta)] = \boldsymbol{X}_0^n.$$

Hence the MVUE for θ^k for $k > n$ does not exist.

Estimating the Modified Power Series Distribution

Let us construct the MVUE of

$$g(\theta) = \boldsymbol{P}_\theta\{X_i = x\} = \frac{a(x)\psi^x(\theta)}{f(\theta)}, \; x \in \boldsymbol{X}_\alpha, \; i = 1, 2, \ldots, n.$$

In this case we have

$$\begin{aligned} g(\theta)f^n(\theta) &= a(x)\psi^x(\theta)f^{n-1}(\theta) \\ &= a(x)\psi^x(\theta)\left(\sum_{x=\alpha}^{\infty} a(x)\psi^x(\theta)\right)^{n-1} \\ &= a(x)\psi^x(\theta)\left(\sum_{s\in(n-1)[\boldsymbol{X}_\alpha]} b(s, n-1)\psi^s(\theta)\right) \\ &= a(x) \sum_{x\in(n-1)[\boldsymbol{X}_\alpha]} b(s, n-1)\psi^{s+x}(\theta) \qquad (17.6) \\ &= \sum_{s\in(n-1)[\boldsymbol{X}_\alpha]+\{x\}} a(x)b(s-x, n-1), \end{aligned}$$

and hence we obtain

$$\boldsymbol{W}[g(\theta)f^n(\theta) = (n-1)[\boldsymbol{X}_\alpha] + \{x\}. \qquad (17.7)$$

On the other hand, we have

$$f^n(\theta) = \left(\sum_{x\in\boldsymbol{X}_\alpha} a(x)\psi^x(\theta)\right)^n = \sum_{s\in(n-1)[\boldsymbol{X}_\alpha]} b(s, n)\psi^s(\theta) \qquad (17.8)$$

and so, from Eqs. (17.7) and (17.8), we obtain

$$\boldsymbol{W}[g(\theta)f^n(\theta)] \subseteq \boldsymbol{W}[f^n(\theta)].$$

From Theorem 17.1, it follows that the MVUE, W_x, of $p_x(\theta) = \boldsymbol{P}_\theta \{X_i = x\}$, exists and $\boldsymbol{E}W_x = p_x(\theta)$. From Eqs. (17.3) and (17.6), we see that

$$W_x = \begin{cases} \frac{a(x)b(S-x,n-1)}{b(S,n)}, & S \in (n-1)[\boldsymbol{X}_\alpha]+\{x\}, \\ 0, & \text{otherwise.} \end{cases} \tag{17.9}$$

Example 17.2. Poisson Distribution. Let us construct now the MVUE of

$$p_x(\theta) = \boldsymbol{P}_\theta\{X_i = x\} = \frac{\theta^x}{x!} e^{-\theta}, \ x \in \boldsymbol{X}_0,$$

using the independent Poisson data $X_1, \ldots, X_n$, with parameter $\theta > 0$. From Eq. (16.7) (see Example 16.7), we have that $b(s,n) = n^s/s!$, and hence, from Eq. (17.9), the MVUE W_x of $p_x(\theta)$ given S and n is

$$W_x = \begin{cases} \binom{S}{x}\left(\frac{1}{n}\right)^x\left(1-\frac{1}{n}\right)^{S-x}, & \text{if } S \geq x, \ x \in \boldsymbol{X}_0, \\ 0, & \text{otherwise.} \end{cases} \tag{17.10}$$

This formula shows us that the MVUE of the Poisson distribution with parameter θ given S has the form of the binomial distribution with parameters S and n,

$$\boldsymbol{E}_\theta W_x = p_x(\theta), \ x = 1, 2, \ldots, S.$$

Example 17.3. Left Truncated Generalized Logarithmic Distribution. Let $X_1, \ldots, X_n$ be i.i.d.,

$$p_x(\theta) = \boldsymbol{P}_\theta\{X_i = x\} = \frac{\binom{\gamma x}{x}[\theta(1-\theta)^{r-1}]^x}{\gamma x f(x)}, \ x \in \boldsymbol{X}_\alpha,$$

$$f(\theta) = \ln\frac{1}{1-\theta} - \sum_{k=1}^{\alpha-1}\frac{1}{\gamma k}\binom{\gamma k}{k}[\theta(1-\theta)^{\gamma-1}]^x,$$

$$\theta \in \Theta = (0,1), \ 0 <| \gamma\theta |< 1,$$

where γ and α are known. In this case, we have the left-truncated generalized logarithmic series distribution (see Example 16.5). The

MVUE W_x of $p_x(\theta)$ given S, obtained by Jani (1977), is

$$W = \begin{cases} \dfrac{\frac{1}{n\gamma x}\binom{\gamma x}{x}(-1)^{x-1}S!s(S-x,n-1)}{(S-x)!s(S,n)}, & \text{if } S \geq \alpha(n-1)+x, \\ 0, & \text{otherwise,} \end{cases} \tag{17.11}$$

where $s(t,n)$ are the modified Stirling numbers of the first kind. The numbers $s(t,n)$ satisfy a simple recurrence relation (see, e.g. Voinov and Nikulin, 1993).

$$s(t,n) = \sum_{x=\alpha}^{t-(n-1)\alpha} \binom{\gamma x}{x} \frac{(-1)^{x-1}t!}{n\gamma x(t-x)!} \cdot s(t-x,n-1),\ t \geq n\alpha,$$

and $s(t,n)$ satisfy the initial conditions

$$s(t,0) = \begin{cases} 1, & \text{if } t = 0, \\ 0, & \text{if } t = 1,2,\ldots . \end{cases}$$

Example 17.4. Negative Binomial Distribution. Let $X_1,\ldots,X_n$ be i.i.d.,

$$p_x(\theta) = \boldsymbol{P}_\theta\{X_i = x\} = \binom{k+x-1}{k-1}\theta^x(1-\theta)^k,$$
$$x \in \boldsymbol{X}_0,\ k \geq 1,\ \theta \in \Theta = (0,1);$$

that is, we have the negative binomial distribution with parameters k and θ (see Example 16.3). In this case, the coefficients $b(s,n) = \binom{kn+s-1}{s}$ and it is easy to see that the MVUE W_x of $p_x(\theta)$ given S is

$$W_x = \begin{cases} \dfrac{\binom{k+x-1}{k-1}\binom{S-x+k(n-1)-1}{S-x}}{\binom{S+kn-1}{S}}, & \text{if } S \geq x, \\ 0, & \text{otherwise;} \end{cases} \tag{17.12}$$

that is, a negative hyper-geometric random variable is the MVUE for a negative binomial distribution.

Example 17.5. Binomial Distribution. One can verify easily by analogy to Example 16.1, that the MVUE given S of the binomial distribution

$$p_x(\theta) = \binom{m}{x}\theta^x(1-\theta)^{m-x},\ 0<\theta<1,\ x \in \boldsymbol{X}_0^m,$$

is a hypergeometric random variable

$$W_x = \begin{cases} \dfrac{\binom{m}{x}\binom{mn-m}{S-x}}{\binom{mn}{S}}, & \text{if } x \le S \le mn, \\ 0, & \text{otherwise.} \end{cases} \tag{17.13}$$

Estimation of Joint Distribution

Now we return to the general case, that is, to the modified power series distribution (16.1). Following the derivation of Eq. (17.9), one obtains that the MVUE, $W_{x,y}$, given S of the joint probability distribution

$$\boldsymbol{P}_\theta\{X_1 = x, X_2 = y\},\ x,y \in \boldsymbol{X}_\alpha$$

is given by

$$W_{x,y} = \begin{cases} \dfrac{a(x)a(y)b(S-x-y,n-2)}{b(S,n)}, & \text{if } S \in (n-2)[\boldsymbol{X}_\alpha]+\{x\}+\{y\}, \\ 0, & \text{otherwise.} \end{cases} \tag{17.14}$$

$$= P\{X := x, X_j = y \mid S\} = W_x(S,n)W_y(S-x,n-1)$$

(see Nikulin and Voinov, 1989).

Since

$$\alpha_1 = \alpha_1(\theta) = \boldsymbol{E}_\theta X_i = \sum_{x\in\boldsymbol{X}_\alpha} x\boldsymbol{P}_\theta\{X_i = x\}$$

from Eq. (17.9), we see that the MVUE $\hat{\alpha}_1$ of α_1 is

$$\hat{\alpha}_1 = \sum_{x\in\boldsymbol{X}_\alpha} xW_x = \sum_{x=2}^{S-(n-1)\alpha} \frac{xa(x)b(S-x,n-1)}{b(S,n)} = \frac{1}{n}S, \tag{17.15}$$

because $\boldsymbol{E}_\theta S = n\boldsymbol{E}_\theta X_1 = n\alpha_1$ and from the completeness of the sufficient statistic.

Estimation of Variance

Now let us calculate the MVUE $\hat{\sigma}^2$ of the variance

$$\sigma^2(\theta) = \mathbf{Var}_\theta X_i = \alpha_2 - \alpha_1^2,$$

where

$$\alpha_2 = \alpha_2(\theta) = \boldsymbol{E}_\theta X_i^2 = \sum_{x=\alpha}^{\infty} \alpha^2 \boldsymbol{P}_\theta\{X_i = x\}.$$

Because $\boldsymbol{E}_\theta W_x = \boldsymbol{P}_\theta\{X_i = x\}$, the MVUE $\hat{\alpha}_2$ of α_2 is

$$\hat{\alpha}_2 = \sum_{x \in \boldsymbol{X}_\alpha} x^2 W_x = \frac{1}{b(S,n)} \sum_{x=\alpha}^{S-(n-1)\alpha} a(x) x^2 b(S-x, n-1) \tag{17.16}$$

The MVUE $\hat{\alpha}_1^2$ of

$$\alpha_1^2 = (\boldsymbol{E}_\theta X_i)^2 = (\boldsymbol{E}_\theta X_i)(\boldsymbol{E}_\theta X_j), \; i \neq j, \tag{17.17}$$

can be calculated as follows. From Eq. (17.16), we have

$$\begin{aligned} \alpha_1^2 &= \left(\sum_{x \in \boldsymbol{X}_\alpha} x \boldsymbol{P}_\theta\{X_i = x\} \right) \left(\sum_{y \in \boldsymbol{X}_\alpha} y \boldsymbol{P}_\theta\{X_j = y\} \right) \\ &= \sum_{x,y \in \boldsymbol{X}_\alpha} xy \boldsymbol{P}_\theta\{X_i = x\} \boldsymbol{P}_\theta\{X_j = y\} \\ &= \sum_{x,y \in \boldsymbol{X}_\alpha} xy \boldsymbol{P}_\theta\{X_i = x, \; X_j = y\}, \end{aligned} \tag{17.18}$$

because X_i and X_j are independent if $i \neq j$. Now using

$$\boldsymbol{E}_\theta W_{x,y} = \boldsymbol{P}_\theta\{X_i = x, X_j = y\}$$

from Eq. (17.14), we obtain

$$\begin{aligned}\hat{\alpha}_1^2 &= \sum_{x,y\in \boldsymbol{X}_\alpha} xyW_{x,y} \\ &= \sum_{y=\alpha}^{S-(n-1)\alpha} ya(y) \sum_{x=\alpha}^{S-(n-2)\alpha-y} \frac{xa(x)b(S-x-y,n-2)}{b(S,n)}.\end{aligned} \tag{17.19}$$

From formula (17.15), we obtain

$$\frac{1}{b(S,n-1)} \sum_{x=\alpha}^{S-(n-2)\alpha} xa(x)b(S-x,n-2) = \frac{1}{n-1}S, \tag{17.20}$$

and hence, from Eqs. (17.19) and (17.20), we have

$$\sum_{x=\alpha}^{S-(n-2)\alpha-y} xa(x)b(S-x-y,n-2) = \frac{S-y}{n-1}\, b(S-y,n-1).$$

Therefore

$$\begin{aligned}\hat{\alpha}_1^2 &= \frac{1}{(n-1)b(S,n)} \sum_{y=\alpha}^{S-(n-1)\alpha} ya(y)b(S-y,n-1)(S-y) \\ &= \frac{S}{(n-1)b(S,n)} \sum_{y=\alpha}^{S-(n-1)\alpha} ya(y)b(S-y,n-1) \\ &\quad - \frac{1}{(n-1)b(S,n)} \sum_{y=\alpha}^{S-(n-1)\alpha} y^2a(y)b(S-y,n-1) \\ &= \frac{1}{n(n-1)}S^2 - \frac{1}{n-1}\,\hat{\alpha}_2.\end{aligned} \tag{17.21}$$

The last equality uses formulas (17.16) and (17.15).

Now combining formulas (17.16) and (17.21), we obtain the MVUE

$\hat{\sigma}^2$ of the variance σ^2:

$$\hat{\sigma}^2 = \hat{\alpha}_2 - \hat{\alpha}_1^2 = \hat{\alpha}_2 - \frac{1}{n(n-1)} S^2 + \frac{1}{n-1}\hat{\alpha}_2 = \frac{n}{n-1}\hat{\alpha}_2 - \frac{1}{n(n-1)} S^2$$

$$= \frac{n}{(n-1)b(S,n)} \sum_{x=\alpha}^{S-(n-1)} x^2 a(x) b(S-x, n-1) - \frac{1}{n(n-1)} S^2. \tag{17.22}$$

It is evident that, in the case of a doubly truncated modified power series distribution, formulas (17.16) to (17.22) remain valid if one replaces the lower limit in the summation by $\max\{\alpha, S-(n-1)\beta\}$, and the upper limit by $\min\{\beta, S-(n-1)\alpha\}$.

Example 17.6. Poisson Distribution. Let us consider again the Poisson case; that is, $X_1, \ldots, X_n$ are i.i.d., and

$$p_x(\theta) = \boldsymbol{P}_\theta\{X_i = x\} = \frac{\theta^x}{x!} e^{-\theta}, \ x \in \boldsymbol{X}_0, \ \theta > 0.$$

From Eq. (17.10), we have that

$$W_x = \begin{cases} \binom{S}{x}(\frac{1}{n})^x(1-\frac{1}{n})^{S-x}, & \text{if } S > x, \\ 0, & \text{otherwise,} \end{cases} \tag{17.23}$$

is the MVUE of $p_x(\theta)$.

From Eqs. (17.14) and (17.17), we obtain that the MVUE $W_{x,y}$ given S of the joint probability

$$\boldsymbol{P}_\theta\{X_1 = x, X_2 = y\} = \boldsymbol{P}_\theta\{X_1 = x\}\boldsymbol{P}_\theta\{X_2 = y\}, \ x, y \in \boldsymbol{X}_0,$$

is

$$W_{x,y} = \begin{cases} \frac{S!}{x!y!(S-x-y)!}\left(\frac{1}{n}\right)^x\left(\frac{1}{n}\right)^y(1-\frac{2}{n})^{S-x-y}, & S \geq x+y, \\ 0, & \text{otherwise.} \end{cases} \tag{17.24}$$

From Eq. (17.24), we have that, if the hypothesis H_0 is true, then for any $x, y \in \boldsymbol{X}_0$

$$\begin{aligned}\boldsymbol{E}_\theta W_{x,y} &= \boldsymbol{E}_\theta(\boldsymbol{P}\{X_i = x, X_j = y \mid S\}) = \boldsymbol{P}_\theta\{X_i = x, X_j = y\} \\ &= \boldsymbol{P}_\theta\{X_i = x\}\boldsymbol{P}_\theta\{X_j = y\} = \frac{\theta^x}{x!}\,e^{-\theta}\cdot\frac{\theta^y}{y!}\,e^{-\theta},\ i \neq j. \quad (17.25)\end{aligned}$$

and from Eq. (17.23) we have

$$\boldsymbol{E}_\theta W_x = \boldsymbol{E}_\theta(\boldsymbol{P}\{X_i = x \mid S\}) = \boldsymbol{P}_\theta\{X_i = x\} = \frac{\theta^x}{x!}\,e^{-\theta},\ i = 1, 2, \ldots, n, \quad (17.26)$$

and at the same time, from Eqs. (17.23) and (17.24) we obtain

$$W_{x,y} \neq W_x \cdot W_y. \quad (17.27)$$

Let us write W_x in formula (17.23) as $W_x(S, 1/n)$, that is, $W_x = W_x(S, 1/n)$. Then from Eq. (17.24) it is easy to see that

$$\begin{aligned}W_{x,y} &= \boldsymbol{P}\{X_i = x, X_j = y \mid S\} \\ &= W_x\Big(S, \frac{1}{n}\Big)W_y\Big(S - x, \frac{1}{n-1}\Big).\end{aligned} \quad (17.28)$$

But one can also show that, if $n \to \infty$ while $S \asymp n$, then for any y, x fixed

$$W_y\Big(S - x, \frac{1}{n-1}\Big) = W_y\Big(S, \frac{1}{n}\Big) + O\Big(\frac{1}{n}\Big),\ (n \to \infty) \quad (17.29)$$

and hence from Eq. (17.28) with this condition we obtain

$$W_{x,y} = W_{x,y}\Big(S, \frac{1}{n}\Big) = W_x\Big(S, \frac{1}{n}\Big)W_y\Big(S, \frac{1}{n}\Big) + O\Big(\frac{1}{n}\Big),\ (n \to \infty). \quad (17.30)$$

This can be done for the general case (17.14).

Finally, it may be of interest to see that formula (17.24) can be obtained directly from generating functions instead of via Theorem 17.1. Let us denote by

$$f(u) = \boldsymbol{E}u^{X_i},\ g(u, v) = \boldsymbol{E}u^{X_i}v^{X_j}\ (i \neq j)$$

the generating functions of X_i and of (X_i, X_j) respectively, where u and

v are positive, X_i is independent of X_j,

$$\boldsymbol{P}_\theta\{X_i = x\} = \frac{\theta^x}{x!}\, e^{-\theta},\ \theta > 0,\ x \in \boldsymbol{X}_0 = \{0, 1, 2, \ldots\}.$$

Then we have

$$f(u) = e^{\theta(u-1)},\ g(u, v) = e^{\theta(u+v-2)} \tag{17.31}$$

Now, let us consider the generating function $g(u, v \mid S)$ of the joint conditional distribution of X_i and X_j given S, where $S = X_1 + \cdots + X_n$ is the sufficient statistic. Of course, for any $\theta > 0$, we have

$$\boldsymbol{E}_\theta\{g(u, v \mid S)\} = g(u, v) = e^{\theta(u+v-2)}. \tag{17.32}$$

Because

$$\boldsymbol{P}_\theta\{S = s\} = \frac{(n\theta)^s}{s!}\, e^{-n\theta},\ s \in \boldsymbol{X}_0,$$

the left-hand side of Eq. (17.32) can be written as

$$\sum_{s=0}^{\infty} g(u, v \mid s) \frac{(n\theta)^s}{s!}\, e^{-n\theta} = e^{\theta(u+v-2)}, \tag{17.33}$$

that is, for any $\theta > 0$ from Eq. (17.33) we have

$$\sum_{s=0}^{\infty} g(u, v \mid s) \frac{(n\theta)^s}{s!} = \sum_{s=0}^{\infty} \frac{(u+v+n-2)^s}{s!}\, \theta^s, \tag{17.34}$$

from which it follows that

$$n^s g(u, v \mid s) = (u+v+n-2)^s,\ s \in \boldsymbol{X}_0,$$

and hence

$$\begin{aligned} g(u, v \mid S) &= (u+v+n-2)^S \\ &= \sum_{\substack{x, y \in \boldsymbol{X}_0 \\ x+y \le S}} \frac{S!}{x!y!(S-x-y)!} \left(\frac{1}{n}\right)^x \left(\frac{1}{n}\right)^y \left(1 - \frac{2}{n}\right)^{S-x-y} u^x v^y. \end{aligned} \tag{17.35}$$

From Eq. (17.35), we compute

$$W_{x,y} = \boldsymbol{P}\{X_i = x, X_j = y \mid S\} = \frac{1}{x!y!}\frac{\partial^{x+y}}{\partial u^x \partial v^y} g(u, v \mid S)\bigg|_{u=0,v=0}$$

$$= \frac{S!}{x!y!(S-x-y)!}\Big(\frac{1}{n}\Big)^n\Big(\frac{1}{n}\Big)^y\Big(1-\frac{2}{n}\Big)^{S-x-y}$$

which is relation (17.34).

Remarks.

In the Poisson case the maximum-likelihood estimator (MLE) of $p_x(\theta)$ is $\hat{p}_x = p_x(\hat{\theta}_n)$, where $\hat{\theta}_n = (1/n)\,S$ is the MLE for θ. Clearly,

$$p_{x,y}(\theta) = \boldsymbol{P}_\theta\{X_i = x, X_j = y\} = p_x(\theta)p_y(\theta), \quad i \neq j,$$

and hence the MLE of $p_x(\theta)$ is $\hat{p}_{x,y} = p_{x,y}(\hat{\theta}_n) = p_x(\hat{\theta}_n)p_y(\hat{\theta}_n)$; that is, $\hat{p}_{x,y}$ is the product of MLEs of $p_x(\theta)$ and $p_y(\theta)$. Notice that $\boldsymbol{E}p_x(\theta) \neq p_x(\theta)$ even though $\boldsymbol{E}\hat{\theta}_n = \theta$. In fact,

$$\boldsymbol{E}p_x(\hat{\theta}_n) = p_x(\theta) + O\Big(\frac{1}{n}\Big).$$

On the other hand, the MVUEs $W_{x,y}$ of $p_{x,y}(\theta)$ do not have the multiplicative property for finite n, even though $\boldsymbol{E}W_{x,y} = p_{x,y}(\theta)$. This is because the independence of the X_is is destroyed by conditioning on S. As $n \to \infty$, the linear relation $S = X_1 + \cdots + X_n$ becomes unimportant. As we can see from Eq. (17.30) the MVUE's have this property asymptotically as $n \to \infty$.

More generally, from Eq. (17.14), it is evident that if $X_1, \ldots, X_n$ is a sample from Eq. (17.1) then

$$W_{x,y} = W_x W_y + O(\frac{1}{n}), \qquad n \to \infty \tag{17.36}$$

18. A CHI-SQUARED TEST FOR MODIFIED POWER SERIES DISTRIBUTIONS

Behavior of the Vector of Frequencies

Here we construct a chi-squared goodness-of-fit test for modified power series distributions. Let us consider the frequencies μ_x, which are defined by formulas (16.26) and (16.27), that is, for any $x \in \boldsymbol{X}_0 = \{0, 1, 2, \ldots\}$

$$\mu_x = \sum_{i=1}^{n} \mu_{ix}, \qquad i = 1, 2, \ldots, n, \tag{18.1}$$

where $\mu_{ix} = 1$, if $X_i = x$ and $\mu_{ix} = 0$ if $X_i \neq x$, and $X_1, \ldots, X_n$ is a sample from Eq. (16.1), that is,

$$\boldsymbol{P}_\theta\{X_i = x\} = p_x(\theta) = \frac{a(x)[\psi(\theta)]^x}{f(\theta)}. \tag{18.2}$$

For simplicity, let us write p_x here instead of $p_x(\theta)$, and $\boldsymbol{E}X_i$ instead of $\boldsymbol{E}_\theta X_i$ etc. From the definition of the random variables μ_{ix}, it follows that

$$\boldsymbol{E}\mu_{ix} = 1 \cdot \boldsymbol{P}_\theta\{X_i = x\} + 0 \cdot \boldsymbol{P}_\theta\{X_i \neq x\} = \boldsymbol{P}_\theta\{X_i = x\} = p_x. \tag{18.3}$$

and hence, for any $x \in \boldsymbol{X}_0$,

$$\boldsymbol{E}\mu_x = np_x. \tag{18.4}$$

In the same way, we find that

$$\textbf{Var}\ \mu_{ix} = \boldsymbol{E}\mu_{ix}^2 - (\boldsymbol{E}\mu_{ix})^2 = p_x - p_x^2 = p_x(1 - p_x). \tag{18.5}$$

and because, for any fixed x, $x \in \boldsymbol{X}_0$, $\mu_{1x}, \mu_{2x}, \ldots, \mu_{nx}$ are i.i.d.; from Eq. (18.5), we obtain

$$\textbf{Var}\ \mu_x = np_x(1 - p_x). \tag{18.6}$$

It is clear that if $x \neq y$ $(x, y \in \boldsymbol{X}_0)$, then

$$\boldsymbol{P}\{X_i = x, X_i = y\} = 0, \qquad i = 1, 2, \ldots, n, \tag{18.7}$$

so we have

$$\boldsymbol{E}\{\mu_{kx}\mu_{ky}\} = 0, \tag{18.8}$$

for all $k = 1, 2, \ldots, n$, when $x \neq y$. Further, let us calculate

$$\mathbf{cov}(\mu_x, \mu_y) = \boldsymbol{E}\{\mu_x \mu_y\} - \boldsymbol{E}\mu_x \boldsymbol{E}\mu_y, \qquad x \neq y. \tag{18.9}$$

We have

$$\boldsymbol{E}\mu_x\mu_y = \boldsymbol{E}\left(\sum_{m=1}^{n} \mu_{mx}\right)\left(\sum_{u=1}^{n} \mu_{uy}\right) = n(n-1)\boldsymbol{E}\mu_{1x}\mu_{2y}, \tag{18.10}$$

from Eq. (18.8) and because $X_1, \ldots, X_n$ are i.i.d.. Also

$$\boldsymbol{E}\mu_{1x}\mu_{2y} = \boldsymbol{E}\mu_{1x}\boldsymbol{E}\mu_{2y} = p_x \cdot p_y, \qquad \text{for any } x, y, \tag{18.11}$$

and hence from Eqs. (18.4), (18.9), (18.10), and (18.11) we obtain

$$\mathbf{cov}(\mu_x, \mu_y) = (n-1)np_x p_y - n^2 p_x p_y = -np_x p_y. \tag{18.12}$$

Now, let us calculate

$$\mathbf{cov}(\mu_x, S) = \boldsymbol{E}\{\mu_x S\} - \boldsymbol{E}\mu_x \cdot \boldsymbol{E}S, \tag{18.13}$$

where $S = X_1 + \cdots + X_n$ is a sufficient statistic for the family (16.1). From the definition of S, it follows that

$$\boldsymbol{E}S = n\boldsymbol{E}X_1 = n\alpha_1, \qquad \mathbf{Var}\, S = n\mathbf{Var}\, X_1 = n\sigma^2. \tag{18.14}$$

Let us now calculate

$$\boldsymbol{E}\{\mu_x S\} = \boldsymbol{E}\Big\{(X_1 + \cdots + X_n)\sum_{i=1}^{n} \mu_{ix}\Big\}. \tag{18.15}$$

But for any $i = 1, 2, \ldots, n$

$$\boldsymbol{E}\{X_i \mu_{ix}\} = xp_x, \qquad x \in \boldsymbol{X}_0, \tag{18.16}$$

and, on the other hand, if $i \neq j$ $(i, j = 1, 2, \ldots, n)$, then

$$\boldsymbol{E}\{X_j \mu_{ix}\} = \boldsymbol{E}X_j \boldsymbol{E}\mu_{ix} = \alpha_1 p_x, \tag{18.17}$$

because $X_1, \ldots, X_n$ are i.i.d. from Eq. (16.1).

Hence from Eqs. (18.15) to (18.17), we obtain that

$$\boldsymbol{E}\{\mu_x S\} = nxp_x + n(n-1)\alpha_1 p_x, \tag{18.18}$$

and the formulas (18.13) to (18.18) give us

$$\mathbf{cov}(\mu_x, S) = np_x(x - \alpha_1) = np_x(x - \boldsymbol{E}X_1). \tag{18.19}$$

Example 18.1. Geometric Distribution. If

$$p_x = \boldsymbol{P}_\theta\{X_i = x\} = \theta(1-\theta)^x,$$

then in formulas (18.12), (18.13), (18.14), (18.18), and (18.19) we can put

$$\alpha_1 = \boldsymbol{E}X_1 = \frac{1-\theta}{\theta}, \qquad \sigma^2 = \mathbf{Var}\ X_1 = \frac{1-\theta}{\theta}, \qquad \theta \in (0,1). \tag{18.20}$$

We follow the work of Park (1973a) further. Let $\varphi_n(\boldsymbol{u}, v)$ denote the characteristic function of the random vector

$$(\mu_0 - \boldsymbol{E}\mu_0, \mu_1 - \boldsymbol{E}\mu_1, \ldots, \mu_k - \boldsymbol{E}\mu_k, S)^T$$

where k is fixed $(k \in \{0, 1, 2, \ldots, s\}$, $\boldsymbol{u} = (u_1, u_2, \ldots, u_k)^T$. Then

$$\begin{aligned} \varphi_n(\boldsymbol{u}, v) &= \boldsymbol{E}\left\{\exp\left[\sum_{x=0}^{k} iu_x(\mu_x - \boldsymbol{E}\mu_x) + ivS\right\}\right. \\ &= \sum_{s=0}^{\infty} \boldsymbol{P}\{S = s\}\left[E\left\{\exp\left[\sum_{x=0}^{k} iu_x(\mu_x - \boldsymbol{E}\mu_x)\right] \mid S = s\right\}\right] e^{ivs}. \end{aligned} \tag{18.21}$$

Limit of the Frequency Vector: Park's Theorem

Let $\psi_n(\boldsymbol{u}, s)$ denote the characteristic function of the random vector

$$(\mu_0 - \boldsymbol{E}\mu_0, \ldots, \mu_k - \boldsymbol{E}\mu_k)^T \text{ given } S = s;$$

that is, $\psi_n(\boldsymbol{u}, s)$ is the characteristic function of the conditional distribution. Then Eq. (18.21) can be written as

$$\varphi_n(\boldsymbol{u}, v) = \sum_{s=0}^{\infty} e^{ivs}\psi_n(\boldsymbol{u}, s)\boldsymbol{P}_\theta\{S = s\},$$

and, by the inverse formula for Eq. (18.21), we obtain

$$\psi_n(\boldsymbol{u}, s) = \frac{1}{2\pi\boldsymbol{P}_\theta\{S = s\}} \int_{-\pi}^{\pi} \varphi_n(\boldsymbol{u}, v)e^{-ivs}\, dv. \tag{18.22}$$

The following slightly generalized version of Park's theorem holds.

Theorem 18.1. *If $n \to \infty$ in such a way that $S \asymp n$, then*

$$\psi_n(\frac{1}{\sqrt{n}}\boldsymbol{u}, s) \to \exp\{\frac{1}{2}Q(\boldsymbol{u})\},$$

where $Q(\boldsymbol{u})$ is the quadratic form in $\boldsymbol{u}$ with covariance matrix $\boldsymbol{\Sigma}_1 = \boldsymbol{\Sigma}_1(\theta)$, given by the formula

$$\boldsymbol{\Sigma}_1 = \boldsymbol{\Sigma}_{11} - \frac{1}{\sigma^2}\boldsymbol{\Sigma}_{12}\,\boldsymbol{\Sigma}_{21}^T, \tag{18.23}$$

and $\boldsymbol{\Sigma}_1$, $\boldsymbol{\Sigma}_{12}$, $\boldsymbol{\Sigma}_{21}$, σ^2 are defined by the $(k+2)\times(k+2)$ matrix $\boldsymbol{\Sigma}=\boldsymbol{\Sigma}(\theta)$,

$$\boldsymbol{\Sigma} = \begin{bmatrix} \boldsymbol{\Sigma}_{11} & \boldsymbol{\Sigma}_{12} \\ \boldsymbol{\Sigma}_{21} & \sigma^2 \end{bmatrix} = [\sigma_{ij}]. \tag{18.24}$$

The elements σ_{ij} *are*

$$\begin{aligned}
\sigma_{ii} &= \frac{1}{n}\,\mathbf{Var}\ \mu_i = p_i(1-p_i), \quad i = 0,1,\ldots,k, \\
\sigma_{ij} &= \frac{1}{n}\,\mathbf{cov}\ (\mu_i,\mu_j) = -p_i p_j \quad i \neq j,\ i,j = 0,2,\ldots,k, \\
\sigma_{i,k+1} &= \frac{1}{n}\,\mathbf{cov}\ (\mu_i, S) = p_i(i-\alpha_1), \quad i = 0,1,2,\ldots,k,\ \alpha_1 = \boldsymbol{E}X_j, \\
\sigma_{k+1,k+1} &= \frac{1}{n}\,\mathbf{Var}\ S = \mathbf{Var}\ X_j = \sigma^2.
\end{aligned} \tag{18.25}$$

The proof of this theorem follows almost exactly the reasoning of Park (1973a) with the help of formulas (18.4) through (18.19) and the result of Scheffe (1947; see also Shanmugam, 1988). According to this result and the asymptotic normality of the statistic $S = X_1 + \cdots + X_n$,

$$\lim_{n\to\infty} \boldsymbol{P}_\theta \left\{ \frac{S-n\alpha_1}{\sigma\sqrt{n}} \le x \right\} = \Phi(x),$$

we have

$$\lim_{n\to\infty} \sqrt{n}\boldsymbol{P}_\theta\{S=s\} = \frac{1}{\sigma}\varphi(0) = \frac{1}{\sigma\sqrt{2\pi}},$$

where $\varphi(x) = \Phi'(x)$, and $\Phi(x)$ is the distribution function of the standard normal law, $\alpha_1 = \boldsymbol{E}_\theta X_1$, $\sigma^2 = \mathbf{Var}_\theta X_1$. Also one obtains the asymptotic normality of the conditional distribution of the frequency vector

$$\frac{1}{\sqrt{n}}(\mu_0 - \boldsymbol{E}\mu_0, \mu_1 - \boldsymbol{E}\mu_1, \ldots, \mu_k - \boldsymbol{E}\mu_k)^T,$$

given S, from the asymptotic normality of the multivariate distribution of the vector

$$\frac{1}{\sqrt{n}}(\mu_0 - \boldsymbol{E}\mu_0, \mu_1 - \boldsymbol{E}\mu_1, \ldots, \mu_k - \boldsymbol{E}\mu_k, S - n\alpha_1)^T$$

with zero mean and the covariance matrix $\boldsymbol{\Sigma}$, which is given by Eq. (18.24).

The matrix $\boldsymbol{\Sigma}_{11}$ in Eq. (18.23) can be expressed as

$$\boldsymbol{\Sigma}_{11} = \boldsymbol{D} - \boldsymbol{p}\boldsymbol{p}^T, \tag{18.26}$$

where

$$\boldsymbol{D} = \begin{bmatrix} p_0 & 0 & \dots & 0 \\ 0 & p_1 & \dots & 0 \\ \vdots & & \ddots & \vdots \\ 0 & 0 & \dots & p_k \end{bmatrix}, \qquad \boldsymbol{p} = (p_0, \dots, p_k)^T. \tag{18.27}$$

If we denote by $\boldsymbol{q}$ the vector with components $q_x = xp_x$, $x = 0, 1, \dots, k$,

$$\boldsymbol{q} = (0, p_1, 2p_2, \dots, kp_k)^T,$$

then

$$\boldsymbol{\Sigma}_{12} = \boldsymbol{\Sigma}_{21} = (\alpha_1 \boldsymbol{p} - \boldsymbol{q})(\alpha_1 \boldsymbol{p} - \boldsymbol{q})^T, \tag{18.28}$$

and the matrix $\boldsymbol{\Sigma}_1$ in Eq. (18.23) can be written as

$$\boldsymbol{\Sigma}_1 = \boldsymbol{D} - \boldsymbol{p}\boldsymbol{p}^T - \frac{1}{\sigma^2}(\alpha_1 \boldsymbol{p} - \boldsymbol{q}_1)(\alpha_1 \boldsymbol{p} - \boldsymbol{q})^T. \tag{18.29}$$

Construction of Grouping Intervals Using the Conditional Distribution

Finally, we are ready to construct a chi-squared test of the hypothesis H_0. First we choose an integer r, where $r \geq 2$. Next, taking into account the observed value, s, of the sufficient statistic S, we choose any nonrandom partition of the sample space $\boldsymbol{X}_0^s = \{0, 1, \dots, s\}$ of X_i given $S = s$ into r nonempty classes $J_1, \dots, J_r$ for grouping the data $X_1, \dots, X_n$. We suppose, for definiteness, that $s \in J_r$. Furthermore, let us define three vectors

$$\boldsymbol{\nu} = (\nu_1 \dots, \nu_r)^T, \qquad \boldsymbol{P} = (P_1, P_2, \dots, P_r)^T, \qquad \boldsymbol{Q} = (Q_1, \dots, Q_r)^T \tag{18.30}$$

with components

$$\nu_i = \sum_{x \in J_i} \mu_x, \qquad P_i = \sum_{x \in J_i} W_x, \qquad Q_i = \sum_{x \in J_i} x W_x, \qquad i = 1, \dots, r, \tag{18.31}$$

where the frequencies μ_x are defined as in Eq. (16.1) and

$$W_x = W_x(S) = \boldsymbol{P}\{X_i = x \mid S\}, \qquad x \in \boldsymbol{X}_0^s,$$

are the MVUE's of the probabilities $p_x(\theta) = \boldsymbol{P}\{X_i = x\}$. It is clear that

P_i is the MVUE for the probability

$$\boldsymbol{P}_\theta\{X_1 \in J_i\} = \sum_{x \in J_i} p_x(\theta), \tag{18.32}$$

and from Eqs. (18.15) and (18.31) we have

$$\sum_{i=1}^{r} Q_i = \frac{1}{n} S.$$

The classes $J_1, \ldots, J_r$ should be chosen by using the joint conditional distribution of the data $X_1, \ldots, X_n$, given S, in such a way that the probabilities

$$P_i = \boldsymbol{P}\{X_1 \in J_i | S\}, \quad i = 1, \ldots, v,$$

will not be small, that is, for some $w > 0$ (see Remark 18.2),

$$\boldsymbol{P}\{\min_{1 \le i \le r} P_i \ge w\} \to 1, \tag{18.33}$$

as $n \to \infty$ with $S \asymp n$. In many cases, one can choose $J_1, J_2, \ldots, J_r$ as intervals in $\boldsymbol{X}_0^s$. For example, take r integers $a_1, a_2, \ldots, a_r$ from $\boldsymbol{X}_0^s$, such that

$$0 \le a_1 < a_2 < \cdots < a_r = s = \mu_1 + 2\mu_2 + \cdots + s\mu_s \tag{18.34}$$

[see Eq. (16.29)] and let $a_0 = -1$. Define

$$J_i = [a_{i-1} + 1, a_i], \qquad i = 1, 2, \ldots, r. \tag{18.35}$$

Then Eq. (18.31) can be written as

$$\nu_i = \sum_{x=a_{i-1}+1}^{a_i} \mu_x, \quad Q_i = \sum_{x=a_{i-1}+1}^{a_i} x W_x, \quad P_i = \sum_{x=a_{i-1}+1}^{a_i} W_x. \tag{18.36}$$

The use of intervals as classes of grouping data in constructing a chi-squared test is traditional but quite unnecessary. The distribution $\boldsymbol{P}_\theta\ \{X_i = x\}$ may have features that cause data to be sparse in some intervals. The combination of such intervals into one or more classes is a natural way to avoid this problem. For example, it is often practical to unify the two ends of a distribution into one class. Notice that here the upper endpoint of the region $\boldsymbol{X}_0^s$ to be partitioned is not ∞ as in $\boldsymbol{X}_0$, but instead the finite random number s, the realization of the sufficient statistic S, based on the data. The partition itself is deterministic apart from the randomness of this value s. An advantage of the present approach to constructing the test is that the conditioning restricts the sample space $\boldsymbol{X}_0$ to the region $\boldsymbol{X}_0^s$ where the data is sitting so that our observations are more concentrated in $J_1, J_2, \ldots, J_r$. Thus the chance is reduced of having only a few observations in particular classes of grouped data.

Limit Conditional Distribution of Grouped Data

Now, using Theorem 18.1, we have the following theorem.

Theorem 18.2. *If the classes $J_1, J_2, \ldots, J_r$ form a partition of $\boldsymbol{X}_0^S$ and $n \to \infty$ so that $S \asymp n$, then the conditional distribution of the vector $n^{-1/2}(\boldsymbol{\nu} - n\boldsymbol{P})$ given S is $AN(\mathbf{0}_r, \boldsymbol{B}_n)$, where*

$$\boldsymbol{B}_n = \boldsymbol{D}_{(0)} - \boldsymbol{P}\boldsymbol{P}^T - \frac{1}{\hat{\sigma}^2}\left[\frac{S}{n}\boldsymbol{P} - \boldsymbol{Q}\right]\left[\frac{S}{n}\boldsymbol{P} - \boldsymbol{Q}\right]^T \tag{18.37}$$

$$\boldsymbol{D}_{(0)} = \begin{bmatrix} P_1 & 0 & \ldots & 0 \\ 0 & P_2 & \ldots & 0 \\ \vdots & \vdots & \ddots & \vdots \\ 0 & 0 & \ldots & P_r \end{bmatrix}, \quad \text{rank } \boldsymbol{B}_n = r - 1, \tag{18.38}$$

and $\hat{\sigma}^2$ is the MVUE of $\sigma^2 = \mathbf{Var}\ X_1$, that is,

$$\hat{\sigma}^2 = \frac{n}{(n-1)b(S,n)} \sum_{x=0}^{S-(n-1)\alpha} x^2 a(x) b(S-x, n-1) - \frac{1}{n(n-1)} S^2, \tag{18.39}$$

as was shown in Eq. (17.22).

This theorem is an extension and unification of the results of Park (1973a) for a geometrical distribution, of Kolchin, Sevastianov, and Chistyakov (1978) for a Poisson distribution, and of Bolshev and Mirvaliev (1978) for a negative binomial distribution. On the other hand, the theorem can be obtained from the results of Nikulin (1973c), because W_x is a consistent estimator of $p_x(\theta)$. The theorem also follows from a central limit theorem of Morris (1975) for multinomial sums.

Remark 18.1. In the case of double truncated (or left-truncated) modified power series distributions, one should replace the lower limit of summation in (18.37) by $\min\{\alpha, S-(n-1)\beta\}$ (or α) and the upper limit by $\max\{\beta, S-(n-1)\alpha\}$.

Examples for Estimation of Variance

1. A Power Series Distribution. Let $X_1, \ldots, X_n$ be a sample from the left-truncated modified generalized logarithmic power series distribution

$$\boldsymbol{P}_\theta\{x_i = x\} = \begin{cases} \frac{\binom{\gamma x}{x}[\theta(1-\theta)^{\gamma-1}]^x}{\gamma x f(\theta)}, & x \in \boldsymbol{X}_\alpha, \\ 0, & \text{otherwise}, \end{cases} \tag{18.40}$$

where

$$\begin{gathered} f(\theta) = \ln\frac{1}{1-\theta} - \sum_{k=1}^{\alpha-1} \frac{1}{k\gamma}\binom{k\gamma}{k}[\theta(1-\theta)^{\gamma-1}]^k, \\ 0<\theta<1, \qquad |\gamma\theta|<1, \end{gathered} \tag{18.41}$$

and the parameters α and γ are given (see Example 16.5). From Eq. (16.11) we have

$$f^n(\theta) = \sum_{s=n\alpha}^{\infty} b(s,n)[\theta(1-\theta)^{\gamma-1}]^s. \tag{18.42}$$

To calculate the MVUE $\hat{\sigma}^2$ of $\sigma^2 = \mathbf{Var}\, X_1$ it is more convenient to use, instead of the coefficients $b(k,n)$, the so-called Stirling numbers of the first kind, $\sigma(k,n)$ (the traditional notation is $s(k,n)$). The Stirling numbers $\sigma(k,n)$ satisfy the equation

$$\begin{aligned} &\sum_{s=\alpha n}^{\infty} (-1)^{s-n}\sigma(s,n)\frac{[\theta(1-\theta)^{\gamma-1}]^s}{s!} \\ &\qquad = \frac{1}{n!}\left[\ln\frac{1}{1-\theta} - \sum_{k=1}^{\alpha-1}\binom{\gamma k}{k}\frac{1}{\gamma k}[\theta(1-\theta)^{\gamma-1}]^k\right]^n, \end{aligned} \tag{18.43}$$

which is the analogue of Eq. (16.11). From Eqs. (16.11) and (18.41), it follows that

$$\sigma(s,n) = (-1)^{s-n}\frac{s!}{n!}\,b(s,n)$$

$$= (-1)^{s-n}\frac{s!}{n!}\sum_{x_1+\cdots+x_n=s} a(x_1)a(x_2)\cdots a(x_n), \qquad (18.44)$$

where (see Example 16.5)

$$a(x) = \frac{1}{\gamma x}\binom{\gamma x}{x}, \qquad x \in \boldsymbol{X}_\alpha.$$

According to Eq. (16.15), it follows, from Eq. (18.44), that the modified Stirling numbers of the first kind satisfy the recurrence relation

$$\sigma(s,t) = \sum_{k=\alpha}^{s-(n-1)\alpha}\binom{\gamma k}{k}\frac{(-1)^{k-1}s!}{n\gamma k(s-k)!}\,\sigma((s-k,n-1), \qquad (18.45)$$

with the initial conditions $\sigma(0,0) = 1$ and $\sigma(s,0) = 0$ if $s > 0$. In this example the MVUE $W_x = W_x(S)$ of p_x is

$$W_x = \begin{cases} \frac{1}{n\gamma x}\binom{\gamma x}{x}\frac{(-1)^{x-1}S!\sigma(S-x,n-1)}{(S-x)!\sigma(S,n)}, & S \geq \alpha(n-1)+x, \\ 0, & \text{if } S < \alpha(n-1)+x,\ x \in \boldsymbol{X}_\alpha, \end{cases} \qquad (18.46)$$

(see Jani, 1977; also Nikulin and Voinov, 1987, 1993).

Now using Eqs. (18.39), (18.44), and (18.45), we see that the MVUE $\hat{\sigma}^2$ of $\sigma^2 = \mathbf{Var}\, X_1$ is

$$\hat{\sigma}^2 = \frac{1}{(n-1)\sigma(S,n)}\sum_{x=\alpha}^{S-(n-1)\alpha}\frac{\gamma}{x}\binom{x\gamma}{x}\frac{(-1)^{x-1}S!}{(S-x)!}\sigma(S-x,n-1) - \frac{1}{n(n-1)}S^2. \qquad (18.47)$$

2. Poisson Distribution. If $X_1,\ldots,X_n$ is a sample from the Poisson distribution with a parameter θ, as is well known, the MVUE of $\theta =$

$\boldsymbol{E}X_i = \mathbf{Var}\ X_i$ is S/n. Let us see how our formulas work in this case. We know [see Eqs. (16.16) and (16.17)] that

$$\boldsymbol{P}_\theta\{S = s\} = \frac{(n\theta)^s}{s!}\, e^{-n\theta} = \frac{b(s,n)\theta^n}{f^n(\theta)}, \qquad s \in \boldsymbol{X}_0,$$

where $b(s,n) = n^s/s!$. Furthermore, using Eqs. (17.23) and (17.15), we have

$$\begin{aligned} \hat{\alpha}_1 &= \sum_{x=0}^{S} xW_x = \sum_{x=0}^{S} x\frac{1}{x!}\,\frac{S!(n-1)^{S-x}}{n^S(S-x)!} \\ &= \sum_{x=0}^{S} x\binom{S}{x}(1-\frac{1}{n})^{S-x}(\frac{1}{n})^x = \frac{1}{n}\,S. \end{aligned} \tag{18.48}$$

Furthermore, from Eq. (18.22) we can compute

$$\hat{\sigma}^2 = \frac{n}{n-1}\,\hat{\alpha}_2 - \frac{1}{n(n-1)}\,S^2 = \frac{1}{n}\,S. \tag{18.49}$$

Construction of the Test Statistic, Its Limit Chi-Squared Law

Now we are prepared to construct a chi-squared-type statistic for testing H_0. As in Chapter 2, we introduce a quadratic form

$$Y_n^2 = \frac{1}{n}\,(\boldsymbol{\nu} - n\boldsymbol{P})^T B_n^-(\boldsymbol{\nu} - n\boldsymbol{P}), \tag{18.50}$$

where $\boldsymbol{B}_n^-$ is a generalized inverse matrix of $\boldsymbol{B}_n$. By direct calculation, one obtains that

$$Y_n^2 = \sum_{i=1}^{r} \frac{(\nu_i - nP_i)^2}{nP_i} + \frac{1}{n\mu}\left(\sum_{i=1}^{r} \frac{Q_i}{P_i}\,\nu_i - S\right)^2, \tag{18.51}$$

where

$$\mu = \hat{\sigma}^2 + (\frac{1}{n}S)^2 - \sum_{i=1}^{r} \frac{Q_i^2}{P_i}. \tag{18.52}$$

Again the statistic Y_n^2 is the sum of a standard Pearson's statistic X^2,

$$X^2 = \sum_{i=1}^{r} \frac{(\nu_i - nP_i)^2}{nP_i}, \tag{18.53}$$

and the square of a linear form of frequencies $\nu_1, \nu_2, \ldots, \nu_r$.

Under H_0 and the hypotheses of Theorem 18.2 we have that Y_n^2 defined by Eq. (18.50) has a limit law which is χ^2_{r-1}. The proof is similar to that for Theorem 12.1, for example. For more information see Nikulin and Voinov (1987, 1993).

Recommendations for Grouping Data

We cite here some recommendations for grouping of data (see Cochran, 1952; Bolshev and Mirvaliev, 1978; Mirvaliev, 1979; Greenwood and Nikulin, 1988; Nikulin and Voinov, 1987, 1993; and Chapter 2). The same general considerations apply as in Section 2 but now, since our grouping intervals are random, depending on the sufficient statistic S, we must take into consideration the observed value of S. The following procedure was developed by Bolshev and colleagues in Steklov Mathematical Institute in Moscow in 1973 to 1975. First, how do we choose w in formula (18.33)? If n is less than 1000, it is recommended that $w = 5/n$, and for $n > 1000$ that $w = 0.005$. Furthermore, let

$$W_M = \max\{W_0, W_1, \ldots, W_s\},$$

where s is the observed value of S, and let $W_u^v = \sum_{x=u}^{v} W_x$, where $0 \le u \le M \le v \le s$. One should choose u and v such that

$$W_u^v \ge w, \qquad W_u^{v-1} < w, \qquad \text{and} \qquad W_{u+1}^v < w. \tag{18.54}$$

Among all such probabilities W_u^v we should choose the probability $W_{u^*}^{v^*}$ for which the difference $W_{u^*}^{v*} - W$ is minimal. With u and v we may construct a *central* grouping interval that will contain all frequencies μ_x for which $u \le x \le v$. The other intervals are defined successively. The interval on the right of the *central* interval contains all frequencies ν_x for which $v+1 \le x \le z$, where z can be found from

$$z > v, \qquad W_{v+1}^z \ge w, \qquad \text{and} \qquad W_{v+1}^{z-1}.$$

The interval on the left of the *central* one is defined similarly. In so doing it is necessary to take into account that if, at some stage, the sum of all remaining ungrouped probabilities is less than w, then the remaining ungrouped points should be united with the preceding interval, if all the probability lies to the left or to the right, or otherwise should be united in one class. It is evident that this grouping rule gives probabilities $P_1, P_2, \ldots, P_r$ to the classes $J_1, J_2, \ldots, J_r$ $(s \in J_r)$ which are almost equal, that is, it makes almost equiprobable classes.

Example. A Test Statistic for the Generalized Negative Binomial Distribution. Let us construct a chi-squared test of the hypothesis H_0 such that $X_1, X_2, \ldots, X_n$ is a sample from the generalized negative binomial distribution (GNBD) (see Examples, Section 16) with unknown parameter θ and fixed, known parameters β and k, that is

$$\boldsymbol{P}_\theta\{X_i = x\} = \frac{k\Gamma(k+\beta x)[\theta(1-\theta)^{\beta-1}]^x}{\Gamma(x+1)\Gamma(k+\beta x - x+1)(1-\theta)^{-k}}, \qquad x \in \boldsymbol{X}_o, \quad (18.55)$$

where $\theta \in \Theta = (0,1)$, $|\theta\beta| < 1$, $k > 0$. It is well known that

$$\boldsymbol{E}X_i = \frac{k\theta}{1-\beta\theta}, \qquad \textbf{Var } X_i = \sigma^2 = \frac{k\theta(1-\theta)}{(1-\beta\theta)^3}. \quad (18.56)$$

If $\beta = 0$ and $k \in \boldsymbol{X}_1 = \{1,2,\ldots\}$ then Eq. (18.55) is the binomial distribution with parameters θ and k. The statistic $S = X_1 + \cdots + X_n$ is sufficient for θ and has the GNBD,

$$\boldsymbol{P}_\theta\{S = s\} = \frac{nk\Gamma(nk+\beta s)[\theta(1-\theta)^{\beta-1}]^s}{\Gamma(s+1)\Gamma(nk+\beta s - s+1)(1-\theta)^{-nk}}, \qquad s \in \boldsymbol{X}_0, \quad (18.57)$$

and hence the MVUE for $\boldsymbol{P}_\theta\{X_i = x\}$ is

$$\boldsymbol{P}\{X_i = x \mid S\} = W_x(S,n) = \frac{a(x)b(S-x,n-1)}{b(S,n)}, \qquad S \geq x, \quad (18.58)$$

where

$$b(s,n) = \frac{nk\Gamma(nk+\beta s)}{\Gamma(s+1)\Gamma(nk+\beta s - s+1)}$$

and (18.59)

$$a(x) = \frac{n\Gamma(n+\beta x)}{\Gamma(x+1)\Gamma(n+\beta x - x + 1)}.$$

To construct our chi-squared test, we need the MVUE for $\sigma^2 = \mathbf{Var}\ X_i$, and, as in Example 17.5, we can obtain it from the MVUE $W_{x,y}$ for the joint distribution

$$\boldsymbol{P}_\theta\{X_1 = x, X_2 = y\} = \boldsymbol{P}_\theta\{X_1 = x\}\boldsymbol{P}_\theta\{X_2 = y\}.$$

From formula (17.14), we have

$$W_{x,y} = \frac{a(x)a(y)b(S-x-y, n-2)}{b(S,n)} = W_x(S,n)W_y(S-x, n-1), \quad (18.60)$$

and, from Eqs. (18.58) and (18.59), we find

$$W_x(S,n) =$$

$$\frac{k(n-1)\Gamma(k+\beta x)\Gamma(nk+\beta S-k-\beta x)\Gamma(nk+\beta S-S+1)\Gamma(S+1)}{n\Gamma(nk+\beta S)\Gamma(k+\beta x-x+1)\Gamma(nk+\beta S+x-S-k-\beta x+1)}. \quad (18.61)$$

Using Eqs. (18.56) and (18.60) we find the MVUE $\hat{\sigma}^2$ for σ^2:

$$\hat{\sigma}^2 = \begin{cases} \frac{(nk+\beta S-S)\ \Gamma(S+1)}{n\Gamma(nk+\beta S)} \sum_{m=0}^{S-1} \beta^m \frac{(m+1)\Gamma(nk+\beta S-m+1)}{\Gamma(S-m)}, & \text{if } S \in \boldsymbol{X}_1, \\ 0 & \text{otherwise.} \end{cases} \quad (18.62)$$

Now we can construct the statistic (18.51) to test H_0 .

Example. A Test Statistic for Stirling's Distribution of the Second Kind. Suppose one wishes to test the hypothesis H_0 according to which the i.i.d. random variables $X_1, X_2, \ldots, X_n$ have Stirling's distribution of the second kind with parameter θ and fixed r, $S_r(\theta)$ (see Singh, 1975). Under this H_0,

$$\boldsymbol{P}_\theta\{X_i = x\} = \frac{\theta^x\Gamma(r+1)}{\Gamma(x+1)(e^\theta - 1)^r}S(x,r), \quad (18.63)$$

where $x \in \boldsymbol{X}_r$, $\theta \in \Theta = [0, +\infty)$, and $r \in \boldsymbol{X}_1$, and

$$S(x,r) = \frac{1}{\Gamma(r+1)} \sum_{k=1}^{r} (-1)^{r-k} \binom{r}{k} k^x \tag{18.64}$$

represents the Stirling numbers of the second kind. The $S(x,r)$ satisfy (see, e.g., Voinov and Nikulin, 1993) the recurrence relations

$$\begin{aligned} rS(x-1,r)+S(x-1,r-1) &= S(x,r), \\ rS(x,r) &= \sum_{i=r-1}^{x-1} \binom{r}{i} S(i,r-1), \\ S(x,r-1) &= \sum_{i=r-1}^{x} (-1)^{x-i} S(i+1,r) \binom{x}{i}. \end{aligned} \tag{18.65}$$

If Y_1 and Y_2 are independent with Stirling distributions $S_m(\theta)$ and $S_r(\theta)$, respectively, then $Y_1 + Y_2$ has distribution $S_{m+r}(\theta)$. It follows that if X_i has the law $S_r(\theta)$ then X_i can be represented as a sum

$$X_i = Y_1 + Y_2 + \cdots + Y_r,$$

where $Y_1, Y_2, \ldots, Y_r$ are independent and each Y_i has law $S_1(\theta)$, which is the Poisson law with parameter θ, truncated at 0:

$$\begin{aligned} \boldsymbol{P}_\theta\{Y_i = 0\} &= 0, \\ \boldsymbol{P}_\theta\{Y_i = x\} &= \frac{\theta^x}{(e^\theta - 1)\Gamma(x+1)}, \qquad x \in \boldsymbol{X}_1, \end{aligned} \tag{18.66}$$

and $\theta > 0$. From Eq. (18.66), we have

$$EY_i = \frac{\theta e^\theta}{e^\theta - 1}$$

and hence

$$\mu = EX_i = \frac{r\theta e^\theta}{e^\theta - 1}. \tag{18.67}$$

Let

$$\mu_k = e(X_i - \mu)^k, \qquad k = 1, 2, 3, \ldots.$$

Then from Eqs. (18.66) and (18.67), one finds that

$$\mu_{k+1} = k\left\{\frac{d\mu_k}{d\theta} + k\frac{d\mu}{d\theta}\mu_{k-1}\right\}. \tag{18.68}$$

Because Eq. (18.63) belongs to the family of modified power series distributions (16.1), the statistic $S = X_1+\cdots+X_n$ is sufficient, and, hence from Eq. (18.63), one obtains

$$\boldsymbol{P}_\theta\{S = s\} = \frac{(nr)!e^\theta}{\Gamma(s+1))(e^\theta - {}_1)^{nr}}S(s, nr), \qquad s \in \boldsymbol{X}_{nr}. \tag{18.69}$$

Using the properties (18.65) of the Stirling numbers, one finds (see, e.g., Theorem 17.1) that, if r is known, then the MVUE $\hat{\mu}$ for $\mu = \boldsymbol{E}X_i$ is the statistic $\hat{\mu} = \bar{X}_n = S/n$, with

$$\boldsymbol{E}\hat{\mu} = \boldsymbol{E}\bar{X}_n = \mu.$$

Furthermore, as is well known

$$W_x = \boldsymbol{P}\{X_i = x \mid S\} = \frac{\binom{S}{x}S(x,r)S(S-x, r(n-1))}{\binom{nr}{r}S(S, nr)}, \tag{18.70}$$

is the MVUE for $P_\theta\{X_i = x\}$, and

$$\boldsymbol{E}W_x = \boldsymbol{P}_\theta\{X_i = x\}, \qquad r \le x \le T - r(n-1). \tag{18.71}$$

Finally, using formula (18.70), one can verify (see, e.g., Voinov and Nikulin, 1993) that

$$\hat{\sigma}^2 = \frac{S}{n} - \frac{S(S-1)S(S-1, nr-1)}{n(nr-1)S(S, nr)}, \qquad S \ge nr, \tag{18.72}$$

is the MVUE for $\sigma^2 = \textbf{Var}\ X_i = \mu_2$,

$$E\hat{\sigma}^2 = \sigma^2.$$

Next we choose numbers $k_1, k_2, \ldots, k_m$ $(2 \le m \le 1 + \log_2 n)$, $k_i \in \boldsymbol{X}_r$, such that

$$r < k_1 < k_2 < \cdots < k_{m-1} < k_m = S = X_1 + \cdots + X_n,$$

and construct grouping intervals

$$J_1 = [r, k_1], \qquad J_2 = [k_1+1, k_2, \ldots, J_m = [k_{m-1}+1, k_m].$$

Let $\boldsymbol{\nu} = (\nu_1, \ldots, \nu_m)^T$ denote the frequency vector obtained by grouping the data $X_1, X_2, \ldots, X_n$ into these intervals and let

$$\boldsymbol{P} = (P_1, \ldots, P_m)^T, \qquad \boldsymbol{Q} = (Q_1, \ldots, Q_m)^T,$$

$$\begin{aligned} P_i &= \sum_{x \in J_i} W_x = P\{X_1 \in J_i \mid S\}, & i &= 1, 2, \ldots, m \\ Q_i &= \sum_{x \in J_i} x W_x, & i &= 1, 2, \ldots, m, \end{aligned} \tag{18.73}$$

where W_x is defined by Eq. (18.70). Our test statistic is

$$Y_n^2 = \sum_{i=1}^m \frac{(\nu_i - nP_i)^2}{nP_i} + \frac{1}{n\gamma}\left(\sum_{i=1}^m \frac{\nu_i Q_i}{P_i} - S\right)^2,$$

where

$$\gamma = \hat{\sigma}^2 + \bar{X}_n^2 - \sum_{i=1}^m \frac{Q_i^2}{P_i},$$

$\hat{\sigma}^2$ and $\bar{X}_n$ are the MVUEs for α^2 and μ. Under H_0,

$$\lim_{n \to \infty} \boldsymbol{P}\{Y_n^2 > x\} = \boldsymbol{P}\{\chi_{m-1}^2 > x\}.$$

Example. A Simple Test for the Poisson Distribution. Suppose one is testing the hypothesis H_0 according to which i.i.d. random variables $X_1, \ldots, X_n$ follow a Poisson law with parameter θ:

$$\boldsymbol{P}\{X_i = x \mid H_0\} = \frac{\theta^x}{x!} e^{-\theta}, \qquad x \in \boldsymbol{X}_0, \qquad \theta > 0. \tag{18.74}$$

If one uses that under H_0,

$$\boldsymbol{P}\{X_1 = x_1, \ldots, X_n = x_n \mid S\} = \frac{S!}{x_1! \ldots x_n!}\left(\frac{1}{n}\right)^S, \tag{18.75}$$

then one can construct a simple test based on the *index of dispersion*

$$X^2 = \frac{n}{S}\sum_{i=1}^{n} X_i^2 - S = n\sum_{i=1}^{n}\frac{\left(X_i - \frac{1}{n}S\right)^2}{S} \tag{18.76}$$

(see e.g., Selby, 1965, Rayner and Best, 1989). This is an example of the Gihman statistic described in Section 7. Under H_0, if $S \approx n$ as $n \to \infty$, we have

$$\mathbf{P}\{X^2 \le x \mid S\} - \mathbf{P}\{\chi^2_{n-1} \le x\} \to 0,$$

and hence the conditional distribution of X^2 is asymptotically chi-squared and

$$\mathbf{E}\{X^2 \mid S\} = n - 1 \quad \text{and} \quad \mathbf{Var}\{X^2 \mid S\} = 2(n-1)\left[1 - \frac{1}{S}\right].$$

Example. A Chi-Squared Test for Rutherford-Chadwick-Geiger Radioactive Decay Data. In this example, we consider the results of the experiment conducted by Rutherford, Geiger, and Chadwick in studies of α-decay. These results are found in detail in the book by Rutherford, Chadwick, and Ellis (1930; see also Marsden and Barrat, 1911); von Mises, 1936). The results of this experiment were used by Cramer (1946), Arley and Buch (1950), and Feller (1948b) to demonstrate that the Poisson distribution can be used as a model for radioactive disintegration. In 1972 to 1976 in the Department of Statistics of the Steklov Mathematical Institute (Moscow), the data of Rutherford et al. were studied once again, this time by Khmaladze, Mirvaliev, and Nikulin under the guidance of Bolshev. Some of these investigations were published in the papers of Bolshev and Mirvaliev (1978), Voinov and Nikulin (1984), and Nikulin (1991). It was shown that the data actually reject rather than support the Poisson law in favor of the binomial distribution. This conclusion does not contradict the exponential model of radioactive disintegration and has a satisfactory explanation if we assume the existence of a "resolving time" of the α-particle counter, as was proposed by Feller (1948b). In Feller's model, the experimenters could not count the α particles themselves. In fact, they observed only the scintillations caused by these particles. Since the duration of scintillation is positive, the α particles hitting the screen during the scintillation are not observed. During a time interval of length $\gamma > 0$

after the beginning of each scintillation, new α particles cannot be counted. In this cxamplc, we consider the statistical analysis of the data of Rutherford made by Bolshev and his group.

Let us describe the problem from the beginning. In the Rutherford experiment, numbers of scintillations, $X_1, \ldots, X_n$, were registered on a screen of sulfurous zinc as a result of penetration into the screen by α particles emitted from a radium source, in n =2608 mutually exclusive time intervals, each with duration 7.5 sec. Let v_x represent the frequency of the event $\{X_i = x\}$, that is, v_x is the number of values of X_i equal to x. Table 18.1 lists the frequencies v_x from Rutherford's data, of the events $\{X_i = x\}$.

Table 18.1

x	0	1	2	3	4	5	6	7	8	9	10	11	12	≥ 13
v_x	57	203	383	525	532	408	273	139	45	27	10	4	2	0

It is evident that v_x and X_j satisfy the relations

$$\sum_{x=0}^{\infty} v_x = n,$$

$$S = \sum_{j=1}^{n} X_j = \sum_{x=0}^{\infty} x v_x = 10094, \tag{18.77}$$

where S is the total number of scintillations registered. From Eq. (18.77), it follows that on average in each time interval

$$\overline{X}_n = \frac{1}{n} \sum_{i=1}^{n} X_i = \frac{1}{n} S = \frac{10094}{2608} = 3.8704 \tag{18.78}$$

particles were observed. According to the exponential model of radioactive decay (hypothesis H_0), we expect the random variables $X_1, \ldots, X_n$ to be mutually independent and have the same Poisson distribution with unknown parameter θ:

$$\boldsymbol{P}\{X_i = x \mid H_0\} = \frac{\theta^x}{x!} e^{-\theta}, \qquad x = 0, 1, \ldots, \qquad \theta > 0, \tag{18.79}$$

where $\theta = \lambda\tau$, λ is unknown, $\lambda > 0$, and $\tau = 7.5$ sec. The statistic S is sufficient for θ and $\overline{X}_n = (1/n)S$ is its MVUE. On the other hand, if H_0 is true,

$$\boldsymbol{E}X_i = \textbf{Var } X_i = \theta. \tag{18.80}$$

The MVUE of θ is

$$s^2 = \frac{1}{n-1}\sum_{i=1}^{n}(X_i - \overline{X}_n)^2. \tag{18.81}$$

Because $\overline{X}_n$ and s^2 are consistent estimators for θ, it follows from Eq. (18.80) that, under H_0,

$$\frac{s^2}{\overline{\overline{X}}} \approx 1 \tag{18.82}$$

for large n (in our case $n = 2,608$). From Table 18.1 we have

$$s^2 = \frac{n}{n-1}\left[\frac{1}{n}\sum_{i=1}^{n} X_i^2 - (\overline{X}_n)^2\right] = \frac{1}{n-1}\left[\sum_{x=0}^{\infty} x^2 v_x - \frac{1}{n}S^2\right] = 3.6756. \tag{18.83}$$

This, together with our value of $\overline{X}$ from Eq. (18.78), gives

$$\frac{s^2}{\overline{\overline{X}}} = \frac{3.6756}{3.8704} = 0.9497 < 1. \tag{18.84}$$

Is the difference from 1 significant? Von Mises (1936) supposed (hypothesis H_1) that $X_1, \ldots, X_n$ follow a binomial distribution with parameters m and p, $0 < p < 1$, $q = 1 - p$,

$$\boldsymbol{P}\{X_i = x \mid H_1\} = \binom{m}{x} p^x q^{m-x}, \qquad x = 0, \ldots, m, \tag{18.85}$$

Under H_1,

$$\boldsymbol{E}X_i = mp, \qquad \textbf{Var } X_i = mpq, \tag{18.86}$$

and, because, for the binomial,

$$\frac{\textbf{Var } X_i}{\boldsymbol{E}X_i} = q < 1, \tag{18.87}$$

the observed value of 0.9497 of the ratio $s^2/\overline{X}_n$ in Eq. (18.84) looks very natural under H_1. Was von Mises correct? This is our question. To test

H_0 let us use the test of Fisher based on the index of dispersion X^2 from the previous example (18.76),

$$(n-1)\frac{s^2}{\overline{X}_n} = X^2 = \sum_{i=1}^{n} \frac{(X_i - \overline{X}_n)^2}{\overline{X}_n}. \tag{18.88}$$

For our data in Table 18.1, we obtain the value of $X^2 = 2475.8$. Moreover, for large n,

$$\mathbf{P}\{X^2 < x \mid S\} \sim \mathbf{P}\{\chi_f^2 < x\}, \tag{18.89}$$

where $f = n-1 = 2608-1 = 2607$. On the other hand,

$$\sqrt{2\chi_f^2} - \sqrt{2f-1} \text{ is } AN(0,1) \text{ as } f \to \infty, \tag{18.90}$$

that is,

$$P\left\{\sqrt{2\chi_f^2} - \sqrt{2f-1} < x\right\} \approx \Phi(x). \tag{18.91}$$

Our value of the statistic (18.90) is

$$\sqrt{2X^2} - \sqrt{2f-1} = -1.834 \tag{18.92}$$

and hence, from Eqs. (18.89) to (18.92), we obtain an approximate P value for the statistic X^2:

$$P = \mathbf{P}\{X^2 < 2475.8 \mid S = 10094\} \approx \Phi(-1.834) = 0.0333. \tag{18.93}$$

From Eq. (18.93), it follows that the hypothesis H_0 is not rejected by the conditional test based on X^2 for any significance level $\alpha < 0.033$. But, if $\alpha > p \approx 0.0333$, for example, if $\alpha = 0.05$, one must reject H_0, since $X^2 < 2475.8$. Notice that in this problem we reject H_0 when X^2 is too small. The statistic S is sufficient under both H_0 and H_1, because the binomial and Poisson distributions both belong to the family (16.1). Following the work of Volodin (1965), Bolshev (1965), and Michel (1979), we can construct a conditional-likelihood ratio test given S for testing H_0 against H_1, which is asymptotically most powerful.

From Eq. (18.79), under H_0, the likelihood function of $X_1, \ldots, X_n$

is

$$\frac{\theta^{X_1}\ldots\theta^{X_n}}{X_1!\ldots X_n!}e^{-\theta\Sigma X_i}=\frac{S!}{X_1!\ldots X_n!}\left(\frac{1}{n}\right)^S\frac{(n\theta)^S}{S}e^{-n\theta}, \tag{18.94}$$

and, under H_1, from Eq. (18.85), the likelihood function is

$$\prod_{i=1}^{n}\binom{m}{X_i}p^{X_i}(1-p)^{m-X_i}=\frac{\binom{m}{X_1}\cdots\binom{m}{X_n}}{\binom{mn}{S}}\binom{mn}{S}p^Sq^{nm-S}. \tag{18.95}$$

The conditional-likelihood function $L=L(S)$ given S is

$$L=\begin{cases}\frac{S!}{X_1!\ldots X_n!}\left(\frac{1}{n}\right)^S, & \text{under } H_0,\\ \frac{\binom{m}{X_1}\cdots\binom{m}{X_1}}{\binom{nm}{S}}, & \text{under } H_1.\end{cases} \tag{18.96}$$

The most powerful conditional test given S of level α $(0\le\alpha<0.5)$ is based on the likelihood ratio statistsic

$$\frac{\binom{m}{X_1}\cdots\binom{m}{X_n}X_1!\cdots X_n!}{\binom{nm}{S}S!\left(\frac{1}{n}\right)^S}, \tag{18.97}$$

where S is given. According to this test, one rejects H_0 if

$$\prod_{i=1}^{n}\binom{m}{X_i}X_i!>c, \tag{18.98}$$

where the constant $c=c(\alpha)$ is chosen so that the significance level of the conditional test is equal to α. The relation (18.98) can be written as

$$\prod_{i=1}^{n}\left\{\left(1-\frac{1}{m}\right)\left(1-\frac{2}{m}\right)\cdots\left(1-\frac{X_i-1}{m}\right)\right\}>c \tag{18.99}$$

where the last factor is equal to 1 if $X_i=0$ or $X_i=1$. Hence H_0 is rejected if

$$\sum_{i=1}^{n}\sum_{j=1}^{X_i-1}\ln\left(1-\frac{j}{m}\right)=-\sum_{i=1}^{n}\sum_{j=1}^{X_j-1}\left(\frac{j}{m}+\frac{j^2}{2m^2}+\frac{j^3}{3m^3}+\cdots\right)>c_1, \tag{18.100}$$

for some c_1. Because

$$-\sum_{i=1}^{n}\sum_{j=1}^{X_i-1}\frac{j}{m}=\frac{1}{2m}\left(\sum_{i=1}^{n}X_i^2-S\right), \qquad (18.101)$$

from Eqs. (18.100) and (18.101), we obtain an approximate version of the best conditional test for large m; namely, H_0 is rejected if

$$-\sum_{i=1}^{n}X_i^2>c_2$$

or

$$\sum_{i=1}^{n}X_i^2<c_3, \qquad (18.102)$$

where $c_3=-c_2$. But the index of dispersion X^2 in the form (18.88) can be written

$$X^2=\sum_{i=1}^{n}\frac{(X_i-\overline{X})^2}{\overline{X}}=\frac{n}{S}\left[\sum_{i=1}^{n}X_i^2-\frac{1}{n}S^2\right] \qquad (18.103)$$

and we reject H_0 if $X^2<c$ for some c. From Eqs. (18.99) to (18.103), we see that the test based on the statistic X^2 is asymptotically equivalent to the test (18.98).

Let us return to Feller's model, mentioned earlier, which specifies that after the beginning of each scintillation, during a time interval of length γ, new α particles cannot be counted. Let γ be a constant, and let $T_1, T_2, \ldots$ be mutually independent exponentially distributed random variables with parameter λ, where $\lambda>0$:

$$\boldsymbol{P}\{T_i>y\}=e^{-\lambda y}, \qquad y>0. \qquad (18.104)$$

If T_i is the time interval between the end of the $i-1$st and the beginning of the ith scintillation, then

$$S_x=T_1+\cdots+T_x+(x-1)\gamma$$

is the registration moment of the xth scintillation. If M is the number of scintillations observed over the interval $(0,\tau)$, then for $\tau>\gamma$, $0\le\ell\le\tau/\gamma$

(ℓ is an integer), we have

$$\boldsymbol{P}\{M \leq \ell\} = \boldsymbol{P}\{S_{\ell+1} > \tau\} = \boldsymbol{P}\{T_1 + \cdots + T_{\ell+1} > \tau - \ell\gamma\}$$
$$= \sum_{k=0}^{\ell} \frac{[\lambda(\tau - \ell\gamma)]^k}{k!} e^{-\lambda(\tau-\ell\gamma)}. \tag{18.105}$$

Approximately

$$\boldsymbol{P}\{M \leq \ell\} = \sum_{k=0}^{\ell} \frac{(\lambda\tau)^k}{k!} e^{-\lambda\tau} + 0(\gamma), \qquad \tau = \text{constant}, \quad \gamma \to 0. \tag{18.106}$$

This is the cumulative Poisson distribution function if $\gamma = 0$.

Bolshev and Mirvaliev (1978) have shown that, in fact,

$$\boldsymbol{P}\{M \leq \ell\} = \sum_{k=0}^{\ell} \binom{m}{k} p^k (1-p)^{m-k} + 0(\gamma^2), \tag{18.107}$$

where

$$m = \frac{\tau}{2\gamma} \text{ and } p = \frac{2\lambda\gamma}{1+\lambda\gamma}, \quad \binom{m}{k} = \frac{\Gamma(m+1)}{\Gamma(k+1)\Gamma(m-k)}, \tag{18.108}$$

and

$$\boldsymbol{P}\{M \leq \ell\} = 1 \text{ if } \ell \geq m = \frac{\tau}{2\gamma},$$

We notice that, in expression (18.107), the binomial distribution appears that reduces to the Poisson if the resolution time $\gamma = 0$. To test the Poisson hypothesis against the binomial alternative, we use a chi-squared test, but first we must estimate the parameters m and p in Eq. (18.108). We can estimate them by the method of moments assuming H_1 [see Eq. (18.86)]:

$$\boldsymbol{E}X_i = mp \quad \text{and} \quad \textbf{Var } X_i = mp(1-p). \tag{18.109}$$

Let $\tilde{m}$ and $\tilde{p}$ denote the estimators of m and p, respectively. Then, from

Eqs. (18.78), (18.83), and (18.108), we have

$$\tilde{m}\tilde{p} = \frac{1}{n}\sum_{i=1}^{n} X_i = \frac{1}{n}S = 3.8704, \qquad (18.110)$$

$$\tilde{m}\tilde{p}(1-\tilde{p}) = 3.6756,$$

and hence

$$\tilde{m} = 76.9 \approx 77 \quad \text{and} \quad \tilde{p} = 0.0503. \qquad (18.111)$$

From Eqs. (18.108) and (18.111), we obtain

$$\tilde{\gamma} = 0.049 \text{ (seconds)}, \qquad (18.112)$$

that is, the "resolution time" is approximately 0.05, and, from Eqs. (18.109) to (18.112), we obtain $\tilde{\lambda} = 0.5294$ (seconds^{-1}), and hence

$$\tilde{\theta} = 7.5\tilde{\lambda} = 3.9705.$$

Rutherford used the maximum-likelihood estimator $\hat{\theta}_n = \overline{X} = 3.8704$. We construct a chi-squared test for H_0 against H_1, using the Neyman-Pearson classes J_1 and J_2, and the MVUE for $\boldsymbol{P}\{X_i = x \mid H_0\}$ and $\boldsymbol{P}\{X_i{=}x\,|\,H_1\}$. One obtains $J_1{=}\{3,4,5,6\}$ and $J_2{=}\{0,1,2,\}\cup\{7,8,\ldots\}$ from

$$J_1 = \left\{x : \overline{X}+0.5-\sqrt{\overline{X}+0.25} \le x \le \overline{X}+0.5+\sqrt{\overline{X}+0.25}\right\},$$

where $\overline{X} = (1/n)S$, $J_2 = \boldsymbol{X}_0\backslash J_1$. Using the technique described in this section, we construct the statistic

$$Y_n^2 = \frac{(\nu_1 - nP_1)^2}{nP_1(1-P_1) - \frac{n}{S}(\overline{X}P_1 - Q_1)^2}, \qquad (18.113)$$

for which, under H_0, as $n \to \infty$, $S \sim n$,

$$\lim_{n\to\infty} \boldsymbol{P}\{Y_n^2 > x \mid H_0\} = \boldsymbol{P}\{\chi_1^2 > x\}. \qquad (18.114)$$

For a test of level α, we choose in Eq. (18.114) the critical value

$$c_\alpha = \left[\Phi^{-1}\left(1-\frac{\alpha}{2}\right)\right]^2$$

and reject H_0 if $Y_n^2 \geq c_\alpha$. Our data gives $Y_n^2 = 5.6644$ and

$$\mathbf{P}\{Y_n^2 > 5.6644 \mid H_0\} < 0.01.$$

We reject H_0 .

Nikulin and Voinov (1983; see also Voinov and Nikulin, 1984) studied the question: is the "resolution time" the main cause of the binomiality? The "resolution time" for i scintillations is on the average equal to $i\tilde{\gamma}$ seconds. During this time, we miss seeing $i\tilde{\gamma}\tilde{\lambda}$ α particles. Hence, in ν_i events each having i scintillations, we miss $i\tilde{\gamma}\lambda\nu_i$ α particles. Correspondingly, we may correct the ν_i by the rounded off values of $\nu_i + \delta_i = \tilde{\nu}_i$, where

$$\delta_i = \begin{cases} 0, & \text{if } i = 0, \\ \tilde{\gamma}\tilde{\lambda}[(i-1)\nu_{i-1} - i\nu_i], & i = 1, 2, \ldots. \end{cases} \tag{18.115}$$

This correction is valid only on the average. The results of such correction of the Rutherford data are shown in Table 18.2.

Table 18.2

$\tilde{\nu}_x$	0	1	2	3	4	5	6	7	8	9	10	11	12	13	14	≥ 15
x	0	198	370	506	519	410	282	154	59	30	13	5	1	1	1	0

From this table, we see that $n = 2606$, $S = \sum_{i=1}^{15} i\tilde{\nu}_i = 10318$. In formula (18.115), we used $\tilde{\gamma} = 0.044$ seconds, $\tilde{\lambda} = 0.528$ seconds^{-1}, the estimators of Bolshev obtained with the help of the MVUE W_x of $p_x(\theta)$ (see Bolshev and Mirvaliev, 1978; Voinov and Nikulin, 1984). The Neyman-Pearson class J_1 in this case is $J_1 = \{2, 3, 4, 5, 6\}$, and the value of the statistic Y_n is

$$Y_n = 1.0229 \quad \text{and} \quad Y_n \sim N(0, 1).$$

Hence, for the data of Rutherford corrected by the method described above, the Poisson hypothesis is not rejected for any $\alpha < 0.153 = 1 - \Phi(1.0229)$; and hence the chi-squared test based on Y^2 does not reject H_0 for any $\alpha < 0.306$.

19. A CHI-SQUARED TEST FOR A DISCRETE MULTIVARIATE EXPONENTIAL FAMILY

In this section, we generalize the results of the previous three sections to the multivariate case. As usual with such extensions, the full story is very long, but essentially the same as for the univariate case. We present here only the bones. More information can be found in Patil (1965), Svensson (1985), Nikulin and Voinov (1993).

Description of the Family

Let $\boldsymbol{Z} = (Z_1, \ldots, Z_k)^T$ be a random vector with values $\boldsymbol{z} = (z_1, \ldots, z_k)^T$ in $\boldsymbol{R}^k$. We consider the problem of constructing chi-square criteria for testing the hypothesis that the law of $\boldsymbol{Z}$ belongs to the family of discrete multivariate distributions,

$$p(\boldsymbol{z}; \boldsymbol{\theta}) = p((z_1, \ldots, z_k); (\theta_1, \ldots, \theta_k)) = \frac{a(\boldsymbol{z})\psi(\boldsymbol{z}, \boldsymbol{\theta})}{f(\boldsymbol{\theta})}, \tag{19.1}$$

where $z_i \in \boldsymbol{X}_0 = \{0, 1, 2, \ldots\}$, $i = 1, 2, \ldots, k$; $a(\boldsymbol{z}) > 0$, $f(\boldsymbol{\theta})$ is a positive, bounded and differentiable function,

$$\boldsymbol{\theta} = (\theta_1, \ldots, \theta_k)^T \in \Theta \subset \boldsymbol{R}_+^k, \qquad \psi(\boldsymbol{z}, \boldsymbol{\theta}) = \theta_1^{z_1} \theta_2^{z_2} \ldots \theta_k^{z_k}.$$

For example, if

$$f(\boldsymbol{\theta}) = (1 + \theta_1 + \cdots + \theta_k)^m$$

then $\boldsymbol{Z}$ has a multinomial distribution. If

$$f(\boldsymbol{\theta}) = -\log(1 - \theta_1 - \theta_2 \cdots - \theta_k)$$

then $\boldsymbol{Z}$ has a so-called multivariate log series distribution. If

$$f(\boldsymbol{\theta}) = \exp(\theta_1 + \cdots + \theta_k)$$

then $\boldsymbol{Z}$ has a multivariate Poisson distribution.

The family (19.1), considered by Patil (1965), is a subfamily of the discrete multivariate exponential family of distributions. We construct a chi-squared test for this family following Patil (1965). One can find examples in the papers of Mirvaliev (1984, 1989).

Let $\boldsymbol{Z}_1, \boldsymbol{Z}_2, \ldots, \boldsymbol{Z}_n$ $(n \geq k+1)$ be independent identically distributed random vectors, $\boldsymbol{Z}_i=(Z_{i1}, \ldots, Z_{ik})^T$, each having distribution (19.1). The family (19.1) has the sufficient statistic

$$\boldsymbol{S} = \boldsymbol{Z}_1 + \boldsymbol{Z}_2 + \cdots + \boldsymbol{Z}_n = (S_1, \ldots, S_k)^T,$$

where $S_j = Z_{1j} + Z_{2j} + \cdots + Z_{nj}$, $j = 1, \ldots, k$. The distribution $g(\boldsymbol{s}; \boldsymbol{\theta})$, $\boldsymbol{s} = (s_1, \ldots, s_k)^T$, $s_i \in \boldsymbol{X}_0$, of the statistic $\boldsymbol{S}$ belongs to the family (19.1) and has the form

$$g(\boldsymbol{s}; \boldsymbol{\theta}) = \frac{b(\boldsymbol{s}; n)\psi(\boldsymbol{s}, \boldsymbol{\theta})}{f^n(\boldsymbol{\theta})}, \tag{19.2}$$

where $\psi(\boldsymbol{s}, \boldsymbol{\theta}) = \theta_1^{s_1} \cdots \theta_k^{s_k}$ and $b(\boldsymbol{s}, n)$ are the coefficients of $\theta_1^{s_1} \cdots \theta_k^{s_k}$ in the series expansion of the function $f^n(\boldsymbol{\theta})$.

Sufficiency and Unbiasedness: Theorem of Patel

Now we formulate a theory of Patel (1978), which is a generalization of Theorem 17.1. This theorem provides a technique for constructing the best unbiased estimator $T = T(\boldsymbol{S})$ in terms of the sufficient statistic $\boldsymbol{S}$ for the value $h(\boldsymbol{\theta})$ of a given function $h(\cdot)$, $\boldsymbol{\theta} \in \boldsymbol{\Theta}$. We use the notation introduced in Sections 16 and 17.

Theorem 19.1. *A necessary and sufficient condition for the existence of the minimum variance unbiased estimator* MVUE *of the value $h(\boldsymbol{\theta})$ of $h(\cdot)$ is*

$$\boldsymbol{W}[h(\boldsymbol{\theta}) f^n(\boldsymbol{\theta})] \subseteq \boldsymbol{W}[f^n(\boldsymbol{\theta})], \tag{19.3}$$

where $\boldsymbol{W}$ is an index set defined in Theorem 17.1. If the MVUE, *T, of $h(\boldsymbol{\theta})$ exists, then the statistic T has the form*

$$T = T(\boldsymbol{S}) = \begin{cases} \frac{c(\boldsymbol{S},n)}{b(\boldsymbol{S},n)}, & \boldsymbol{S} \in \boldsymbol{W}[h(\boldsymbol{\theta}) f^n(\boldsymbol{\theta})], \\ 0 & \textit{otherwise}, \end{cases} \tag{19.4}$$

where the numbers $c(\boldsymbol{s}, n)$ are the coefficients of $\psi(\boldsymbol{s}, \boldsymbol{\theta})$ in the series expansion of the function $G(\boldsymbol{\theta}) = h(\boldsymbol{\theta}) f^n(\boldsymbol{\theta})$ in powers of $\psi(\boldsymbol{s}, \boldsymbol{\theta})$, and the $b(\boldsymbol{s}, n)$ are defined following Eq. (19.2).

From this theorem, it follows that

$$W_{\boldsymbol{z}} = \begin{cases} \frac{a(\boldsymbol{z})b(\boldsymbol{S}-\boldsymbol{z},n-1)}{b(\boldsymbol{S},n)}, & \boldsymbol{S} \in (n-1)[\boldsymbol{J}]+\{\boldsymbol{z}\}, \\ 0 & \text{otherwise}, \end{cases} \tag{19.5}$$

is the best unbiased estimator for $p(\boldsymbol{z};\boldsymbol{\theta}) = \boldsymbol{P}(\boldsymbol{Z} = \boldsymbol{z};\boldsymbol{\theta})$, given by Eq. (19.1), where

$$\boldsymbol{J} = \{\boldsymbol{J} = (j_1,\ldots,j_k)^T, \quad j_\ell = 0,1,\ldots,S_\ell; \quad \ell = 1,2,\ldots,k\}. \tag{19.6}$$

Example. The Multinomial Distribution. If $f(\boldsymbol{\theta}) = (1+\theta_1+\cdots+\theta_k)^m$, that is, if $\boldsymbol{Z}_i$ has a multinomial distribution, then

$$a(\boldsymbol{z}) = \frac{m!}{z_1!\ldots z_k!(m-\sum_{i=1}^{k} z_i)!} = \binom{m}{z_1,\ldots,z_k}, \tag{19.7}$$

$$g(\boldsymbol{s};\theta) = \frac{b(\boldsymbol{s},n)\theta_1^{S_1}\theta_2^{S_2}\ldots\theta_k^{S_k}}{f^n(\boldsymbol{\theta})}, \tag{19.8}$$

$$b(\boldsymbol{s},n) = \binom{mn}{s_1,s_2,\ldots,s_k}. \tag{19.9}$$

Using this formula and from Theorem 19.1, we can obtain the best unbiased estimator $W_{\boldsymbol{z}_1,\boldsymbol{z}_2,\ldots,\boldsymbol{z}_i}$ for $\boldsymbol{P}\{\boldsymbol{Z}_1 = \boldsymbol{z}_1\boldsymbol{Z}_2 = \boldsymbol{z}_2,\ldots,\boldsymbol{Z}_i = \boldsymbol{z}_i;\theta\}$:

$$\begin{aligned} \boldsymbol{W}_{\boldsymbol{z}_1,\boldsymbol{z}_2,\ldots\boldsymbol{z}_i} &= \frac{a(\boldsymbol{z}_1)a(\boldsymbol{z}_c)\ldots a(\boldsymbol{z}_i)b(\boldsymbol{S}-\boldsymbol{z}_1-\boldsymbol{z}_2-\cdots-\boldsymbol{z}_i,n)}{b(\boldsymbol{S},n)} \\ &= \prod_{j=1}^{i}\binom{m}{z_{j1},\ldots,z_{jk}}\binom{m(n-i)}{S_1-z_{11}-\cdots-z_{i1},\ldots,S_k-z_{1k}-\cdots-z_{ik}}. \end{aligned} \tag{19.10}$$

for any $i \leq n$. Let us consider the case $k = 3$, that is, let $\boldsymbol{Z}_1,\boldsymbol{Z}_2,\ldots,\boldsymbol{Z}_n$ be a sample from the trinomial distribution, $\boldsymbol{Z}_i = (Z_{i1},Z_{i2},Z_{i3})$:

$$\begin{aligned} \boldsymbol{P}\{\boldsymbol{Z}_i = \boldsymbol{z};\boldsymbol{\theta}\} &= \boldsymbol{P}\{Z_{i1} = z_1, Z_{i2} = z_2, Z_{i3} = z_3;\boldsymbol{\theta}\} \\ &= \frac{m!}{z_1!z_2!(m-z_1-z_2)!}\theta_1^{z_1}\theta_2^{z_2}(1-\theta_1-\theta_2)^{m-z_1-z_2}, \end{aligned}$$

$0 \leq z_i \leq m, \quad z_1+z_2 \leq m_1, \quad m \in \{1,2,\ldots\}.$

The statistic $\boldsymbol{S} = (S_1, S_2, S_3)^T$ is sufficient for $\boldsymbol{\theta} = (\theta_1, \theta_2, \theta_3)^T$, $\theta_3 = 1 - \theta_1 - \theta_2$,

$$S_j = \sum_{i=1}^{n} Z_{ji}.$$

The density function of $\boldsymbol{S}$ is

$$g(\boldsymbol{s}; \boldsymbol{\theta}) = \frac{\Gamma(mn+1)}{s_1! s_2! (mn - s_1 - s_2)} \theta_1^{s_1} \theta_2^{s_2} (1 - \theta_1 - \theta_2)^{mn - s_1 - s_2}, \tag{19.11}$$
$$0 \leq s_i \leq mn, \quad s_1 + s_2 \leq mn.$$

The best unbiased estimator $W_{\boldsymbol{z}}$ of $\boldsymbol{P}\{\boldsymbol{Z}_i = \boldsymbol{z} : \theta\}$ is

$$W_{\boldsymbol{z}} = \binom{m}{z_1, z_2} \frac{\binom{m(n-1)}{S_1 - z_1, S_2 - z_2}}{\binom{mn}{S_1, S_2}}, \qquad \binom{m}{a, b} = \frac{m!}{a! b! (m - a - b)!} \tag{19.12}$$

Construction of a Test Statistic

Let $\boldsymbol{j} = (j_1, \ldots, j_k)^T$, $j_l \in \boldsymbol{X}_0$, and let

$$\delta_{\boldsymbol{j}}(\boldsymbol{Z}_i) = \begin{cases} 1, & \text{if } \boldsymbol{Z}_i = \boldsymbol{j}, \quad i = 1, 2, \ldots, n, \\ 0, & \text{otherwise,} \end{cases}$$

$$\boldsymbol{J} = \{\boldsymbol{j} = (j_1, \ldots, j_k)^T; \qquad j_l = 0, 1, \ldots, s_l; \qquad l = 1, 2, \ldots, k\} \subset \boldsymbol{R}^k,$$

and $\boldsymbol{s} = (s_1, \ldots, s_k)^T$ is the observed value of the sufficient statistic $\boldsymbol{S}$. Then the frequency $\nu_{\boldsymbol{j}}$, with which the sequence of random vectors $\boldsymbol{Z}_1, \boldsymbol{Z}_2, \ldots, \boldsymbol{Z}_n$ takes value $\boldsymbol{j}$, can be written

$$\nu_{\boldsymbol{j}} = \nu(j_1, \ldots, j_k) = \sum_{i=1}^{n} \delta_{\boldsymbol{j}}(\boldsymbol{Z}_i). \tag{19.13}$$

The linear condition relating the frequencies $\nu_{\boldsymbol{j}}$ can be expressed in the form

$$\sum_{\boldsymbol{J}\in \boldsymbol{J}} \nu_{\boldsymbol{J}} = \sum_{j_1=0}^{s_1} \sum_{j_2=0}^{s_2} \cdots \sum_{j_k=0}^{s_k} \nu(j_1, \ldots, j_k) = n$$

$$\sum_{\boldsymbol{J}\in \boldsymbol{J}} j_r \nu_{\boldsymbol{J}} = \sum_{j_1=0}^{s_1} \sum_{j_2=0}^{s_2} \cdots \sum_{j_k=0}^{s_k} j_r \nu_{\boldsymbol{J}} = s_r, \qquad (r = 1, 2, \ldots, k).$$

From Eq. (19.3), one can see that

$$\begin{aligned}
&\boldsymbol{E}\nu_{\boldsymbol{J}} = np_{\boldsymbol{J}} = np(\boldsymbol{J}; \boldsymbol{\theta}), \qquad \boldsymbol{J} \in \boldsymbol{J},\\
&\mathbf{Var}\ \nu_{\boldsymbol{J}} = np_{\boldsymbol{J}}(1 - p_{\boldsymbol{J}})\\
&\mathbf{Cov}\ (\nu_{\boldsymbol{I}}, \nu_{\boldsymbol{J}}) = -np_{\boldsymbol{I}} p_{\boldsymbol{J}}, \qquad \boldsymbol{I} \neq \boldsymbol{J}\\
&\mathbf{Cov}\ (\nu_{\boldsymbol{J}}, S_i) = np_{\boldsymbol{J}}(j_i - \boldsymbol{E}Z_{1i}), \qquad i = 1, 2, \ldots, k,\\
&\mathbf{Var}\ S_i = n\mathbf{Var}\ Z_{1i}, \qquad i = 1, 2, \ldots, k,\\
&\mathbf{Cov}\ (S_i, S_j) = n\mathbf{Cov}\ (Z_{1i}, Z_{1j}), \qquad i \neq j; \qquad i, j = 1, 2, \ldots, k),
\end{aligned} \tag{19.14}$$

where the probabilities $p_{\boldsymbol{J}}$ are defined by Eq. (19.1): $p_{\boldsymbol{J}} = p(\boldsymbol{J}; \boldsymbol{\theta})$.

We partition the collection $\boldsymbol{J}$, which represents the sample space of the random vector $\boldsymbol{Z}_i$ given $\boldsymbol{S}$ into $r \geq 2$ mutually disjoint classes $J_1, J_2, \ldots, J_r$ such that $J_1 \cup J_2 \cup \cdots \cup J_r = \boldsymbol{J}$ and let

$$\begin{aligned}
\boldsymbol{\mu} = (\mu_1, \ldots, \mu_r)^T, \quad \boldsymbol{P} = (P_1, \ldots, P_r)^T,\\
\boldsymbol{Q}_i = (Q_{i1}, \ldots, Q_{ir})^T, \quad i = 1, \ldots, k,
\end{aligned}$$

be vectors whose components are defined as

$$\mu_m = \sum_{\boldsymbol{J}\in J_m} \nu_{\boldsymbol{J}}, \qquad P_m = \sum_{\boldsymbol{J}\in J_m} p_{\boldsymbol{J}}, \qquad Q_{im} = \sum_{\boldsymbol{J}\in J_m} j_i p_{\boldsymbol{J}},$$

$i = 1, 2, \ldots, k$; $m = 1, 2, \ldots, r$.

Consider the vector

$$\boldsymbol{U} = \frac{1}{\sqrt{n}}\begin{pmatrix}\boldsymbol{\mu} - \boldsymbol{E}\boldsymbol{\mu} \\ \boldsymbol{S} - \boldsymbol{E}\boldsymbol{S}\end{pmatrix}$$

$$= \frac{1}{\sqrt{n}}(\mu_1 - \boldsymbol{E}\mu_1, \ldots, \mu_r - \boldsymbol{E}\mu_r, S_1 - \boldsymbol{E}S_1, \ldots, S_k - \boldsymbol{E}S_k)^T.$$

Using the formula (19.4), the covariance matrix $\boldsymbol{\Sigma}$ of the components of this vector can be written as

$$\boldsymbol{\Sigma} = \begin{bmatrix}\boldsymbol{\Sigma}_{11} & \boldsymbol{\Sigma}_{12} \\ \boldsymbol{\Sigma}_{21} & \boldsymbol{\Sigma}_{22}\end{bmatrix}_{(r+s)\times(r+s)},$$

where

$$\boldsymbol{\Sigma}_{11} = \boldsymbol{D} - \boldsymbol{P}\boldsymbol{P}^T,$$

$\boldsymbol{D}$ is the diagonal matrix with elements $P_1, P_2, \ldots, P_r$ on the main diagonal,

$$\boldsymbol{\Sigma}_{22} = \textbf{Var } \boldsymbol{Z}_1 = \begin{bmatrix} \textbf{Var } Z_{11} & \textbf{Cov }(Z_{11}, Z_{12}) & \ldots & \textbf{Cov }(Z_{11}, Z_{1k}) \\ \textbf{Cov }(Z_{12}, Z_{11}) & \textbf{Var } Z_{12} & \ldots & \textbf{Cov }(Z_{12}, Z_{1k}) \\ \vdots & \vdots & \ldots & \vdots \\ \textbf{Cov }(Z_{1k}, Z_{11}) & \textbf{Cov }(Z_{1k}, Z_{12}) & \ldots & \textbf{Var } Z_{1k} \end{bmatrix}$$

$$= \frac{1}{n}\textbf{Var } \boldsymbol{S}$$

is the dispersion matrix of the random vector $\boldsymbol{Z}_1$, and $\boldsymbol{\Sigma}_{12} = [\sigma_{ij}]$ is the matrix with elements

$$\sigma_{ij} = Q_{ji} - P_i \boldsymbol{E} Z_{1j}, \qquad i = 1, \ldots, r; j = 1, \ldots, k; \text{ and } \boldsymbol{\Sigma}_{21} = \boldsymbol{\Sigma}_{12}^T.$$

From the structure of the vector $\boldsymbol{U}$ it follows by the multivariate Lindeberg-Levy central limit theorem that the vector $\boldsymbol{U}$ is $AN(\mathbf{0}_{r+k}, \boldsymbol{\Sigma})$, where $\mathbf{0}_m = \underbrace{(0, \ldots, 0)}^T$.

Now using the results of Holst (1981), Sethuraman (1961), and Sweeting (1989) one can see, from the asymptotic normality of the vector $\boldsymbol{U}$, that the following asymptotic result holds.

Theorem 19.2. *If the number n of observations $\boldsymbol{Z}_1, \boldsymbol{Z}_2, \ldots, \boldsymbol{Z}_n$ goes to infinity in such a way that $S_i \asymp n$ $(i = 1, \ldots, k)$ and the classes*

$J_1, J_2, \ldots, J_r$ remain fixed, then the asymptotic distribution of the conditional law of

$$\frac{1}{\sqrt{n}}\,(\boldsymbol{\mu}-\boldsymbol{E}\boldsymbol{\mu}) = \frac{1}{\sqrt{n}}\,(\mu_1-\boldsymbol{E}\mu_1,\ldots,\mu_k-\boldsymbol{E}\mu_k)^T,$$

given $\boldsymbol{S}$, is the same as for the unconditioned $1/(\sqrt{n})(\boldsymbol{\mu}-\boldsymbol{E}\boldsymbol{\mu})$ and it is $AN(\mathbf{0}_k, \boldsymbol{B})$, where

$$\boldsymbol{B} = \boldsymbol{D}-\boldsymbol{P}\boldsymbol{P}^T - \boldsymbol{\Sigma}_{12}\,\boldsymbol{\Sigma}_{22}^{-}\,\boldsymbol{\Sigma}_{21}, \qquad \textit{rank } \boldsymbol{B} = r-1,$$

and $\boldsymbol{\Sigma}_{22}^{-}$ is a generalized inverse of the matrix $\boldsymbol{\Sigma}_{22}$, and in general $\boldsymbol{B} = \boldsymbol{B}(\boldsymbol{\theta})$ depends on $\boldsymbol{\theta}$.

Corollary 19.1. The statement of the theorem holds for the vector $\frac{1}{\sqrt{n}}\,(\boldsymbol{\mu}-n\widehat{\boldsymbol{P}})$ with the matrix $\widehat{\boldsymbol{B}}$ instead of $\boldsymbol{B}$, where

$$\widehat{\boldsymbol{B}} = \widehat{\boldsymbol{D}}-\widehat{\boldsymbol{P}}\widehat{\boldsymbol{P}}^T - \widehat{\boldsymbol{\Sigma}}_{12}\,\widehat{\boldsymbol{\Sigma}}_{22}^{-}\,\widehat{\boldsymbol{\Sigma}}_{21},$$

$\widehat{\boldsymbol{P}}, \widehat{\boldsymbol{\Sigma}}_{ij}$ are consistent estimators of $\boldsymbol{P}$ and $\boldsymbol{\Sigma}_{ij}$, respectively, $\widehat{\boldsymbol{D}}$ is the diagonal matrix with the elements of the vector $\widehat{\boldsymbol{P}}$ on the main diagonal, and $\widehat{\boldsymbol{B}}$ is a consistent estimator of $\boldsymbol{B}$. To obtain these consistent estimators, one can use the conditional distribution $W_{\boldsymbol{Z}}$ given by Eq. (19.5) or the maximum-likelihood estimator $\hat{\boldsymbol{\theta}}_n$ of $\boldsymbol{\theta}$.

From Corollary 1 and the results of Puri and Russell (1984), we obtain the following corollary.

Corollary 19.2. If H_0 is true, then the statistic

$$Y^2 = \frac{1}{n}\,(\boldsymbol{\mu}-n\widehat{\boldsymbol{P}})^T\widehat{\boldsymbol{B}}^{-}(\boldsymbol{\mu}-n\widehat{\boldsymbol{P}})$$

has in the limit as $n \to \infty$, with $S_i \asymp n$, a chi-square distribution with $r-1$ degrees of freedom.

Remarks. For a matrix $\widehat{\boldsymbol{B}}$ of rank $r-1$, a generalized inverse $\widehat{\boldsymbol{B}}^{-}$ is defined as follows:

> *The matrix obtained by crossing out any row and column, for example, the last row and column, of $\widehat{\boldsymbol{B}}$, is invertible. To this inverse one adds a last row and column of 0s to obtain $\widehat{\boldsymbol{B}}^{-}$.*

Using Theorem 1 of Portnoy (1977), or Michel (1979), one sees that Corollary 19.2 is in some cases, a generalization of results of Nikulin (1973a,b) where the maximum-likelihood estimator was used and of Nikulin and Voinov (1987, 1993), where an MVUE was used. Mirvaliev (1989) studied the problem of constructing chi-squared goodness-of-fit tests for multinomial, and multidimensional Poisson, multinomial, negative multinomial, and multidimensional binomial, hypergeometric, and negative hypergeometric distributions.

The technique of Michel (1979) is useful for studying the power of these tests. More about the asymptotic efficiency of conditional tests for exponential functions is in the work of Michel (1979), Witting and Nölle (1970), Steck (1957).

20. THE CHI-SQUARED TEST FOR A CONTINUOUS EXPONENTIAL FAMILY

In this section, we consider the hypothesis, H_0, which states that i.i.d. $X_1, \ldots, X_n$ follow a probability distribution with density

$$f(x;\theta) = \begin{cases} \frac{1}{b(\theta)} c(x) \exp\{xQ(\theta)\}, & \alpha \leq x \leq \beta, \\ 0, & \text{otherwise}, \end{cases} \tag{20.1}$$

where $b(\theta) > 0$, $\theta \in \Theta \subset \boldsymbol{R}^1$, $c(x) > 0$ if $\alpha \leq x \leq \beta$, and the parameters α and β are known. The family (20.1) is a continuous exponential family of rank one. The same Rao-Blackwell approach, which was developed in Sections 16 to 18 is used for this continuous family.

Sufficient Statistic and Minimum Variance Unbiased Estimator for the Density

The statistic $S = X_1 + \cdots + X_n$ is sufficient for the parameter θ. The probability density function of S is

$$g(s;\theta) = \frac{b(s,n,\alpha,\beta)}{b^n(\theta)} \exp\{sQ(\theta)\}, \qquad n\alpha \leq s \leq n\beta \tag{20.2}$$

(see Lehmann, 1991; Nikulin and Voinov, 1989).

The functions $b(s, n, \alpha, \beta)$ can be found from the relation

$$\int_{n\alpha}^{n\beta} g(s;\theta)\,ds = 1.$$

It can be shown that the functions $b(s, n, \alpha, \beta)$ satisfy the recurrence relation

$$b(s,n,\alpha,\beta) = \int_{\ell(s)}^{L(s)} c(x)b(s-x,n-1,\alpha,\beta)\,dx, \tag{20.3}$$

where $\ell(s) = \max\{\alpha, s-(n-1)\beta\}$, $L(s) = \min\{\beta, s-(n-1)\alpha\}$. It is easily verified (see, for example, a technique proposed by Karakostas, 1985), or Voinov and Nikulin, 1989) that the MVUE $\hat{f}(x)$ of $f(x;\theta)$ is

$$\hat{f}(x) = \frac{c(x)b(S-x,n-1,\alpha,\beta)}{b(S,n,\alpha,\beta)}, \qquad \ell(S) \le x \le L(S). \tag{20.4}$$

Using the same technique as in Section 16, we find

$$\boldsymbol{E}_\theta\{X_1 \mid S\} = \int_{\ell(S)}^{L(S)} \frac{xc(x)b(S-x,n-1,\alpha,\beta)}{b(S,n,\alpha,\beta)}\,dx = \frac{1}{n}S = \bar{X}_n, \tag{20.5}$$

and the MVUE, $\hat{f}(x_1, x_2)$, of the joint probability density $f(x_1, x_2; \theta)$ of X_1 and X_2 is

$$\hat{f}(x_1,x_2) = \frac{c(x_1)c(x_2)b(S-x_1-x_2,n-2,\alpha,\beta)}{b(S,n,\alpha,\beta)}, \qquad \alpha \le x_i \le \beta, \tag{20.6}$$
$$(n-2)\alpha + x_1 + x_2 \le S \le n\beta.$$

The MVUE, $\widehat{\sigma^2}$, of $\sigma^2 = \mathbf{Var}_\theta X_1$ is

$$\widehat{\sigma^2} = \frac{n}{n-1}\int_{\ell(S)}^{L(S)} \frac{x^2c(x)b(S-x,n-1,\alpha,\beta)}{b(S,n,\alpha,\beta)}\,dx - \frac{1}{n(n-1)}S^2. \tag{20.7}$$

Following Section 13, we choose a vector $\boldsymbol{p} = (p_1, \ldots, p_r)^T$ such that $\boldsymbol{p}^T\mathbf{1}_r = 1$, $p_i > 0$, and define quantiles $x_0, x_1, \ldots, x_{r-1}, x_r$ by the conditions

$$\int_\alpha^{x_i} \hat{f}\,dx = p_1 + p_2 + \cdots + p_i, \qquad i = 1, \ldots, r-1, \tag{20.8}$$

$$x_0 = \alpha, \quad x_r = \min\{\beta, S-(n-1)\alpha\}.$$

Let $\boldsymbol{\nu} = (\nu_1, \ldots, \nu_r)^T$ be the vector of frequencies for $x_1, \ldots, X_n$ falling in the intervals

$$[x_0, x_1), [x_1, x_2), \ldots, [x_{r-1}, x_r].$$

Clearly, $\boldsymbol{E\nu} = \boldsymbol{E}\{\boldsymbol{\nu} \mid S\} = n\boldsymbol{p}$. Let $\boldsymbol{\Sigma}$ be the conditional covariance matrix, given S, of the vector

$$\boldsymbol{Z} = \frac{1}{\sqrt{n}}(\boldsymbol{\nu}-n\boldsymbol{p}).$$

Using the results of, for example, Holst (1981), Sethuraman (1961), or Sweeting (1989), one is able to show that for $n \to \infty$, with $S \asymp n$, the conditional distribution of $\boldsymbol{Z}$ is asymptotically normal with

$$\boldsymbol{E}\{\boldsymbol{Z} \mid S\} = \mathbf{0}_r, \quad \boldsymbol{E}\{\boldsymbol{Z}\boldsymbol{Z}^T \mid S\} = \boldsymbol{\Sigma} + o(\mathbf{1}_{r\times r}),$$

where [see Eq. (18.37)]

$$\boldsymbol{\Sigma} = \boldsymbol{P} - \boldsymbol{p}\cdot\boldsymbol{p}^T - \frac{1}{\widehat{\sigma^2}}\left[\frac{1}{n}S\boldsymbol{p} - \boldsymbol{q}\right]\left[\frac{1}{n}S\boldsymbol{p} - \boldsymbol{q}\right]^T, \tag{20.9}$$

$$\operatorname{rank} \boldsymbol{\Sigma} = r-1, \qquad \boldsymbol{p} = (p_1, \ldots, p_r)^T, \qquad \boldsymbol{q} = (q_1, \ldots, q_r)^T,$$

$$\boldsymbol{P} = \begin{bmatrix} p_1 & 0 & \cdots & 0 \\ 0 & p_2 & \cdots & 0 \\ \vdots & \vdots & \ddots & 0 \\ 0 & 0 & \cdots & p_r \end{bmatrix}, \qquad p_i = \int_{x_{i-1}}^{x_i} \hat{f}(x)\,dx, \qquad q_i = ip_i.$$

The Test Statistic

We are now prepared to construct a chi-squared type statistic Y_n^2 for testing the hypothesis H_0, as was done in Sections 13 to 15,

$$Y_n^2 = \frac{1}{n}(\boldsymbol{\nu}-n\boldsymbol{p})^T\boldsymbol{\Sigma}^-(\boldsymbol{\nu}-n\boldsymbol{p}). \tag{20.10}$$

Direct calculation shows that

$$Y_n^2 = \sum_{i=1}^{r} \frac{(\nu_i - np_i)^2}{np_i} + \frac{1}{n\mu}\left(\sum_{i=1}^{r} i\nu_i - S\right)^2, \tag{20.11}$$

where

$$\mu = \widehat{\sigma^2} + \left(\frac{1}{n}S\right)^2 - \sum_{i=1}^{r} i^2 p_i. \tag{20.12}$$

Example. Exponential distribution with location and scale. Let $X_1, \ldots, X_n$ be i.i.d. with probability density function, under H_0,

$$f(x;\theta) = \begin{cases} \theta \exp\{-\theta(x-\alpha)\}, & x \geq \alpha \\ 0, & \text{otherwise,} \end{cases} \tag{20.13}$$

which belongs to the family (20.1) with

$$c(x) \equiv 1, \qquad b(\theta) = \frac{1}{\theta}\exp(-\theta\alpha), \qquad Q(\theta) = -\theta, \qquad \beta = +\infty,$$

where α is known. The probability density function $g(s;\theta)$ of the sufficient statistic $S = X_1 + \cdots + X_n$ is

$$g(s;\theta) = \begin{cases} \frac{1}{\Gamma(n)}\theta^n \exp(n\theta\alpha)\exp(-\theta s)(s-n\alpha)^{n-1}, & s \geq n\alpha, \\ 0, & \text{otherwise,} \end{cases}$$

that is,

$$b(s,n,\alpha,\beta) = b(s,n,\alpha) = \frac{1}{\Gamma(n)}(s-n\alpha)^{n-1}.$$

From Eqs. (20.4) and (20.7), it follows that

$$\hat{f}(x) = \frac{n-1}{(s-n\alpha)^{n-1}}(S-x-(n-1)\alpha)^{n-2}, \quad \alpha \leq x \leq S-(n-1)\alpha. \tag{20.14}$$

The MVUE $\widehat{\boldsymbol{E}_\theta X_1}$ of $\boldsymbol{E}_\theta X_1 = \alpha + 1/\theta$ is

$$\widehat{\boldsymbol{E}_\theta X_1} = \int_{\alpha}^{S-(n-1)\alpha} \frac{x[S-x-(n-1)\alpha]^{n-2}\Gamma(n)}{(S-n\alpha)^{n-1}\Gamma(n-1)}\,dx$$

$$= \frac{n-1}{(S-n\alpha)^{n-1}} \int_{\alpha}^{S-(n-1)\alpha} x[S-x-(n-1)\alpha]^{n-2}\,dx$$

$$= \frac{1}{n}S = \bar{X}_n,$$

$$\widehat{\sigma^2} = \frac{n\Gamma(n)}{(n-1)(S-n\alpha)^{n-1}} \int_{\alpha}^{S-(n-1)\alpha} \frac{x^2(S-x-(n-1)\alpha)^{n-2}}{\Gamma(n-1)}\,dx - \frac{S^2}{n(n-1)}$$

$$= \frac{1}{n(n+1)}S^2 + \frac{n\alpha^2 - 2\alpha S}{n+1}. \tag{20.15}$$

As $n \to \infty$,

$$\widehat{\sigma^2} = \left(\frac{1}{n}S - \alpha\right)^2 + o(1).$$

Let us choose equiprobable cells; that is,

$$p_1 = p_2 = \cdots = p_r = \frac{1}{r}.$$

Then

$$\int_{\alpha}^{x_i} \hat{f}(x)\,dx = p_1 + \cdots + p_i = \frac{i}{r} = q_i, \qquad i = 1, \ldots, r.$$

Substituting into this equation the MVUE $\hat{f}$, which is given by Eq. (20.14), gives

$$\frac{n-1}{(s-n\alpha)^{n-1}} \int_{\alpha}^{x_i} (S-x-(n-1)\alpha)^{n-2}\,dx = \frac{i}{r}. \tag{20.16}$$

If, in Eq. (20.16), we put $S-x-(n-1)\alpha = y$, then we obtain $dy = -dx$, and hence

$$\frac{(n-1)}{(S-n\alpha)^{n-1}} \int_{S-(n-1)\alpha-x_i}^{S-n\alpha} y^{n-2}\,dy = \frac{i}{r},$$

$$\frac{i}{r} = 1 - \frac{[S-(n-1)\alpha - x_i]^{n-1}}{[S-n\alpha]^{n-1}},$$

from which we find

$$x_i = [S-(n-1)\alpha] - (S-n\alpha)(1-i/r)^{\frac{1}{n-1}}, \qquad i = 1,\ldots,r. \qquad (20.17)$$

Using the quantiles x_i given by Eq. (20.17), we define the vector of frequencies $\boldsymbol{\nu}$. From Eqs. (20.10) to (20.12), taking into account that

$$p_i = \frac{1}{r}, \qquad q_i = ip_i = \frac{i}{r}, \qquad i = 1,\ldots,r,$$

we obtain the statistic

$$\begin{aligned} Y_n^2 &= \frac{1}{nr}\sum_{i=1}^{r}(\nu_i r - n)^2 + \frac{1}{n\mu}\left(\sum_{i=1}^{r} i\nu_i - S\right)^2 \\ &= X^2 + \frac{1}{n\mu}\left(\sum_{i=1}^{r} i\nu_i - S\right)^2, \end{aligned}$$

where

$$X^2 = \frac{r}{n}\sum_{i=1}^{r}\nu_i^2 - n, \quad \mu = \left(\frac{1}{n}S - \alpha\right)^2 + \left(\frac{1}{n}S\right)^2 + (r+1)(2r+1).$$

An important point is that we can work with the conditional distribution in this problem owing to the characterization of Prokhorov (1965) of the family of exponential distributions by their conditional distributions. See also Dzhaparidze and Nikulin (1982a), and Nikulin and Voinov (1993).

21. A CHI-SQUARED TEST FOR NATURAL EXPONENTIAL FAMILIES WITH POWER VARIANCE FUNCTION

A Morris Family of Distributions

Suppose that we have i.i.d. random variables $X_1, X_2, \ldots, X_n$ and the distribution function of X_1 belongs to the family $\{F(x;\theta),\ \theta \in \Theta \subseteq \boldsymbol{R}^1\}$,

where

$$F(x;\theta)=\boldsymbol{P}\{X_1\leq x\}=\int_{-\infty}^{x}\exp\{-u\theta+\psi(\theta)\}\,dG(u),\qquad x\in\boldsymbol{R}^1,\tag{21.1}$$

where $G(x)$ is taken to be a cumulative distribution function not depending on the *natural parameter* θ. Morris, 1982, 1983, Barndorff-Nielsen, 1978, and Bar-Lev and Enis, 1986) have studied this class of distributions and called $\{F(x;\theta);\ \theta\in\Theta\}$ a *natural exponential family* (NEF) *generated by* G. The *cumulant generating function* $\psi(\theta)$ gives Eq. (21.1) unit probability, and one can easily verify that

$$\mu=\boldsymbol{E}_\theta X_i=\frac{d\psi(\theta)}{d\theta},\qquad v^2=\mathbf{Var}_\theta X_i=\boldsymbol{E}_\theta(X_i-\mu)^2=-\frac{d^2\psi(\theta)}{d\theta^2},\tag{21.2}$$

for any $\theta\in\Theta_0=$ interior of Θ. Let $V(\mu)=\mathbf{Var}\ X_i$ denote the variance of X_i expressed in terms of μ. The function $V(\mu)$ on its domain $\Omega\equiv\psi(\Theta)$ is called the *power variance function* (PVF), if $V(\mu)$ has the form

$$V(\mu)=c\mu^\beta,\qquad \mu\in\Omega,\tag{21.3}$$

for some constants $c>0$ and $\beta\geq 0$, which are called the *scale and power parameters*, respectively. Following Barndorff-Nielsen (1978), Hougaard (1986), Morris (1982), and Bar-Lev and Enis (1986), a class (21.1) of NEFs with $V(\mu)$ in the form (21.3) is called the class of NEF-PVF distributions. For example, if $\beta=0$, we have the family of normal distributions. If $\beta=1$, we obtain the family of Poisson distributions, and, for $\beta=2$, we obtain the family of gamma distributions. As was shown by Hougaard (1986), and Bar-Lev and Enis (1986), for $\beta\in(0,1)$, no NEF-PVF distributions exist, but for any $\beta\in(1,2)$ we obtain NEF-PVF distributions that are compound Poisson distributions, generated by gamma distributions. When $\beta>2$, we obtain the subclass of NEF-PVF distributions *generated by stable distributions on* $\boldsymbol{R}^1_+$, which are unimodal and absolutely continuous with respect to Lebesgue measure on $\boldsymbol{R}^1_+$. We consider here the subclass of all NEFs on the positive numbers. As is well, known a stable distribution on $\boldsymbol{R}^1_+$ has a stable index $\alpha\in(0,1)$, and, as was shown by Pollard (1946), its probability density $g_\alpha(x)$ is given by the formula

$$g_\alpha(x)=-\frac{1}{\pi x}\sum_{k=1}^{\infty}\sin(\pi\alpha k)\,\frac{\Gamma(\alpha k+1)}{\alpha k},\qquad x>0.\tag{21.4}$$

Suppose the generating distribution $G(x)$ in Eq. (21.1) is stable on $\boldsymbol{R}^1_+$ with the density $g(x) = 1/\sigma g_\alpha(x/\sigma)$, where σ is known and $\sigma > 0$. Then, the distribution $F(x;\theta)$, $\theta \in \Theta = [0, \infty)$, is unimodal and absolutely continuous with respect to Lebesgue measure on $\boldsymbol{R}^1_+$ with density

$$f(x,\theta) = \begin{cases} g(x)\exp\{-x\theta + \psi(\theta)\}, & x > 0, \\ 0, & x \leq 0, \end{cases} \tag{21.5}$$

where $\psi(\theta) = (\sigma\theta)^\alpha$ and $\theta \in \Theta = [0, \infty)$. The power variance function is

$$V(\mu) = c\mu^\beta, \tag{21.6}$$

where

$$\begin{aligned} \mu &= \boldsymbol{E}_\theta X_1 = \alpha\sigma^\alpha\theta^{\alpha-1}, \quad && c = (1-\alpha)(\alpha\sigma^\alpha)^{1/(\alpha-1)}, \\ \beta &= \frac{2-\alpha}{1-\alpha} > 2, && 0 < \alpha < 1 \end{aligned}$$

The Laplace transform $\mathcal{L}(s)$ of the random variable X_1 is

$$\mathcal{L}(s) = \boldsymbol{E}_\theta\{\exp(-sX_1)\} = \exp\{-\sigma^\alpha[(\theta+s)^\alpha - \theta^\alpha]\}. \tag{21.7}$$

Sufficiency and Unbiasedness: Theorem of Washio, Morimoto, and Ikeda

This theorem gives us the UMVUEs that we need to construct a chi-squared test. In view of the factorization theorem of Neyman, the family (21.5) possesses the complete sufficient statistic $S_n = X_1 + X_2 + \cdots + X_n$ for the parameter θ. As was shown by Bar-Lev and Enis (1986) and Hougaard (1986) the density $f_n(x;\theta)$ of S_n is

$$f_n(x;\theta) = \begin{cases} n^{-1/\alpha}g(n^{-1/\alpha}x)\exp\{-x\theta + \psi(n^{1/\alpha}\theta)\}, & x > 0, \\ 0, & \text{otherwise.} \end{cases} \tag{21.8}$$

A theorem of Washio, Morimoto, and Ikeda (1956) gives us a general expression for the UMVUE of a function $u(\theta)$ (see also Morris, 1982; Zaks, 1971; Voinov and Nikulin, 1989; Nikulin, 1991). The characteristic function $\varphi_n(t,\theta)$ of S_n is

$$\varphi_n(t,\theta) = \exp\{n\psi(\theta) - n\psi(\theta - it)\}. \tag{21.9}$$

Theorem (Washio, Morimoto, Ikeda). *Assume that the following conditions are satisfied:*

1. *$u(z)$, $z=\theta+it$, is analytic in a strip whose real part is Θ_0, the interior of Θ, $t \in \boldsymbol{R}$.*
2. *the characteristic function $\varphi_n(t,\theta)$ of the complete sufficient statistic S_n is such that*

$$\lim_{|t|\Rightarrow\infty} \varphi_n(-t,\theta)u(\theta+it)=0$$

uniformly in any subinterval of Θ_0 and

$$\int_{-\infty}^{+\infty} |\varphi_n(-t,\theta)u(\theta+it)|dt<\infty, \qquad \theta\in\Theta_0.$$

Then the essentially unique minimum quadratic risk unbiased estimator (UMVUE) $\hat{u}=T(S_n)$ of $u(\theta)$ is

$$T(S_n)=\begin{cases} \frac{n^{\alpha}W(S_n)}{g(s^{-1/\alpha}S_n)}, & \text{if } g(n^{-1/\alpha}S_n)>0, \\ 0, & \text{otherwise,}\end{cases} \tag{21.10}$$

where

$$W(x)=\Im^{-1}\{u(\theta)e^{-n\psi(\theta)}\}=\frac{1}{2\pi i}\int_{\theta-i\infty}^{\theta+i\infty} u(z)\exp[zx-n\psi(z)]\,dz. \tag{21.11}$$

Using this theorem, one can find (see also Zacks, 1971; Voinov and Nikulin, 1993) the UMVUEs $\hat{v}^2$ for $\mathbf{Var}\,X_1$ and $\hat{f}(x)$ for the density $f(x;\theta)$. They are

$$\hat{v}^2=\frac{\beta\alpha}{g(S_n n^{-1/\alpha})\Gamma(1-\alpha)}\int_0^{S_n} g(yn^{-1/\alpha})(S_n-y)^{1-\alpha}\,dy. \tag{21.12}$$

and

$$\hat{f}(x)=\begin{cases} \left(\frac{n}{n-1}\right)^{1-\alpha} g(x)g\left[(S_n-x)(n-1)^{-1/\alpha}\right]\frac{1}{g(S_n n^{-1/\alpha})}, & 0<x<S_n, \\ 0, & \text{otherwise.}\end{cases} \tag{21.13}$$

A Chi-Squared Test Statistic

We can now construct a chi-squared test for testing the hypothesis, H_0, that the X_i have distribution (21.5). Taking into account the observed value s of the sufficient statistic S_n, we choose the vector $\boldsymbol{p}=(p_1,\ldots,p_r)^T$ such that $p_1+\cdots+p_r=1$, $p_i>0$, $r\geq 2$, and construct classes

$$\boldsymbol{J}_1=(0,x_1],\qquad \boldsymbol{J}_2=(x_1,x_2],\ldots,\qquad \boldsymbol{J}_r=(x_{r-1}x_r],\qquad x_r=s,$$

for grouping the data $X_1,\ldots,X_n$, where $x_1,\ldots,x_{r-1}$ are quantiles of the estimated distribution defined by

$$\int_0^{x_i}\hat{f}(x)\,dx=p_1+\cdots+p_i,\qquad i=1,\ldots,r-1.$$

Furthermore, define the vectors $\boldsymbol{\nu}=(\nu_1,\ldots,\nu_r)^T$, $\boldsymbol{q}=(q_1,\ldots,q_r)^T$, where

$$q_i=\int_{\boldsymbol{J}_i}x\hat{f}(x)\,dx,\qquad i=1,\ldots,r,$$

and $\boldsymbol{\nu}$ is a vector of frequencies arising as the result of grouping the data $X_1,\ldots,X_n$ into the classes $\boldsymbol{J}_1,\ldots,\boldsymbol{J}_r$. By the same argument as in Section 20, we have asymptotic normality for the vector of frequencies.

Theorem. *If $n\Rightarrow\infty$ so that $S_n=O(n)$, then the conditional distribution of the vector $1/\sqrt{n}(\boldsymbol{\nu}-n\boldsymbol{p})$, given S_n is $AN(\boldsymbol{O}_r,\boldsymbol{B})$, where*

$$\boldsymbol{B}=\boldsymbol{P}-\boldsymbol{p}\boldsymbol{p}^T-\frac{1}{\hat{v}^2}\left[\frac{1}{n}S_n\boldsymbol{p}-\boldsymbol{q}\right]\left[\frac{1}{n}S_n\boldsymbol{p}-\boldsymbol{q}\right]^T.$$

$\boldsymbol{P}$ is a diagonal matrix with elements $p_1,\ldots,p_r$ on the main diagonal, rank $\boldsymbol{B}=r-1$ and $\hat{v}^2$, given by Eq. (21.12) is the UMVUE of **Var** *X_1.*

As in previous sections, we construct the test statistic

$$Y_n^2=\frac{1}{n}(\boldsymbol{\nu}-n\boldsymbol{p})^T\boldsymbol{B}^-(\boldsymbol{\nu}-n\boldsymbol{p}),$$

where $\boldsymbol{B}^-$ is a generalized inverse matrix of $\boldsymbol{B}$. An alternate form is

$$Y_n^2=\sum_{i=1}^r\frac{(\nu_i-np_i)^2}{np_i}+\frac{1}{n\kappa}\left(\sum_{i=1}^r\frac{q_i}{p_i}\nu_i-S_n\right)^2,$$

where

$$\kappa = \hat{v}^2 + \left(\frac{1}{n}S_n\right)^2 - \sum_{i=1}^{r} \frac{q_i^2}{p_i}.$$

From the above theorem the limiting distribution of Y_n^2 under H_0 will be chi-squared with $r-1$ degrees of freedom.

CHAPTER 4

Some Additional Examples

22. A TEST FOR HOMOGENEITY

A Problem of Several Samples

In a study of the effect of diet on cholesterol level, we have r groups of patients. Each group has been on a different diet for several weeks. At this point, the cholesterol level of each individual is recorded and these numbers are our data: $\xi_{i1}, \ldots, \xi_{in_i}$, $i = 1, \ldots, r$. There are n_i patients in the ith group. We assume that it is known that, in the entire population, cholesterol level has a distribution belonging to a parameterized family $\{F_{\boldsymbol{\theta}}, \boldsymbol{\theta} \in \boldsymbol{\Theta}\}$ of continuous distributions, with densities $f_{\boldsymbol{\theta}}$, and that, for each of our r groups, the distribution is $F_{\boldsymbol{\theta}_i}$, $i = 1, \ldots, r$, for some $\boldsymbol{\theta}_i \in \boldsymbol{\Theta}$. We wish to test the hypothesis H_0 that all values $\boldsymbol{\theta}_i$ are the same, $\boldsymbol{\theta}_1 = \cdots = \boldsymbol{\theta}_r = \boldsymbol{\theta}_0$ where $\boldsymbol{\theta}_0$ is unknown. We also consider, separately, an alternative hypothesis H_1 that $\boldsymbol{\theta}_i$ differs from group to group. Suppose $\boldsymbol{\Theta} \subset \boldsymbol{R}^s$. There are many approaches to this famous problem. For two samples, the problem was studied by Murthy and Gafarian (1970), Chase (1972), and Nukulin and Yusas (1976, 1977). We use the technique of this guide to construct a quadratic statistic that has, asymptotically, a chi-squared distribution, central if homogeneity exists, that is, if H_0 holds, and noncentral under the alternative hypothesis of nonhomogeneity, following Bolshev and Nikulin (1975).

For each $i = 1, \ldots, r$, we regard our observations $\xi_{i,1}, \ldots, \xi_{i,n_i}$ as being independent random variables with the same distribution $F_{\boldsymbol{\theta}_i}$. We begin by forming grouping intervals. Since, as in Chapter 2, we have a composite hypothesis, the grouping intervals will depend on $\boldsymbol{\theta}_0$, which must be estimated. We choose a number, k, of grouping intervals

$k > 2$. If $\boldsymbol{\Theta} \subset \boldsymbol{R}^s$, $s > 1$, then k should be more than $s + 1$. Choose $\boldsymbol{p} = (p_1, \ldots, p_k)^T$ so that $p_i \geq 0$ and $\sum_{i=1}^k p_i = 1$. A good choice would be $p_i = 1/k$, $i = 1, \ldots, k$. Set

$$x_j(\boldsymbol{\theta}) = F_{\boldsymbol{\theta}}^{-1}(p_1 + \cdots + p_j), \quad j = 1, \ldots, k-1.$$
$$x_0(\boldsymbol{\theta}) = -\infty, \quad x_k(\boldsymbol{\theta}) = \infty.$$

Suppose for the family F_θ estimators exist satisfying Eq. (10.16), for estimating $\boldsymbol{\theta}_0$, computed using all the observations ξ_{ij}. There are $N = n_1 + \cdots + n_r$ observations. Then

$$\sqrt{N}(\bar{\boldsymbol{\theta}}_N - \boldsymbol{\theta}_0) = \frac{1}{N^{1/2}} \sum_{i,j} \boldsymbol{I}^{-1}(\boldsymbol{\theta}_0)\boldsymbol{\Lambda}(\xi_{ij}) + o_p(1)$$

where $\boldsymbol{I}(\theta)$ is the Fisher information matrix,

$$\boldsymbol{\Lambda}(\xi_{ij}) = \boldsymbol{\Lambda}(\boldsymbol{\theta}_0; \xi_{ij}) = \textbf{grad } \log f_{\boldsymbol{\theta}}(\xi_{ij})|_{\boldsymbol{\theta}=\boldsymbol{\theta}_0},$$

and $o_p(1)$ is a random vector that goes to $\mathbf{0}_s$ in probability as $N \to \infty$. We assume that, as N increases, each of $n_1, \ldots, n_r$ increases in such a way that $c < n_i/N < C$, $i = 1, \ldots, r$, where c and C are positive constants.

Let $\boldsymbol{\mu}_i = (\mu_{i1}, \ldots, \mu_{ik})^T$ be the vector obtained by grouping the data $\xi_{i1}, \ldots, \xi_{in_i}$, using the intervals

$$(-\infty, x_1(\bar{\boldsymbol{\theta}}_N)), \qquad [x_1(\bar{\boldsymbol{\theta}}_N), x_2(\bar{\boldsymbol{\theta}}_N)), \ldots, [x_{k-1}(\bar{\boldsymbol{\theta}}_N), \infty)$$

and let $\tilde{\boldsymbol{\mu}}_i = (\mu_{i1}, \ldots, \mu_{i,n-1})^T$, $\tilde{\boldsymbol{p}} = (p_1, \ldots, p_{k-1})^T$. If H_0 is true, then

$$\boldsymbol{\nu}_i = n_i^{-1/2}(\tilde{\boldsymbol{\mu}}_i - n_i\tilde{\boldsymbol{p}}), \quad i = 1, \ldots, r$$

is asymptotically normal as $N \to \infty$ with mean $\mathbf{0}_{k-1}$ and covariance matrix

$$\boldsymbol{\Xi}_i = \boldsymbol{D} - \tilde{\boldsymbol{p}}\tilde{\boldsymbol{p}}^T - \frac{n_i}{N}\boldsymbol{W}^T\boldsymbol{I}^{-1}\boldsymbol{W}$$

where $\boldsymbol{D}$ is a diagonal matrix with elements $p_1, \ldots, p_{k-1}$ on the diagonal and $\boldsymbol{W} = \boldsymbol{W}(\theta_0)$ is a $s \times (k-1)$ matrix with elements

$$w_{ij}(\boldsymbol{\theta}_0) = \int_{x_{j-1}(\boldsymbol{\theta}_0)}^{x_j(\boldsymbol{\theta}_0)} \frac{\partial}{\partial\theta_i} f_{\boldsymbol{\theta}}(x)\bigg|_{\boldsymbol{\theta}=\boldsymbol{\theta}_0} dx.$$

The vector $\boldsymbol{\nu} = (\boldsymbol{\nu}_1^T, \ldots, \boldsymbol{\nu}_r^T)^T$, as $\min n_i \to \infty$ is asymptotically normal with mean a $\mathbf{0}$-vector of length $r(k-1)$ and nondegenerate covariance matrix $\boldsymbol{U}$ consisting of blocks $\boldsymbol{U}_{ij}$, $i, j = 1, \ldots, r$, where

$$\boldsymbol{U}_{ij} = \begin{cases} \boldsymbol{\Xi}_i, & \text{if } i = j, \\ -\frac{\sqrt{n_i n_j}}{n} \boldsymbol{W}^T \boldsymbol{I}^{-1} \boldsymbol{W}, & \text{if } i \neq j. \end{cases}$$

If the inverse matrix $\boldsymbol{U}^{-1}$ exists then the quadratic form

$$Y^2(\boldsymbol{\theta}_0) = \boldsymbol{\nu}^T \boldsymbol{U}^{-1}(\boldsymbol{\theta}_0)\boldsymbol{\nu}$$

has, in the limit as $N \to \infty$, a chi-squared distribution with $r(k-1)$ degrees of freedom.

The quadratic form $Y^2(\boldsymbol{\theta}_0)$ cannot be used to test the hypothesis H_0 because it depends on $\boldsymbol{\theta}_0$. But by choice of the estimator $\bar{\boldsymbol{\theta}}_N$ we have that $Y^2(\bar{\boldsymbol{\theta}}_N)$ and $Y^2(\boldsymbol{\theta}_0)$ have asymptotically the same distribution as $N \to \infty$, so we can use $Y^2(\bar{\boldsymbol{\theta}}_N)$, which is asymptotically distributed like $\chi^2_{r(k-1)}$, as our test statistic.

Now suppose that H_1 holds; that is, not all $\boldsymbol{\theta}_i$ are the same. Let $\boldsymbol{\theta}^* = (n_1\boldsymbol{\theta}_1 + \cdots + n_r\boldsymbol{\theta}_r)/N$, and $c_{il} = (\boldsymbol{\theta}_i - \boldsymbol{\theta}^*)\boldsymbol{W}_l(\boldsymbol{\theta}^*)$, $\boldsymbol{W}_l(\boldsymbol{\theta}^*) = (w_{il}, \ldots, w_{sl})^T$, the lth column of matrix $\boldsymbol{W}$, $i = 1, \ldots, r$, $l = 1, \ldots, k$. Then, since $\bar{\boldsymbol{\theta}}_N - \boldsymbol{\theta}^* = o_p(1)$ under H_1,

$$\boldsymbol{P}\{x_{\ell-1}(\bar{\boldsymbol{\theta}}_N) < \xi_{ij} \leq x_\ell(\bar{\boldsymbol{\theta}}_N)\} = p_\ell + c_{i\ell} + o(1).$$

As in Section 10, we conclude that $Y^2(\bar{\boldsymbol{\theta}}_N)$ has a limiting noncentral chi-squared distribution with noncentrality parameter

$$\kappa(\boldsymbol{\theta}^*) = \sum_{i=1}^{r} \sum_{\ell=1}^{k} n_i c_{i\ell}^2 / p_\ell.$$

Location and Scale Families

Now we consider the problem of homogeneity of two samples in the case when the parametric family of distributions depends only on location and scale parameters.

Let us suppose that we have two independent samples

$$\xi_{11}, \xi_{12}, \ldots, \xi_{1n_1} \qquad \text{and} \qquad \xi_{21}, \xi_{22}, \ldots, \xi_{2n_2},$$

such that

$$\boldsymbol{P}\{\xi_{1j} \le x\} = G\Big[\frac{x - m_1}{\sigma_1}\Big] \quad \text{and} \quad \boldsymbol{P}\{\xi_{2j} \le x\} = G\Big[\frac{x - m_2}{\sigma_2}\Big], \quad x \in \boldsymbol{R}^1,$$

where $|m_i| < \infty$, $\sigma_i > 0$, m_i and σ_i are unknown. We suppose that

$$g(x) = G'(x) > 0,$$

for all $x \in \boldsymbol{R}^1$. We wish to test the hypothesis H_0 according to which

$$m_1 = m_2 \qquad \text{and} \qquad \sigma_1 = \sigma_2,$$

that is, under H_0,

$$\xi_{ij} \sim G\Big(\frac{x - m}{\sigma}\Big),$$

for some m and σ, $|m| < \infty$, $\sigma > 0$, for all i, j. Let $\hat{m}_n$ and $\hat{\sigma}_n$ be the maximum-likelihood estimators for m and σ obtained from all $N = n_1 + n_2$ data $\xi_{11}, \ldots, \xi_{1n_1}, \xi_{21}, \ldots, \xi_{2n_2}$ under H_0, that is,

$$L(\hat{m}_n, \hat{\sigma}_n) = \sup_{m,\sigma} \prod_{i=1}^{2} \prod_{j=1}^{n_i} g\Big(\frac{\xi_{ij} - m}{\sigma}\Big).$$

As before, we choose a number k of grouping intervals $k > 2$, a vector $\boldsymbol{p} = (p_1, \ldots, p_k)^T$ of probabilities $p_i > 0$, $p_1 + \cdots + p_k = 1$ and define the quantiles $x_0, x_1, \ldots, x_{k-1}, x_k$:

$$x_k = -x_0 = +\infty, \qquad x_i = G^{-1}(p_1 + \cdots + p_i), \qquad i = 1, 2, \ldots, k-1.$$

Let $\boldsymbol{\mu}_1 = (\mu_{11}, \ldots, \mu_{1k})^T$ and $\boldsymbol{\mu}_2 = (\mu_{21}, \ldots, \mu_{2k})^T$ be the vectors obtained by grouping the samples using the intervals

$$(\hat{m}_n + \hat{\sigma}_n x_{\ell-1}; \hat{m}_n + \hat{\sigma}_n x_\ell], \qquad \ell = 1, 2, \ldots, k.$$

If H_0 is true, then

$$\boldsymbol{\nu}_i = \frac{1}{\sqrt{n_i}}(\tilde{\boldsymbol{\mu}}_i - n_i \tilde{\boldsymbol{p}}), \qquad i = 1, 2,$$

as $\min n_i \to \infty$ is asymptotically normal with mean $\mathbf{0}_{k-1}$ and covariance matrix

$$\boldsymbol{\Sigma}_i = \mathbf{D} - \tilde{\mathbf{p}}\tilde{\mathbf{p}}^T - \frac{n_i}{n}\mathbf{W}^T \mathbf{I}^{-1}\mathbf{W}, \qquad i = 1, 2,$$

where

$$\mathbf{W}^T = \|\mathbf{u} \vdots \mathbf{v}\|, \qquad \mathbf{u} = (u_1, \dots, u_{k-1})^T, \qquad \mathbf{v} = (v_1, \dots, v_{k-1})^T,$$
$$u_i = g(x_i) - g(x_{i-1}), \qquad v_i = x_i g(x_i) - x_{i1} y(x_{i-1}), \qquad i = 1, \dots, k-1.$$

The elements of I_{ij} $(i, j = 1, 2)$ of the Fisher information matrix $\mathbf{I} = \|I_{ij}\|$ are

$$I_{11} = \int_{-\infty}^{\infty} \left[\frac{g'(x)}{g(x)}\right] g(x)\,dx,$$
$$I_{22} = \int_{-\infty}^{\infty} x^2 \left[\frac{g'(x)}{g(x)}\right]^2 g(x)\,dx - 1,$$
$$I_{12} = I_{21} = \int_{-\infty}^{\infty} x \left[\frac{g'(x)}{g(x)}\right]^2 g(x)\,dx.$$

One can see that the matrix $\boldsymbol{\Sigma}_i$ does not depend on the parameters and is nonsingular $(i = 1, 2)$. Under H_0, the vector $\boldsymbol{\nu} = (\boldsymbol{\nu}_1, \boldsymbol{\nu}_2)^T = (\nu_{11}, \dots, \nu_{1,k-1}, \nu_{21}, \dots, \nu_{2,k-1})^T$, as $\min n_i \to \infty$ is asymptotically normal with mean $\mathbf{0}_{2(k-1)}$ and covariance matrix

$$\mathbf{U} = \left\| \begin{matrix} \boldsymbol{\Sigma}_1 & -\frac{\sqrt{n_1 n_2}}{N}\mathbf{W}^T \mathbf{I}^{-1} W \\ -\frac{\sqrt{n_1 n_2}}{N}\mathbf{W}^T \mathbf{I}^{-1}\mathbf{W} & \boldsymbol{\Sigma}_2 \end{matrix} \right\|.$$

The matrix $\mathbf{U}$ is nonsingular and its inverse is

$$\mathbf{U}^{-1} = \left\| \begin{matrix} \mathbf{C} & \vdots & 0 \\ \cdots & & \cdots \\ 0 & \vdots & \mathbf{C} \end{matrix} \right\| + \frac{1}{N} \left\| \begin{matrix} n_1 \mathbf{M} & \sqrt{n_1 n_2}\mathbf{M} \\ \sqrt{n_1 n_2}\mathbf{M} & n_2 \mathbf{M} \end{matrix} \right\|,$$

where (see Sections 12 and 13)

$$\mathbf{C} = \mathbf{D}^{-1} + \frac{1}{p_k}\mathbf{1}_{k-1} \cdot \mathbf{1}_{k-1}^T,$$

$$\mathbf{M} = \frac{1}{\lambda_1\lambda_2 - \lambda_3^2}\left[\lambda_2 \mathbf{y}\mathbf{y}^T - \lambda_3(\mathbf{z}\mathbf{y}^T + \mathbf{y}\mathbf{z}^T) + \lambda_1 \mathbf{z}\mathbf{z}^T\right],$$

$$\boldsymbol{y} = \boldsymbol{D}^{-1}\boldsymbol{u} - \frac{u_k}{p_k}\mathbf{1}_{k-1}, \qquad \boldsymbol{z} = \boldsymbol{D}^{-1}\boldsymbol{v} - \frac{v_k}{p_k}\mathbf{1}_{k-1},$$

$$\lambda_1 = I_{11} - \sum_{i=1}^{k}\frac{u_i^2}{p_i}, \qquad \lambda_2 = I_{22} - \sum_{i=1}^{k}\frac{v_i^2}{p_i}, \qquad \lambda_3 = I_{12} - \sum_{i=1}^{k}\frac{u_i v_i}{p_i}.$$

Under H_0, the statistic $Y_2^2 = \boldsymbol{\nu}^T \boldsymbol{U}^{-1}\boldsymbol{\nu}$ has, in the limit as $\min n_i \to \infty$, a chi-squared distribution with $2(k-1)$ degrees of freedom:

$$\lim_{\min n_i \to \infty} \boldsymbol{P}\{Y_n^2 \le x | H_0\} = \boldsymbol{P}\{\chi^2_{2(k-1)} \le x\}.$$

The statistic Y_n^2 has the representation

$$\begin{aligned} Y_n^2 = {} & \sum_{i=1}^{k}\frac{(\mu_{1i} - n_1 p_i)^2}{n_1 p_i} + \sum_{i=1}^{k}\frac{(\mu_{2i} - n_2 p_i)^2}{n_2 p_i} \\ & + \frac{1}{N(\lambda_1\lambda_2 - \lambda_3^2)}\left[\lambda_1\Big(\sum_{i=1}^{k}\frac{v_i\mu_i}{p_i}\Big)^2 + \lambda_2\Big(\sum_{i=1}^{k}\frac{u_i\mu_i}{p_i}\Big)^2 \right. \\ & \left. - 2\lambda_3\Big(\sum_{k=1}^{k}\frac{v_i\mu_i}{p_i}\Big)\Big(\sum_{i=1}^{k}\frac{u_i\mu_i}{p_i}\Big)\right], \end{aligned}$$

where $\mu_i = \mu_{1i} + \mu_{2i}$, $i = 1, \dots, k$.

Normal Families

If there exists a sufficient statistic, then it is reasonable to use other methods for testing H_0, since we have a parametric problem. As an example, let us consider the problem of two samples for normal families, that is, we suppose that $G(x) = \Phi(x)$, where $\Phi(x)$ is the distribution function of the standard normal law. Let $\Phi[(x - m_1)/\sigma_1]$ and $\Phi[(x - m_2)/\sigma_2]$ be the distribution functions of ξ_{1j} and ξ_{2j}, respectively. In this case, we have a sufficient statistic $\boldsymbol{T} = (\bar{\xi}_1, \bar{\xi}_2, s_1^2, s_2^2)^T$, where

$$\bar{\xi}_i = \frac{1}{n_i}\sum_{j=1}^{n_i}\xi_{ij}, \qquad s_i^2 = \frac{1}{f_i}\sum_{i=1}^{n_i}(\xi_{ij} - \bar{\xi}_i)^2, \qquad f_i = n_i - 1,$$

are the best unbiased estimators for m_i and σ_i^2, $i = 1, 2$. The components $\bar{\xi}_1, \bar{\xi}_2, s_1^2, s_i^2$ of the sufficient statistic $\boldsymbol{T}$ are independent and

$$\bar{\xi}_i = m_i + \frac{1}{n_i}\zeta_i\sigma_i, \qquad s_i^2 = \frac{1}{f_i}\sigma_i^2\chi_{f_i}^2, \qquad i = 1, 2,$$

where ζ_1 and ζ_2 are standard normal random variables and $\zeta_1, \zeta_2, \chi_{f_1}^2, \chi_{f_2}^2$ are independent.

Let

$$a_i = \sqrt{\frac{2}{(f_i}}\,\frac{\Gamma\Big((f_i + 1)/2\Big)}{\Gamma\Big(f_i/2\Big)}, \qquad i = 1, 2,$$

and consider the statistic

$$Z^2 = \frac{(f_1 + f_2)(f_1 + f_2 - 2)}{(f_1 + f_2 - 1)(f_1 s_1^2 + f_2 s_2)} - \left[\frac{(\bar{\xi}_1 - \bar{\xi}_2)^2}{\frac{1}{n_1} + \frac{1}{n_2}} + \frac{\left(\frac{s_1}{a_1} - \frac{s_2}{a_2}\right)^2}{\frac{1}{a_1^2} + \frac{1}{a_2^2} - 2}\right].$$

If H_0 is true, that is, if $m_1 = m_2$ and $\sigma_1 = \sigma_2$, then

$$Z^2 = \frac{f_1 + f_2 - 2}{f_1 + f_2 - 1}\left[t^2 + \frac{f_1 + f_2}{\frac{1}{a_1^2} + \frac{1}{a_2^2} - 2}\left(\frac{1}{a_1\sqrt{f_1}}\sqrt{\beta} - \frac{1}{a_2\sqrt{f_2}}\sqrt{1 - \beta}\right)^2\right],$$

where t and β are independent random variables, t has a Student distribution with $f_1 + f_2$ degress of freedom and β has a beta distribution with parameters $f_1/2$ and $f_2/2$, from which it follows that

$$\lim_{\min n_i \to \infty} \boldsymbol{P}\{Z^2 \leq x | H_0\} = \boldsymbol{P}\{\chi_2^2 \leq x\}.$$

The power function of the test based on the statistic Z^2 for a wide class of close alternatives (see, for example, Sections 9 to 13) can be expressed in terms of a noncentral chi-squared distribution with two degrees of freedom and the parameter of noncentrality

$$\lambda = \frac{(f_1 + f_2)(f_1 + f_2 - 2)}{(f_1 + f_2 - 1)(f_1\sigma_1^2 + f_2\sigma_2^2)}\left[\frac{(m_1 - m_2)^2}{\frac{1}{n_1} + \frac{1}{n_2}} + \frac{(\sigma_1 - \sigma_2)^2}{\frac{1}{a_1^2} + \frac{1}{a_2^2} - 2}\right].$$

23. A TEST FOR A PARAMETER DIFFERENCE AMONG BINOMIAL DISTRIBUTIONS

A Problem of Comparison with a Control Group

A laboratory is running tests on k groups of mice. One group, say the kth, is the control group that is subjected to no treatment or a treatment with known effect. Each mouse is observed to be in one of two states: healthy or unhealthy. In group k, the probability of "success," that is, a healthy mouse, is known and equal to p_k. The probability of success for a mouse in group i, p_i, $i = 1, \ldots, k-1$, is not known. We wish to test the hypothesis that all p_i, or some subset of them, are the same and equal to p_k.

To model this situation, we let $\mu_1, \ldots, \mu_k$ be independent, binomially distributed random variables, where μ_i represents the number of healthy mice in group i,

$$\boldsymbol{P}\{\mu_i = m\} = \binom{n_i}{m} p_i^m (1-p_i)^{n_i - m}, \quad m = 0, \ldots, n_i; \quad i = 1, \ldots, k,$$

The probabilities $p_1, \ldots, p_{k-1}$ are unknown elements of (0,1).

Construction of a Test Statistic

One could use various tests. Let us show how one can use chi-squared technique. As in Section 22, the test is for homogeneity. However there are several important differences between the two situations, for example, here there is no binning of data, and one group is viewed differently. We will push further in this example and decide in which groups inhomogeneity lies.

We put $\xi_i = \mu_i / n_i$, $i = 1, \ldots, k$. If $(x_i - p_i)(n_i / p_i q_i)^{1/2} = O(1)$ as $n_i \to \infty$, $q_i = 1 - p_i$, by the de Moivre-Laplace theorem

$$\boldsymbol{P}\{\xi_1 \le x_1, \ldots, \xi_k \le x_k\} \sim \prod_{i=1}^{k} \Phi\left(n_i^{1/2} \frac{x_i - p_i}{(p_i q_i)^{1/2}}\right),$$

where $\Phi(x)$ is the standard normal distribution function. We put

$\eta_i = \xi_i - \xi_k$, $i = 1, \ldots, k-1$, and let

$$\Delta_i = \boldsymbol{E}\eta_i = p_i - p_k, \quad \sigma_i^2 = \frac{1}{n_i} p_i q_i, \quad i = 1, \ldots, k.$$

The random vector $\boldsymbol{\eta} = (\eta_1, \ldots, \eta_{k-1})^T$ is asymptotically normally distributed as $\min(n_1, \ldots, n_k) \to \infty$, with parameters

$$\boldsymbol{E}\boldsymbol{\eta} = \boldsymbol{\Delta} = (\Delta_1, \ldots, \Delta_{k-1})^T, \quad \boldsymbol{E}(\boldsymbol{\eta} - \boldsymbol{\Delta})(\boldsymbol{\eta} - \boldsymbol{\Delta})^T = \boldsymbol{\Sigma},$$

where

$$\boldsymbol{\Sigma} = \begin{bmatrix} \sigma_1^2 & 0 & \ldots & 0 \\ 0 & \sigma_2^2 & \ldots & 0 \\ & & \ddots & \vdots \\ 0 & & \ldots & \sigma_{k-1}^2 \end{bmatrix} + \sigma_k^2 \mathbf{1}_{k-1} \mathbf{1}_{k-1}^T,$$

$\mathbf{1}_{k-1} = (1, \ldots, 1)^T$.

The matrix $\boldsymbol{\Sigma}$ is nonsingular and

$$\boldsymbol{\Sigma}^{-1} = \begin{bmatrix} \frac{1}{\sigma_1^2} & 0 & \ldots & 0 \\ 0 & \frac{1}{\sigma_2^2} & & 0 \\ \vdots & & \ddots & \\ 0 & 0 & \ldots & \frac{1}{\sigma_{k-1}^2} \end{bmatrix} - \frac{1}{\sum_{i=1}^{k} \sigma_i^2} \begin{bmatrix} \frac{1}{\sigma_1^2} & \frac{1}{\sigma_1\sigma_2} & \ldots & \frac{1}{\sigma_1\sigma_{k-1}} \\ \frac{1}{\sigma_2\sigma_1} & \frac{1}{\sigma_2^2} & \ldots & \frac{1}{\sigma_2\sigma_{k-1}} \\ \vdots & & & \\ \frac{1}{\sigma_{k-1}\sigma_1} & \frac{1}{\sigma_{k-1}\sigma_2} & \ldots & \frac{1}{\sigma_{k-1}^2} \end{bmatrix}$$

From the asymptotic normality of the vector $\boldsymbol{\eta}$ it follows (see Chapter 1) that the quadratic form

$$Y^2 = (\boldsymbol{\eta} - \boldsymbol{\Delta})^T \boldsymbol{\Sigma}^{-1} (\boldsymbol{\eta} - \boldsymbol{\Delta}) \tag{23.1}$$

has, asymptotically as $\min(n_1, \ldots, n_k) \to \infty$, a chi-squared distribution with $k-1$ degrees of freedom.

The quadratic form (23.1) can be written as

$$Y^2 = \sum_{i=1}^{k-1} \left(\frac{\eta_i - \Delta_i}{\sigma_i} \right)^2 - \left(\sum_{i=1}^{k} \frac{1}{\sigma_i^2} \right)^{-1} \left[\sum_{i=1}^{k-1} \frac{\eta_i - \Delta_i}{\sigma_i^2} \right]. \tag{23.2}$$

It follows from the general theory in Section 9 that the limiting distribution of the quadratic form Y^2 does not change if the unknown parameters σ_i^2 are replaced by the unbiased estimators

$$\hat{\sigma}_i^2 = \frac{\xi_i(1-\xi_i)}{n_i - 1}, \qquad i = 1, \ldots, k.$$

Statistical Analysis when the Null Hypothesis is Rejected

Let α be a given level, $0 < \alpha < 1/2$, and let $\tilde{c}_\alpha$ be the upper α quantile of a chi-squared distribution with $k-1$ degrees of freedom, that is,

$$\boldsymbol{P}\{\chi_{k-1}^2 \geq \tilde{c}_\alpha\} = \alpha, \qquad \tilde{c}_\alpha = \chi_{k-1,\alpha}^2$$

as in Eq. (1.18). Then the probability $\boldsymbol{P}\{Y^2 \leq \tilde{c}_\alpha\}$ is approximately equal to $1-\alpha$ and the set of values of the vector $\boldsymbol{\Delta}$ for which Y^2 satisfies the inequality $Y^2 \leq \tilde{c}_\alpha$ form a confidence region whose confidence level is near $1-\alpha$.

Let I denote the collection of all ordered subsets of the set $\{1, \ldots, k-1\}$ and let $\boldsymbol{I} = \{i_1, \ldots, i_r\}$, $1 \leq i_1 < \cdots < i_r \leq k-1$, denote an element of I. We consider the set of hypotheses of the form

$$H_{\boldsymbol{I}} : \Delta_{i_1} = \Delta_{i_2} = \cdots = \Delta_{i_r} = 0, \quad \boldsymbol{I} = (i_1, \ldots, i_r) \in I.$$

We say that the hypothesis $H_{\boldsymbol{I}}$ is *not contradicted* by a given experiment if, in the $k-1$ dimensional space of vectors $\boldsymbol{\Delta}$, the hyperplane defined by $\Delta_{i_1} = \Delta_{i_2} = \cdots = \Delta_{i_r} = 0$ has a nonempty intersection with the interior of the ellipsoid defined by the inequality $Y^2 \leq \tilde{c}_\alpha$.

The final step is to identify the collection of hypotheses $H_{\boldsymbol{I}}$ that are not contradicted by the given experiment. Some of the hypotheses are consequences of others; that is, if the set $\boldsymbol{J} \subset \boldsymbol{I}$, then $H_{\boldsymbol{I}}$ implies $H_{\boldsymbol{J}}$. Hence, we need list only those $H_{\boldsymbol{I}}$ for which the set $\boldsymbol{I} = (i_1, \ldots, i_r)$ is maximal. Let us consider the hypothesis $H_{\boldsymbol{I}}$, where $\boldsymbol{I} = (k-r, k-r+1, \ldots, k-1)$, the string of r adjacent integers ending with $k-1$. Under the hypothesis, the quadratic form (23.2) is

$$\begin{aligned}Y^2 = &\sum_{i=1}^{k-r-1}\left(\frac{\eta_i - \Delta_i}{\sigma_i}\right)^2 + \sum_{i=k-r}^{k-1}\left(\frac{\eta_i}{\sigma_i}\right)^2 \\ &- \left[\sum_{j=1}^{k-r-1} c_j(\eta_j - \Delta_j) + \sum_{j=k-r}^{k-1} c_j\eta_j\right]^2,\end{aligned} \tag{23.3}$$

where

$$c_j = \sigma_j^{-2}\left(\sum_{i=1}^{k}\sigma_j^2\right)^{-1/2}, \quad j = 1, \ldots, k-1.$$

One can see that the smallest value of the statistic obtained from Eq. (23.3) by replacing each σ_i by $\hat{\sigma}_i$ is attained at the point

$$\Delta_i = \Delta_i^* = \eta_i - c_i\hat{\sigma}_i^2\left(1 - \sum_{j=1}^{k-r-1} c_j^2\hat{\sigma}_j^2\right)^{-1}\sum_{j=k-r}^{k-1} c_j\eta_j, \quad i = 1, \ldots, k-r-1$$

and this minimum value is

$$Y_*^2 = \sum_{i=k-r}^{k-1}(\eta_i/\hat{\sigma}_i)^2 - \left(\sum_{i=k-r}^{k-1} c_i\eta_i\right)^2\left(1 + \sum_{i=1}^{k-r-1} c_i^2\hat{\sigma}_i^2\right)^{-1}. \tag{23.4}$$

If $r = k-1$, the last factor should be 1. The hypothesis H_I should be rejected if $Y_*^2 \geq \tilde{c}_\alpha$.

Numerical Example

Let us consider a numerical example. Let $k = 4, n_1 = n_2 = n_3 = n_4 = 100$ and suppose the observed values of μ_i, $i = 1, \ldots, 4$ are $\mu_1 = 20$, $\mu_2 = 50$, $\mu_3 = 60$, $\mu_4 = 40$. Then $\xi_1 = 0.2$, $\xi_2 = 0.5$, $\xi_3 = 0.6$, $\xi_4 = 0.4$, $\eta_1 = -0.2$, $\eta_2 = 0.1$, $\eta_3 = 0.2$. Suppose we estimate σ_i^2 by $\hat{\sigma}_i^2 = \xi_i(1-\xi_i)/n_i$, $i = 1, 2, 3, 4$. Then $\hat{\sigma}_1^2 = 0.0016$, $\hat{\sigma}_2^2 = 0.0025$, $\hat{\sigma}_3^2 = 0.0024$, $\hat{\sigma}_4^2 = 0.0024$, and $\sum_{i=1}^{n} 1/\hat{\sigma}_i^2 = 22,300/12$, so that the c_j appearing in Eq. (23.3) are $c_1 = 125(3/223)^{1/2}$, $c_2 = 80(3/223)^{1/2}$, $c_3 = (250/3)(3/223)^{1/2}$. The statistic Y^2 has, approximately, a χ_3^2 distribution and for $\alpha = 0.05$, $\tilde{c}_\alpha = 7.815$.

Consider $H_{1,2,3}$. Y_*^2 of Eq. (23.4) has the value $Y_*^2 = 45.665$. Since $P\{\chi_3^2 \geq 45.665) < 10^{-7}$, this hypothesis will be rejected by any chi-squared criterion whose level is no less than 10^{-7}.

For $H_{2,3}$, $Y_*^2 = 14.541$. Since $P\{\chi_3^2 \geq 14.541\} = 0.00225$, this hypothesis will be rejected if the level is no less than 0.00225.

For $H_{1,3}$, $Y_*^2 = 40.898$. Again $P\{\chi_3^2 > 40.898\} < 10^{-7}$, so $H_{1,3}$ will be rejected if $\alpha \geq 10^{-7}$.

For $H_{1,2}$, $Y_*^2 = 25.824 > \tilde{c}_{.05} = 7.815$, so $H_{(1,2)}$ is rejected if $\alpha = 0.05$.

For $H_{(1)}$, $Y_*^2 = 19.159$, again more than 7.815, so we reject $H_{(1)}$ if $\alpha = 0.05$.

For $H_{(2)}$, $Y_*^2 = 3.448$ which is smaller than $\tilde{c}_{0.05} = 7.815$, so $H_{(2)}$ is not rejected; we obtain the estimators $\Delta_1^* = -0.249$, $\Delta_3^* = 0.151$ for $p_1 - p_4$ and $p_3 - p_4$.

For $H_{(3)}$, $Y_*^2 = 14.258$ so $H_{(3)}$ is rejected at level $\alpha = 0.05$.

Our conclusion is that only hypothesis $H_{(2)}$ agrees with the experimental results, that is, $p_2 = p_4$. If this is correct, then it is reasonable to estimate p_4 by $(\mu_2 + \mu_4)/(n_2 + n_4)$, which in this example equals 0.45. Using the above estimates Δ_1^*, Δ_3^* we obtain the estimates 0.201 for p_1 and 0.601 for p_3.

Remark. To follow this approach in the general case, we must test

$$\sum_{r=1}^{k-1} \binom{k-1}{r} = 2^{k-1} - 1$$

hypotheses. In practice, however, one can restrict oneself to no more than $k-1$ tests. To do this it is necessary to compute the ratios $\eta_i^2/\hat{\sigma}_i^2$, $i = 1, 2, \ldots, k-1$, and order them:

$$\left(\frac{\eta_1}{\hat{\sigma}_1}\right)^2 \geq \left(\frac{\eta_2}{\hat{\sigma}_2}\right)^2 \geq \cdots \geq \left(\frac{\eta_{k-1}}{\hat{\sigma}_{k-1}}\right)^2$$

(the numbering can be assigned after the ordering). Then one successively tests the hypotheses

$$H_r = H_{(k-r, k-r+1, \ldots, k-1)}, \qquad r = k-1, k-2, \ldots .$$

If, at any step, one finds that

$$\left(\frac{\eta_m}{\hat{\sigma}_m}\right)^2 > \left(\frac{\eta_{m+1}}{\hat{\sigma}_{m+1}}\right)^2 = \cdots = \left(\frac{\eta_{m+t}}{\hat{\sigma}_{m+t}}\right)^2 > \left(\frac{\eta_{m+t+1}}{\sigma_{m+t+1}}\right)^2$$

and the hypothesis $H_{(m,m+1,\ldots,k-1)}$ has been rejected, then next one tests the hypothesis

$$H_{(m+t,m+t+1,\ldots,k-1)} \qquad \text{and not} \qquad H_{(m+1,m+2,\ldots,k-1)}.$$

24. STATISTICAL EVALUATION OF RATINGS IN SPORTS COMPETITIONS SUCH AS SKATING OR GYMNASTICS

The Problem

We consider the problem of evaluating the performance of a committee of judges whose task is to rate a list of contestants in an athletic event, let's say a figure skating competition. Suppose there are J judges and I competing skaters. Each judge indicates his rating of each contestant independently, always using the same scale. A higher rating indicates a better performance. Let x_{ij} denote the rating of skater i by judge j, $i = 1, \ldots, I$; $j = 1, \ldots, J$. In our model the rating x_{ij} is a realization of the random variable

$$X_{ij} = a_i + b_j + Y_{ij} \tag{24.1}$$

where a_i is the "true" competence of skater i, b_j is the "systematic error" or "judge effect" of judge j, and Y_{ij} is a "random error." We assume that the random errors Y_{ij} are independent with the same normal law, $N(0, \sigma^2)$, where σ^2 is unknown, and that the a_is and b_js are unknown constants. One would like to order the contestants according to the true values of the a_i and to evaluate the judges by ordering them by their absolute sysematic errors, $|b_j|$. These numbers are not available, but there are several statistical questions here. One might argue that the final interpretation of performances should be based on a statistical analysis of the ratings. However, in practice competing sportsmen are usually ordered according to the ratings x_{ij}.

If we measure the systematic error of each judge relative to the whole group of judges, we have the additional condition on the b_js,

$b_1 + \cdots + b_J = 0$, and our model can be written

$$\begin{aligned} X_{ij} &= a_i + b_j + Y_{ij}, \quad i = 1, 2, \ldots, I; \quad j = 1, 2, \ldots, J-1, \\ X_{iJ} &= a_i - b_1 - b_2 - \cdots - b_{J-1} + Y_{iJ}. \end{aligned} \tag{24.2}$$

The best skater corresponds to the largest value of a_i. We define the "best" judge to be the one with the smallest value of $|b_j|$.

The questions we will study are: how to detect nonobjectivity on the part of some judge, and to correct for this; how to test for equality of subsets of the a_is and of the b_js; how to test for uniform precision among the judges. The essential ideas in this section come from Bolshev (1969b). For a more detailed version see Nikulin (1984a).

Gauss-Markov Regression Model

The model (24.2) is a linear Gauss-Markov regression model. It can be written as

$$\boldsymbol{X} = \boldsymbol{A}\boldsymbol{\Theta} + \boldsymbol{Y}$$

where

$$\begin{aligned} \boldsymbol{X} &= (X_{11}, \ldots, X_{1J}, X_{21}, \ldots, X_{2J}, \ldots, X_{I1}, \ldots, X_{IJ})^T, \\ \boldsymbol{Y} &= (Y_{11}, \ldots, Y_{1J}, Y_{21}, \ldots, Y_{2J}, \ldots, Y_{I1}, \ldots, Y_{IJ})^T, \\ \boldsymbol{\Theta} &= (a_1, \ldots a_I, b_1, \ldots b_{J-1})^T, \end{aligned} \tag{24.3}$$

and $\boldsymbol{A}$ is the "matrix of regression coefficients" of dimension $IJ \times (I+J-1)$, which we avoid writing explicitly here. The best unbiased estimator of $\boldsymbol{\Theta}$ is the least squares estimator, with the familiar form

$$\widehat{\boldsymbol{\Theta}} = (\boldsymbol{A}^T \boldsymbol{A})^{-1} \boldsymbol{A}^T \boldsymbol{X}.$$

With patience, one can work out $\widehat{\boldsymbol{\Theta}}$ in terms of $\boldsymbol{X}$. Since $\widehat{\boldsymbol{\theta}}$ is a linear transformation of $\boldsymbol{X}$, it is normally distributed and, from Eq. (24.3), one has

$$\hat{a}_i = \frac{1}{J} \sum_{j=1}^{J} X_{ij}, \qquad \hat{b}_j = \frac{1}{I} \sum_{i=1}^{I} X_{ij} - \frac{1}{IJ} \sum_{i=1}^{I} \sum_{j=1}^{J} X_{ij}. \tag{24.4}$$

Also

$$\boldsymbol{E}\hat{a}_i = a_i, \qquad \textbf{Var } \hat{a}_i = \frac{\sigma^2}{J}, \qquad i = 1, \ldots, I$$
$$\textbf{cov } (\hat{a}_i, \hat{a}_j) = \boldsymbol{E}(\hat{a}_i - a_i)(\hat{a}_j - a_j) = 0, \qquad i \neq j,$$

so that $\hat{a}_1, \ldots, \hat{a}_I$ are independent. Furthermore,

$$\boldsymbol{E}\hat{b}_j = b_j, \qquad \textbf{Var } \hat{b}_j = \frac{J-1}{JI}\sigma^2, \qquad j = 1, \ldots, J$$
$$\textbf{cov } (\hat{b}_i, \hat{b}_j) = -\frac{\sigma^2}{IJ}, \qquad i \neq j,$$

and for each i, j

$$\textbf{cov } (\hat{a}_i, \hat{b}_j) = 0,$$

so the vectors $\hat{\boldsymbol{a}} = (\hat{a}, \ldots, \hat{a}_I)$, $\hat{\boldsymbol{b}} = (\hat{b}_1, \ldots, \hat{b}_J)$ are independent, and

$$\hat{\boldsymbol{a}} \sim N_I(\boldsymbol{a}, \frac{\sigma^2}{J}\boldsymbol{I}_I), \qquad \hat{\boldsymbol{b}} \sim N_{J-1}(\boldsymbol{b}, \frac{\sigma^2}{IJ}\boldsymbol{\Sigma}). \tag{24.5}$$

Let $\boldsymbol{\Delta}_{ij}$ denote the vector of apparent errors

$$\delta_{ij} = X_{ij} - \hat{a}_i - \hat{b}_j, \qquad \boldsymbol{\Delta} = (\delta_{11}, \ldots, \delta_{IJ})^T.$$

The statistic $S^2 = \boldsymbol{\Delta}^T\boldsymbol{\Delta}$ together with $\hat{\boldsymbol{a}}, \hat{\boldsymbol{b}}$ is a sufficient set of statistics for our model. Under our assumptions, the vector $\boldsymbol{\Delta}$ has an IJ-dimensional normal law, which is degenerate of order $(I-1)(J-1)$, and $\boldsymbol{E}\delta_{ij} = 0$, all i, j;

$$\boldsymbol{E}\delta_{ij}\delta_{rs} = \begin{cases} (1 - \frac{I+J-1}{IJ})\sigma^2, & i = r,\ j = s, \\ -\frac{1}{J}(1 - \frac{1}{I})\sigma^2, & i = r,\ j \neq s, \\ -\frac{1}{I}(1 - \frac{1}{J})\sigma^2, & i \neq r,\ j = s, \\ \frac{1}{IJ}\sigma^2, & i \neq r,\ j \neq s. \end{cases}$$

The law of the statistic S^2 is

$$\sigma^2\chi^2_{(I-1)(J-1)},$$

Test for a Nonobjective Rating

Let

$$\Lambda = \frac{IJ}{(J-1)(I-1)} \frac{\max_{ij} \delta_{ij}}{S^2}.$$

Bolshev (1969b) showed that for $z \geq 0$,

$$\boldsymbol{P}\{\Lambda \geq z\} \leq IJ\{1 - I_z(\frac{1}{2}, \frac{1}{2}(IJ - I - J))\},$$

where $I_z(x, y)$ denotes Euler's incomplete beta function.

Let i_0, j_0 be an index i, j where $\max_{ij} \delta_{ij}^2$ is attained. We shall regard the estimate $X_{i_0 j_0}$ as *nonobjective* if $\Lambda \geq z$ where $z = z(\alpha)$ is the solution of the equation

$$1 - \frac{\alpha}{IJ} = I_z(\frac{1}{2}, \frac{1}{2}(IJ - I - J)),$$

where α has a prescribed significance level, $0 < \alpha < 1$. If the estimate $X_{i_0 j_0}$ is regarded as nonobjective, the estimates (24.2) of a_i and b_j should be corrected as follows:

$$\hat{\hat{a}}_i = \hat{a}_i \text{ if } i \neq i_0, \qquad \hat{\hat{a}}_{i_0} = \hat{a}_{i_0} - \frac{I}{(I-1)(J-1)} \delta_{i_0 j_0};$$

$$\hat{\hat{b}}_j = \hat{b}_j + \frac{\delta_{i_0 j_0}}{(I-1)(J-1)}, \text{ if } j \neq j_0, \qquad \hat{\hat{b}}_{j_0} = \hat{b}_{j_0} - \frac{1}{I-1} \delta_{i_0 j_0}.$$

The rating $X_{i_0 j_0}$ should be replaced by

$$\hat{\hat{X}}_{i_0 j_0} = X_{i_0 j_0} + \frac{IJ}{(I-1)(J-1)} (\hat{a}_{i_0} + \hat{b}_{j_0} - X_{i_0 j_0}).$$

More details are in the paper of Bolshev (1969b). Mirvaliev (1978) carried the computations further to include the case of two or more nonobjective observations, but recommended that one keep in mind that the linear model may be inadequate.

Tests for Indistinguishable Contestants and Judges

Next we construct tests of the hypotheses that certain subsets of the a_is are equal, meaning that the contestants' performance is indistinguishable in this model from the data. From Eq. (24.5), we have the following exact distributional relations:

$$\frac{J}{\sigma^2}\sum_{i=1}^{I}(\hat{a}_i - a_i)^2 \sim \chi_I^2, \tag{24.6}$$

$$\frac{IJ}{\sigma^2}(\hat{\boldsymbol{b}} - \boldsymbol{b})^T \boldsymbol{\Sigma}^{-1}(\hat{\boldsymbol{b}} - \boldsymbol{b}) \sim \chi_{J-1}^2, \tag{24.7}$$

and hence also

$$\frac{IJ}{\sigma^2}\sum_{j=1}^{J}(\hat{b}_j - b_j)^2 \sim \chi_{J-1}^2. \tag{24.8}$$

We recall from Eq. (24.3) that

$$\frac{S^2}{\sigma^2} \sim \chi_{(I-1)(J-1)}^2 \tag{24.9}$$

From Eqs. (24.5) and (24.9) we have

$$J(I-1)(J-1)\sum_{i=1}^{I}\frac{(\hat{a}_i - a_i)^2}{IS^2} \sim F_{I,(I-1)(J-1)}, \tag{24.10}$$

and

$$I(I-1)\sum_{j=1}^{J}\frac{(\hat{b}_j - b_j)^2}{S^2} \sim F_{J-1,(I-1)(J-1)}, \tag{24.11}$$

where $F_{p,q}$ is the F distribution with p and q degrees of freedom (see, e.g., Kendall and Stuart, 1963). Relations (24.10) and (24.11) allow us to construct confidence ellipsoids for the vectors $\boldsymbol{a}$ and $\boldsymbol{b}$.

Let $\boldsymbol{I} = (i_1, \ldots, i_k)$, $2 \leq k \leq I$, denote a subset of the integers $(1, \ldots, I)$, and let $\boldsymbol{a}(\boldsymbol{I})$ denote the corresponding subset $\boldsymbol{a}(\boldsymbol{I}) = (a_{i_1}, \ldots, a_{i_k})$ of the a_is. Suppose we wish to test the hypothesis $H_{\boldsymbol{a}(\boldsymbol{I})}$ that the elements of the set $\boldsymbol{a}(\boldsymbol{I})$ are equal,

$$H_{\boldsymbol{a}(\boldsymbol{I})} : a_{i_1} = \cdots = a_{i_k}.$$

Statement (24.6) with the sum restricted to the set $\boldsymbol{I}$ is

$$\frac{J}{\sigma^2}\sum_{i\in \boldsymbol{I}}(\hat{a}_i - a_i)^2 \sim \chi_k^2.$$

We use the notation

$$X_{\boldsymbol{I}}^2 = \min_{\alpha} \sum_{i\in \boldsymbol{I}}(\hat{a}_i - \alpha)^2.$$

It is easy to see that in fact

$$X_{\boldsymbol{I}}^2 = \sum_{i\in \boldsymbol{I}}\Big(\hat{a}_i - \frac{1}{k}\sum_{i\in \boldsymbol{I}} \hat{a}_i\Big)^2. \tag{24.12}$$

By the same kind of argument used in Section 9, if $H_{\boldsymbol{a}(\boldsymbol{I})}$ is true then

$$\frac{J}{\sigma^2}X_{\boldsymbol{I}}^2 \sim \chi_{k-1}^2.$$

Consequently, under $H_{\boldsymbol{a}(\boldsymbol{I})}$ as in Eq. (24.10) we have

$$\frac{(I-1)(J-1)X_{\boldsymbol{I}}^2}{(k-1)S^2} \sim F_{k-1,(I-1)(J-1)}. \tag{24.13}$$

A test of level α, $0 < \alpha < 0.5$, based on the statistic (24.12), will reject $H_{\boldsymbol{a}(\boldsymbol{I})}$ if

$$X_{\boldsymbol{I}}^2 > \frac{(k-1)}{J(J-1)(I-1)}S^2 F_{k-1,(I-1)(J-1q)}(\alpha).$$

Similarly, to test the hypothesis $H_{\boldsymbol{b}(\boldsymbol{I})}$,

$$H_{\boldsymbol{b}(\boldsymbol{I})} : b_{i_1} = \cdots = b_{i_k},$$

where now $z \leq k \leq J-1$. We use the information from Eq. (24.5) that the vector of estimates $\hat{\boldsymbol{b}}(\boldsymbol{I}) = (\hat{b}_i, \ldots, \hat{b}_{i_k})$ is normally distributed with

$$\boldsymbol{E}\hat{\boldsymbol{b}}(\boldsymbol{I}) = \boldsymbol{b}(\boldsymbol{I})$$

and covariance matrix

$$\frac{\sigma^2}{IJ}(J\boldsymbol{I}_k - \mathbf{1}_k \cdot \mathbf{1}_k^T)$$

with inverse

$$\frac{I}{\sigma^2}(\mathbf{I}_k + \frac{1}{J-k}\mathbf{1}_k \cdot \mathbf{1}_k^T).$$

It follows that

$$\frac{I}{\sigma^2}\sum_{i\in\mathbf{I}}(\hat{b}_i - b_i)^2 + \frac{I}{\sigma^2(J-k)}\left[\sum_{i\in\mathbf{I}}(\hat{b}_i - b_i)\right]^2 \sim \chi_k^2.$$

We use the notation

$$Y_{\mathbf{I}}^2 = \min_{\beta}\left\{\sum_{i\in\mathbf{I}}(\hat{b}_i - \beta)^2 + \frac{1}{J-k}\left[\sum_{i\in\mathbf{I}}(\hat{b}_i - \beta)\right]^2\right\}. \tag{24.14}$$

As with Eq. (24.12), one can check that

$$Y_{\mathbf{I}}^2 = \sum_{i\in\mathbf{I}}(\hat{b}_i - \beta^*)^2 + \frac{1}{J-k}\left[\sum_{i\in\mathbf{I}}(\hat{b}_i - \beta^*)\right]^2,$$

where $\beta^* = (1/k)\sum_{i\in\mathbf{I}}\hat{\beta}_i$. Again, by the method of Section 9, if $H_{\mathbf{b}(\mathbf{I})}$ is true, then

$$\frac{I}{\sigma^2}Y_{\mathbf{I}}^2 \sim \chi_{k-1}^2,$$

and, similarly to Eq. (24.13), we have

$$\frac{I(I-1)(J-1)Y_{\mathbf{I}}^2}{(k-1)S^2} \sim F_{k-1,(I-1)(J-1)}. \tag{24.15}$$

A test at level α, $0 < \alpha < 0.5$, based on the statistic (24.15) will reject $H_{\mathbf{b}(\mathbf{I})}$ if

$$Y_{\mathbf{I}}^2 > \frac{k-1}{I(I-1)(J-1)}S^2 F_{k-1,(I-1)(J-1)}(\alpha).$$

Test for a Less Precise Judge

Finally, we describe a test of the hypothesis, H_{σ^2}, that all the Y_{ij} appearing in Eq. (24.1) have the same unknown variance, σ^2, against the alternative hypothesis, H, that for some one unknown $j = j_0$, Y_{ij_0} has a larger variance, that is,

$$H_{\sigma^2} : \mathbf{Var}\ Y_{ij} = \sigma^2, \quad i = 1, \ldots, I; \quad j = 1, \ldots, J,$$

$$H : \begin{cases} \mathbf{Var}\ Y_{i,j_0} > \sigma^2, & i = 1, \ldots, I, \\ \mathbf{Var}\ Y_{ij} = \sigma^2, & i = 1, \ldots, I, \quad j \neq j_0, \end{cases}$$

If H_{σ^2} is rejected when tested against H, we will conclude that one of the judges is less precise than the others in the sense that his judgements are more variable. In addition, we will be able to identify the index, j_0 of one such judge. It is possible that this judge has produced gross errors and there are outlying observations. Acceptance of the hypothesis H_{σ^2} against the alternative H affirms our belief on the basis of the data that the members of the committee of experts have the same qualification in the sense of variability.

If H_{σ^2} is true, the error of X_{ij} as an estimate of a_i is

$$Y_{ij} = X_{ij} - a_i - b_j,$$

where $Y_{ij} \sim N_1(0, \sigma^2)$, whereas the "apparent error" of the estimate X_{ij} is

$$\delta_{ij} = X_{ij} - \hat{a}_i - \hat{b}_j.$$

The statistic S_j^2 defined by

$$S_j^2 = \sum_{i=1}^{I} \delta_{ij}^2, \quad j = 1, \ldots, J,$$

will be our estimate of the precision of the jth judge, and the relative precision will be estimated by

$$\Lambda_j = \frac{J}{J-1} \frac{S_j^2}{S}, \quad j = 1, \ldots, J.$$

The following result of Bolshev (1969b) gives us essential information about the distribution of Λ_j.

Theorem 24.1. *If H_{σ^2} is true, then, for each Λ_j, $j = 1, \ldots, J$, and $0 < z < 1$,*

$$\boldsymbol{P}\{\Lambda_j \le z \mid H_{\sigma^2}\} = I_z(\frac{I-1}{2}, \frac{1}{2}(I-1)(J-2)), \tag{24.16}$$

where I_z is Euler's incomplete beta function.

The Chauvenet test for the identification of the index j_0 for which $\mathbf{Var}\, Y_{ij_0} > \sigma^2$ is as follows: Since, under H_{σ^2},

$$\boldsymbol{P}\{\max_j \Lambda_j \ge z \mid H_{\sigma^2}\} \le J\boldsymbol{P}\{\Lambda_j \ge z \mid H_{\sigma^2}\},$$

we have, from Eq. (24.16), that

$$\boldsymbol{P}\{\max \Lambda_j \ge z \mid H_{\sigma^2}\} \le J\left[1 - I_z(\frac{I-1}{2}, \frac{(I-1)(J-2)}{2})\right].$$

The hypothesis H_{σ^2} will, accordingly, be rejected in favor of H by a test of level α based on Λ_j, $j = 1, \ldots, J$, if

$$\max_{1 \le j \le J} \Lambda_j > z_\alpha \tag{24.17}$$

where z_α is the solution of the equation

$$I_z\left(\frac{I-1}{2}, \frac{(J-2)(I-1)}{2}\right) = 1 - \frac{\alpha}{J}.$$

If H_{σ^2} is rejected, the judge with index j_0 at which the maximum in Eq. (24.17) is attained is deemed to be less precise than the others, having a greater variability of judgement.

APPENDIX

Multivariate Normal and Chi-Squared Distributions: Some Definitions and Basic Properties

Here we define the multivariate normal and chi-squared distributions and present background material, some of which is familiar to the student of introductory probability theory. Expanded versions can be found, for example, in Feller (1968), Rao (1973), and Serfling (1980).

Univariate Normal Distribution

A random variable ξ is said to have a *normal distribution with parameters μ and σ^2* if, for any $x \in R$, the probability of the event $\{\xi \leq x\}$ is given by

$$\boldsymbol{P}\{\xi \leq x\} = \frac{1}{\sqrt{2\pi}\sigma} \int_{-\infty}^{x} e^{-(u-\mu)^2/2\sigma^2}\, du \equiv \Phi\left(\frac{x-\mu}{\sigma}\right). \tag{A.1}$$

where $|\mu| < \infty$, $\sigma > 0$, and

$$\boldsymbol{E}\xi = \int_{-\infty}^{\infty} x\, d\Phi(x) = \mu, \quad \textbf{Var}\ \xi = \boldsymbol{E}(\xi - \mu)^2 = \sigma^2,$$

that is, μ and σ^2 appearing in Eq. (A.1) are the mean and variance of ξ. The notation $\xi \sim N_1(\mu, \sigma^2)$ will indicate that the random variable ξ has a normal distribution with mean μ and variance σ^2.

The characteristic function $f_\xi(t) = \boldsymbol{E}e^{it\xi}$ of $\xi \sim N_1(\mu, \sigma^2)$ is given by

$$f_\xi(t) = \boldsymbol{E}e^{it\xi} = \exp\{it\mu - \frac{1}{2}\sigma^2 t\}, \quad |t| < \infty. \tag{A.2}$$

If $\xi \sim N_1(0,1)$, we say that ξ has a standard normal distribution, and in this case from (A.1), we have

$$\boldsymbol{P}\{\xi \le x\} = \Phi(x) = \frac{1}{\sqrt{2\pi}} \int_{-\infty}^{x} e^{-t^2/2}\, dt, \quad \boldsymbol{E}\xi = 0, \quad \mathbf{Var}\ \xi = 1.$$

We say that a sequence $\{\xi_n\}$, $n = 1, 2, \ldots$ of random variables is *asymptotically normal with normalizing sequences* $\{\mu_n\}$ *and* $\{\sigma_n\}$, if there exist sequences of real numbers $\{\mu_n\}$ and $\{\sigma_n\}$ such that $\sigma_n > 0$ for all n sufficiently large and

$$\lim_{n\to\infty} \boldsymbol{P}\left\{\frac{1}{\sigma_n}(\xi_n - \mu_n) \le x\right\} = \Phi(x), \quad \text{all } x \in \boldsymbol{R}^1.$$

This convergence will be denoted by

$$\xi_n \text{ is } AN(\mu_n, \sigma_n^2)$$

or by

$$\frac{\xi_n - \mu_n}{\sigma_n} \overset{P}{\Rightarrow} \xi,$$

where ξ represents a random variable with distribution $N_1(0,1)$; that is, $\xi \sim N_1(0,1)$. The normalizing sequences need not consist of the means and variances of the sequence $\{\xi_n\}$. Indeed, we do not speak about the existence of $\boldsymbol{E}\xi_n$ and $\mathbf{Var}\ \xi_n$.

Chi-squared Distribution

Let $\xi_1, \ldots, \xi_n$ be independent standard normal random variables, and let

$$\chi_n^2 = \xi_1^2 + \cdots + \xi_n^2.$$

The distribution of the random variable χ_n^2 is called *the chi-squared distribution with n degrees of freedom* and is given by

$$\boldsymbol{P}\{\chi_n^2 \le x\} = \frac{1}{2^{n/2}\Gamma(\frac{n}{2})} \int_0^x t^{\frac{n}{2}-1} e^{-\frac{t}{2}}\, dt, \quad x \ge 0. \tag{A.3}$$

From the definition of χ_n^2, it can be seen that $\boldsymbol{E}\chi_n^2 = n$, **Var** $\chi_n^2 = 2n$, the characteristic function $f_{\chi_n^2}(t)$ of χ_n^2 has the form

$$f_{\chi_n^2}(t) = \boldsymbol{E}e^{it\chi_n^2} = (1-2it)^{-n/2}.$$

Fisher's Approximation

From the central limit theorem, we have that the sequence $\{\chi_n^2\}$ is asymptotically normal with normalizing sequences

$$\mu_n = \boldsymbol{E}\chi_n^2 = n \quad \text{and} \quad \sigma_n = \sqrt{\mathbf{Var}\chi_n^2} = \sqrt{2n};$$

that is, in the notation introduced above, χ_n^2 is $AN(n, 2n)$. Fisher showed that $\sqrt{2\chi_n^2}$ is $AN(\sqrt{2n-1}, 1)$.

Multivariate Characteristic Function

Let $\boldsymbol{\eta} = (\eta_1, \ldots, \eta_r)^T \in \boldsymbol{R}^r$ denote a random vector. The characteristic function of $\boldsymbol{\eta}$ is

$$f_{\eta}(\boldsymbol{w}) = Ee^{i\boldsymbol{w}^T\eta}, \quad \boldsymbol{w} \in \boldsymbol{R}^r.$$

Note that $f_{\eta}(\boldsymbol{w}) = f_{\boldsymbol{w}^T\eta}(1)$. We state formally the useful fact that the distribution of $\boldsymbol{\eta}$ is determined by the distributions of its linear combinations.

Lemma A.1. Suppose the random vector $\boldsymbol{\eta} = (\eta_1, \ldots, \eta_r)^T$ takes values in $\boldsymbol{R}^r$, and let $\boldsymbol{z} = (z_1, \ldots, z_r)^T$ be an arbitrary element of $\boldsymbol{R}^r$. The distributions of the random variables

$$\boldsymbol{z}^T\boldsymbol{\eta} = \sum_{i=1}^{r} z_i\eta_i, \quad \boldsymbol{z} \in \boldsymbol{R}^r,$$

uniquely determine the distribution of $\boldsymbol{\eta}$, and $f_{\boldsymbol{z}^T\boldsymbol{\eta}}(t) = f_{\boldsymbol{\eta}}(t\boldsymbol{z})$.

Multivariate Normal Distribution

Definition A.1. *If for any $\boldsymbol{z} \in \boldsymbol{R}^r$, $\boldsymbol{z}^T\boldsymbol{z} \neq 0$, the (scalar) random variable $\boldsymbol{z}^T\boldsymbol{\eta}$ has a normal distribution, then the vector $\boldsymbol{\eta}$ is said to have a normal* *distribution in* $\boldsymbol{R}^r$.

If $\boldsymbol{\eta}$ has a normal distribution in $\boldsymbol{R}^r$, then $\boldsymbol{E}\boldsymbol{\eta} = (\mu_1, \ldots, \mu_r)^T = \boldsymbol{\mu}$ and **Var** $\xi = \boldsymbol{\Sigma} = [\sigma_{ij}]$ exist, and $\boldsymbol{\Sigma}$ is positive semidefinite where

$$\sigma_{ij} = \boldsymbol{E}(\eta_i - \mu_i)(\eta_j - \mu_j), |\sigma_{ij}| < \infty, \quad i, j = 1, \ldots, r,$$
$$\mu_i = \boldsymbol{E}\eta_i, \quad |\mu_i| < \infty, \quad i = 1, \ldots, r; \quad \boldsymbol{\mu} = (\mu_1, \ldots, \mu_r)^T.$$

We write $\boldsymbol{\eta} \sim N_r(\boldsymbol{\mu}, \boldsymbol{\Sigma})$, and as we will see shortly, the distribution is determined by $\boldsymbol{\mu}$ and $\boldsymbol{\Sigma}$.

Let $m =$ rank $\boldsymbol{\Sigma}$. If $m = r$, we say that the distribution of $\boldsymbol{\eta}$ is *nondegenerate*, whereas, if $m < r$, then $\boldsymbol{\eta}$ has a *degenerate multivariate distribution*, which is concentrated on an m-dimensional subspace $\boldsymbol{R}^m$ of $\boldsymbol{R}^r$. Now we derive the distribution of $\boldsymbol{\eta}$ from its indirect definition. This is a useful exercise for understanding the material in Chapters 1 and 2. Consider the vector

$$\boldsymbol{\zeta} = (\zeta_1, \ldots, \zeta_r)^T = \boldsymbol{B}\boldsymbol{\eta},$$

where $\boldsymbol{B}$ is any $r \times r$ matrix, and let $\boldsymbol{z} = (z_1, \ldots, z_r)^T$ belonging to $\boldsymbol{R}^r$. Then

$$\boldsymbol{z}^T\boldsymbol{\zeta} = \boldsymbol{z}^T\boldsymbol{B}\boldsymbol{\eta} = (\boldsymbol{z}^T\boldsymbol{B})\boldsymbol{\eta} = (\boldsymbol{B}^T\boldsymbol{z})^T\boldsymbol{\eta},$$

where

$$\boldsymbol{\eta} \sim N_r(\boldsymbol{\mu}, \boldsymbol{\Sigma}),$$

is a normal random vector. Hence, $\boldsymbol{\zeta}$ has a multivariate normal distribution. Then

$$\boldsymbol{E}\boldsymbol{\zeta} = \boldsymbol{E}(\boldsymbol{B}\boldsymbol{\eta}) = \boldsymbol{B}\boldsymbol{\mu},$$

and

$$\begin{aligned}\textbf{Var}\,\boldsymbol{\zeta} &= \boldsymbol{E}(\boldsymbol{\zeta} - \boldsymbol{E}\boldsymbol{\zeta})(\boldsymbol{\zeta} - \boldsymbol{E}\boldsymbol{\zeta})^T,\\ &= \boldsymbol{E}\{\boldsymbol{B}(\boldsymbol{\eta} - \boldsymbol{\mu})(\boldsymbol{\eta} - \boldsymbol{\mu})^T\boldsymbol{B}^T\} = \boldsymbol{B}\boldsymbol{\Sigma}\boldsymbol{B}^T,\end{aligned}$$

where $\boldsymbol{\Sigma}$ is the covariance matrix of $\boldsymbol{\eta}$. The existence of **Var** $\boldsymbol{\zeta}$ follows from the fact that $\boldsymbol{\zeta}$ has a multivariate normal distribution. Thus, if

$$\boldsymbol{\eta} \sim N_r(\boldsymbol{\mu}, \boldsymbol{\Sigma}),$$

then

$$\zeta = \boldsymbol{B\eta} \sim N_r(\boldsymbol{B\mu}, \boldsymbol{B\Sigma B}^T).$$

Let us select a matrix $\boldsymbol{B}$ such that

1. $\det \boldsymbol{B} \neq 0$, that is, rank $\boldsymbol{B} = r$.
2. The matrix $\boldsymbol{B\Sigma B}^T$ has the diagonal form

$$\boldsymbol{B\Sigma B}^T = \left[\begin{array}{ccccc|ccc} 1 & 0 & \cdot & \cdot & 0 & 0 & \cdot \cdot & 0 \\ 0 & 1 & & & 0 & 0 & & 0 \\ \cdot & \cdot & & & \cdot & & & \cdot \\ 0 & & & 1 & 0 & 0 & \cdot \cdot & 0 \\ 0 & 0 & \cdot & 0 & 1 & 0 & \cdot \cdot & 0 \\ \hline 0 & 0 & & 0 & 0 & 0 & & 0 \\ \cdot & & & & \cdot & & & \cdot \\ \dot{0} & 0 & \cdot & \cdot & \dot{0} & 0 & \cdot \cdot & \dot{0} \\ \multicolumn{5}{c|}{\underbrace{\hphantom{0000000000}}_{m}} & \multicolumn{3}{c}{\underbrace{\hphantom{000000}}} \end{array}\right] \left.\vphantom{\begin{array}{c}1\\0\\\cdot\\0\\0\end{array}}\right\} m ,$$

$$= \left[\begin{array}{c|c} \boldsymbol{I}_m & \boldsymbol{0}_{m\times(r-m)} \\ \hline \boldsymbol{0}_{(r-m)\times m} & \boldsymbol{0}_{(r-m)(r-m)} \end{array}\right],$$

where $\boldsymbol{I}_m$ is the unit matrix of *rank* m, $\mathbf{0}_{u\times v}$ - the null matrix of dimension $u \times v$.

Clearly the rank of $\boldsymbol{B\Sigma B}^T$ is also $m \leq r$. Such a $\boldsymbol{B}$ exists but is not unique. Furthermore, if $\boldsymbol{z} = (z_1, \ldots, z_r)^T$ is any element of $\boldsymbol{R}^r$ and $f_{\boldsymbol{z}^T\boldsymbol{B\eta}}(t)$ is the characteristic function of the random variable $\boldsymbol{z}^T\boldsymbol{B\eta}$, then we have

$$f_{\boldsymbol{z}^T \boldsymbol{B}\boldsymbol{\eta}}(t) = \boldsymbol{E}\exp\{it(\boldsymbol{z}^T\boldsymbol{B}\boldsymbol{\eta})\} = \boldsymbol{E}\exp\{it(\boldsymbol{z}^T\boldsymbol{\zeta})\} = f_{\boldsymbol{z}^T\boldsymbol{\zeta}}(t), \quad \text{(A.5)}$$

where $f_{\boldsymbol{z}^T\boldsymbol{\zeta}}(t)$ is the characteristic function of $\boldsymbol{z}^T\boldsymbol{\zeta}$, and

$$\boldsymbol{z}^T\boldsymbol{\zeta} \sim N_1(\boldsymbol{z}^T\boldsymbol{B}\boldsymbol{\mu}, \boldsymbol{z}^T\boldsymbol{B}\boldsymbol{\Sigma}\boldsymbol{B}^T\boldsymbol{z});$$

that is, from Definition A.1 and Lemma A.1, $\boldsymbol{\zeta} \sim N_r(\boldsymbol{B}\boldsymbol{\mu}, \boldsymbol{B}\boldsymbol{\Sigma}\boldsymbol{B}^T)$. From Eqs. (A.3) and (A.5), we have

$$f_{\boldsymbol{z}^T\boldsymbol{B}\boldsymbol{\eta}}(t) = f_{\boldsymbol{z}^T\boldsymbol{\zeta}}(t) = e^{it\boldsymbol{z}^T\boldsymbol{B}\boldsymbol{\mu} - \frac{1}{2}t^2\boldsymbol{z}^T\boldsymbol{B}\boldsymbol{\Sigma}\boldsymbol{B}^T\boldsymbol{z}}.$$

Also

$$f_{\boldsymbol{\zeta}}(\boldsymbol{z}) = f_{\boldsymbol{z}^T\boldsymbol{\zeta}}(1),$$

and, from (A.4), it follows that

$$f_{\boldsymbol{\zeta}}(\boldsymbol{z}) = e^{i\boldsymbol{z}^T\boldsymbol{B}\boldsymbol{\mu} - \frac{1}{2}\boldsymbol{z}^T\boldsymbol{B}\boldsymbol{\Sigma}\boldsymbol{B}^T\boldsymbol{z}} \quad \text{(A.6)}$$

is the characteristic function of $\boldsymbol{\zeta} = \boldsymbol{B}\boldsymbol{\eta}$, $\boldsymbol{\eta} \sim N_r(\boldsymbol{\mu}, \boldsymbol{\Sigma})$.

Let us write

$$\boldsymbol{B}\boldsymbol{\mu} = \boldsymbol{b} = (b_1, \ldots, b_r)^T.$$

Because $\boldsymbol{B}$ satisfies conditions 1 and 2, we have

$$f_{\boldsymbol{\zeta}}(\boldsymbol{z}) = f_{\boldsymbol{B}\boldsymbol{\eta}}(\boldsymbol{z}) = \left(\prod_{j=1}^{m} \exp(iz_jb_j - \frac{1}{2}z_j^2)\right)\left(\prod_{j=m+1}^{r} \exp(iz_jb_j)\right). \quad \text{(A.7)}$$

In other words, if $\boldsymbol{\Sigma}$ is degenerate, that is, rank $\Sigma = m < r$, then there exists an $m \times m$ matrix $\boldsymbol{B}$ satisfying conditions 1 and 2, such that the first m coordinates $\zeta_1, \ldots, \zeta_m$ of the vector $\boldsymbol{\zeta} = (\zeta_1, \ldots, \zeta_r)^T = \boldsymbol{B}\boldsymbol{\eta}$ are independent normal random variables

$$\mathbf{Var}\,\zeta_j = 1, \qquad \boldsymbol{E}\zeta_j = b_j\cdot, \qquad j = 1, \ldots, m,$$

and the remaining coordinates $\zeta_{m+1}, \ldots, \zeta_r$ are degenerate, that is,

$$\boldsymbol{P}(\zeta_j = b_j) = 1, \qquad j = m+1, \ldots, r.$$

If $m = r$, that is, if the matrix $\boldsymbol{\Sigma}$ is nondegenerate, we say that η has a nondegenerate normal law, $N_r(\boldsymbol{\mu}, \boldsymbol{\Sigma})$. In this case, from Eq. (A.7), the characteristic function of $\boldsymbol{\zeta}$ has the form

$$f_{\boldsymbol{\zeta}}(\boldsymbol{z}) = \prod_{j=1}^{r} \exp(iz_j b_j - \frac{1}{2} z_j^2).$$

From the form of the Fourier transform, we see that the density function $p_{\boldsymbol{\zeta}}(\cdot)$ of $\boldsymbol{\zeta}$ has the form

$$p_{\boldsymbol{\zeta}}(\boldsymbol{x}) = p_{\boldsymbol{\zeta}}(\boldsymbol{x}; \boldsymbol{B}\boldsymbol{\mu}, \boldsymbol{I}_r) = \frac{1}{(2\pi)^{r/2}} \exp\{-\frac{1}{2}(\boldsymbol{x} - \boldsymbol{b})^T(\boldsymbol{x} - \boldsymbol{b})\},$$
$$\boldsymbol{x} = (x_1, \ldots, x_r)^T \in \boldsymbol{R}, \qquad \boldsymbol{b} = \boldsymbol{E}\boldsymbol{\zeta} = \boldsymbol{B}\boldsymbol{\mu}, \qquad \boldsymbol{\mu} = \boldsymbol{E}\eta$$

where $\boldsymbol{I}_r$ is the unit matrix of rank r. To calculate the density, $p_{\boldsymbol{\eta}}(\cdot)$, of the random vector $\boldsymbol{\eta}$, when rank $\boldsymbol{\Sigma} = r$, we change variables in the differential $p_{\boldsymbol{\zeta}}(\boldsymbol{x})\, dx_1 \ldots dx_r$ by the formula $\boldsymbol{x} = \boldsymbol{B}\boldsymbol{y}$, $\boldsymbol{y} \in \boldsymbol{R}^r$. Since

$$\boldsymbol{B}\boldsymbol{\Sigma}\boldsymbol{B}^T = \boldsymbol{I}_r,$$

we have

$$\boldsymbol{\Sigma} = \boldsymbol{B}^{-1}(\boldsymbol{B}^{-1})^T = (\boldsymbol{B}^T\boldsymbol{B})^{-1},$$

and it follows that det $\boldsymbol{\Sigma} = |\det \boldsymbol{B}|^{-2}$, that is,

$$|\det \boldsymbol{B}| = \frac{1}{(\det\, \boldsymbol{\Sigma})^{1/2}}.$$

Det $\boldsymbol{\Sigma} \neq 0$, since rank $\boldsymbol{\Sigma} = r$. Hence

$$\begin{aligned} p_{\boldsymbol{\eta}}(\boldsymbol{y}) &= p_{\boldsymbol{\eta}}(\boldsymbol{y}; \boldsymbol{\mu}, \boldsymbol{\Sigma}) \\ &= \frac{1}{(2\pi)^{r/2}(det\boldsymbol{\Sigma})^{1/2}} \exp\{-\frac{1}{2}(\boldsymbol{y} - \boldsymbol{\mu})^T \boldsymbol{B}^T\boldsymbol{B}(\boldsymbol{y} - \boldsymbol{\mu})\} \\ &= \frac{1}{(2\pi)^{r/2}(det\boldsymbol{\Sigma})^{1/2}} \exp\{-\frac{1}{2}(\boldsymbol{y} - \boldsymbol{\mu})^T \boldsymbol{\Sigma}^{-1}(\boldsymbol{y} - \boldsymbol{\mu})\}. \qquad \text{(A.8)} \end{aligned}$$

The likelihood function $L(\boldsymbol{\mu}, \boldsymbol{\Sigma})$ of the vector $\boldsymbol{\eta}$ is

$$L(\boldsymbol{\mu}, \boldsymbol{\Sigma}) = \frac{1}{(2\pi)^{r/2}(\det\ \boldsymbol{\Sigma})^{1/2}} \exp(-\frac{1}{2}(\boldsymbol{\eta} - \boldsymbol{\mu})^T \boldsymbol{\Sigma}^{-1}(\boldsymbol{\eta} - \boldsymbol{\mu})). \tag{A.9}$$

A bit of algebra will show us that the quadratic form in the exponent in Eq. (A.9) is χ_r^2. In fact,

$$\begin{aligned}(\boldsymbol{\eta} - \boldsymbol{\mu})^T \boldsymbol{\Sigma}^{-1}(\boldsymbol{\eta} - \boldsymbol{\mu}) &= (\boldsymbol{\eta} - \boldsymbol{\mu})^T \boldsymbol{B}^T \boldsymbol{B}(\boldsymbol{\eta} - \boldsymbol{\mu}), \\ &= [\boldsymbol{B}(\boldsymbol{\eta} - \boldsymbol{\mu})]^T [\boldsymbol{B}(\boldsymbol{\eta} - \boldsymbol{\mu})] = (\boldsymbol{\zeta} - \boldsymbol{b})^T(\boldsymbol{\zeta} - \boldsymbol{b}).\end{aligned}$$

Since

$$\boldsymbol{\zeta} = \boldsymbol{B}\boldsymbol{\eta} \sim N_r(\boldsymbol{b}, \boldsymbol{I}_r),$$

that is, the coordinates, $\zeta_1 - b_1, \ldots, \zeta_r - b_r$, of the vector $\boldsymbol{\zeta} - \boldsymbol{b}$ are independent standard normal random variables, we have, $\boldsymbol{\zeta} - \boldsymbol{b} \sim N_r(\mathbf{0}_r, \boldsymbol{I}_r)$ and

$$(\boldsymbol{\eta} - \boldsymbol{\mu})^T \boldsymbol{\Sigma}^{-1}(\boldsymbol{\eta} - \boldsymbol{\mu}) = \chi_r^2. \tag{A.10}$$

Hence, if $\boldsymbol{\eta} \sim N_r(\boldsymbol{\mu}, \boldsymbol{\Sigma})$ and rank $\boldsymbol{\Sigma} = r$, the density function of the vector $\boldsymbol{\eta}$ is

$$\begin{aligned}p_{\boldsymbol{\eta}}(\boldsymbol{y}) &= p_{\boldsymbol{\eta}}(\boldsymbol{y}; \boldsymbol{\mu}, \boldsymbol{\Sigma}) \\ &= \frac{1}{(2\pi)^{r/2}(\det \boldsymbol{\Sigma})^{1/2}} \exp(-\frac{1}{2}(\boldsymbol{y} - \boldsymbol{\mu})\boldsymbol{\Sigma}^{-1}(\boldsymbol{y} - \boldsymbol{\mu})), \qquad \boldsymbol{y} \in \boldsymbol{R}^r\end{aligned}$$

and Eq. (A.10) holds.

If rank $\boldsymbol{\Sigma} = m < r$ then

$$(\boldsymbol{\zeta} - \boldsymbol{B}\boldsymbol{\mu})^T(\boldsymbol{\zeta} - \boldsymbol{B}\boldsymbol{\mu}) = \chi_m^2.$$

In fact,

$$(\boldsymbol{\zeta} - \boldsymbol{B}\boldsymbol{\mu})^T(\boldsymbol{\zeta} - \boldsymbol{B}\boldsymbol{\mu}) = (\boldsymbol{\eta} - \boldsymbol{\mu})^T \boldsymbol{B}^T \boldsymbol{B}(\boldsymbol{\eta} - \boldsymbol{\mu}).$$

We cannot use $\boldsymbol{\Sigma}^{-1}$, since it doesn't exist. Instead, we use the matrix $\boldsymbol{C} \equiv \boldsymbol{B}^T \boldsymbol{B}$, which has the property

$$\boldsymbol{\Sigma}\boldsymbol{C}\boldsymbol{\Sigma} = \boldsymbol{\Sigma}. \tag{A.11}$$

Any matrix $\boldsymbol{C}$ that satisfies Eq. (A.11) is called a *generalized inverse matrix* of $\boldsymbol{\Sigma}$. If $\boldsymbol{\Sigma}$ is degenerate, as in the present case, it possesses many generalized inverses. Any generalized inverse of the matrix $\boldsymbol{\Sigma}$ will be denoted by $\boldsymbol{\Sigma}^-$. With this notation we can write

$$(\boldsymbol{\nu}-\boldsymbol{\mu})^T\boldsymbol{\Sigma}^-(\boldsymbol{\eta}-\boldsymbol{\mu}) = (\boldsymbol{\eta}-\boldsymbol{\mu})^T\boldsymbol{B}^T\boldsymbol{B}(\boldsymbol{\eta}-\boldsymbol{\mu}) = [\boldsymbol{B}(\boldsymbol{\eta}-\boldsymbol{\mu})]^T[\boldsymbol{B}(\boldsymbol{\eta}-\boldsymbol{\mu})]$$
$$= (\boldsymbol{\zeta}-\boldsymbol{b})^T(\boldsymbol{\zeta}-\boldsymbol{b}) = \sum_{i=1}^{m}(\zeta_i - b_i)^2 = \chi_m^2.$$

In other words, if $\boldsymbol{\eta} \sim N_r(\boldsymbol{\mu}, \boldsymbol{\Sigma})$ where rank $\boldsymbol{\Sigma} = m < r$, then Eq. (A.10) holds in the sense that

$$(\boldsymbol{\eta}-\boldsymbol{\mu})^T\boldsymbol{\Sigma}^-(\boldsymbol{\eta}-\boldsymbol{\mu}) = \chi_m^2. \tag{A.12}$$

We say that a sequence $\boldsymbol{X}^{(i)}$, $i = 1, 2, \ldots$ of random vectors in $\boldsymbol{R}^r$ is *asymptotically normal* with parameters $\boldsymbol{\mu}$ and $\boldsymbol{\Sigma}$ if $\boldsymbol{X}^{(i)}$ converge in distribution and the limit distribution is a normal distribution $N_r(\boldsymbol{\mu}, \boldsymbol{\Sigma})$ with mean vector $\boldsymbol{\mu}$ and covariance matrix $\boldsymbol{\Sigma}$, and we write

$$\boldsymbol{X}^{(i)} \text{ is } AN_r(\boldsymbol{\mu}, \boldsymbol{\Sigma}).$$

Noncentral Chi-squared Distribution

The noncentral chi-squared distribution is constructed as follows. Let $\xi_1, \ldots, \xi_n$ be independent normal random variables,

$$\boldsymbol{E}\xi_i = c_i, \qquad \mathbf{Var}\,\xi_i = 1, \qquad i = 1, \ldots, n,$$

and consider the random variable $\xi_1^2 + \cdots + \xi_n^2$. It can be computed (Cramer, 1946; Kendall and Stuart, 1963) that

$$\boldsymbol{P} = \left\{\sum_{i=1}^{n}\xi_i^2 \le x\right\} = \begin{cases} \int_0^x g_n(t,\lambda)\,dt, & x \ge 0, \\ 0, & x < 0, \end{cases} \tag{A.13}$$

where

$$g_n(t,\lambda) = \frac{e^{-t/2}e^{-\lambda/2}}{2^{\frac{n}{2}}\Gamma(\frac{n}{2})}t^{\frac{n}{2}-1}\left(1 + \frac{1}{n}\frac{t\lambda}{2} + \frac{1}{n(n+2)}\frac{1}{2!}\left(\frac{t\lambda}{2}\right)^2 + \cdots\right)$$

and

$$\lambda = \sum_{i=1}^{n} c_i^2$$

is the *parameter of noncentrality*, $\lambda \ge 0$. We write

$$\chi_n^2(\lambda) := \sum_{i=1}^{n} \xi_i^2. \tag{A.14}$$

The random variable $\chi_n^2(\lambda)$ is said to have a *noncentral chi-squared distribution with n degrees of freedom and parameter of noncentrality λ.*

One can see from the definition of $\chi_n^2(\lambda)$ that

$$\boldsymbol{E}\chi_n^2(\lambda) = n + \lambda, \qquad \mathbf{Var}\ \chi_n^2(\lambda) = 2n + 4\lambda.$$

If all $c_i = 0$, then $\xi_i \sim N(0,1)$, $\lambda = 0$, and

$$\chi_n^2(0) = \sum_{i=1}^{n} \xi_i^2 = \chi_n^2$$

has the distribution of a chi-squared random variable with n degrees of freedom. From Eq. (A.13) in the case $\lambda = 0$, we obtain [see Eq. (A.3)]

$$\boldsymbol{P}\{\chi_n^2 \le x\} = \begin{cases} \frac{1}{2^{n/2}\Gamma(n/2)} \int_0^x t^{n/2-1} e^{-t/2}\, dt, & x > 0, \\ 0, & x < 0. \end{cases}$$

The characteristic function of ξ_i^2 in Eq. (A.14) is

$$\begin{aligned} \boldsymbol{E} e^{it\xi_i^2} &= \frac{1}{\sqrt{2\pi}} \int_{-\infty}^{\infty} e^{ity - \frac{1}{2}(y-c_i)^2}\, dy \\ &= \frac{1}{\sqrt{1-2it}}\, e^{itc_i^2/(1-2it)}, \qquad i = 1, \ldots, n, \end{aligned}$$

hence the characteristic function of $\chi_n^2(\lambda) = \xi_1^2 + \cdots + \xi_n^2$ is

$$\boldsymbol{E} e^{it\chi_n^2(\lambda)} = \prod_{i=1}^{n} \left[\frac{1}{\sqrt{1-2it}}\, e^{itc_i^2/(1-2it)} \right] = \frac{1}{(1-2it)^{n/2}}\, e^{it\lambda/(1-2it)}. \tag{A.15}$$

Because in Eq. (A.15) the factor $(1-2it)^{-n/2}$ is the characteristic function of χ_n^2, we have

$$\chi_n^2(\lambda) = \chi_n^2 + \rho(\lambda),$$

where $\rho(\lambda)$ is independent of χ_n^2, and $\boldsymbol{E} e^{it\rho(\lambda)} = e^{it\lambda/(1-2it)}$. Furthermore,

$$\boldsymbol{E}\chi_n^2 = n, \qquad \mathbf{Var}\chi_n^2 = 2n, \qquad \boldsymbol{E}[\chi_n^2 - n]^3 = 8n,$$
$$\boldsymbol{E}\rho(\lambda) = \lambda, \qquad \mathbf{Var}\rho(\lambda) = 4\lambda, \qquad \boldsymbol{E}[\rho(\lambda) - \lambda]^3 = 24\lambda,$$
$$\boldsymbol{E}\chi_n^2(\lambda) = n + \lambda, \qquad \mathbf{Var}\chi_n^2(\lambda) = 2(n + 2\lambda),$$
$$\boldsymbol{E}[\chi_n^2(\lambda) - (n + \lambda)]^3 = 8(n + 3\lambda).$$

REFERENCES

Agresti, A. (1984). *Analysis of Ordinal Categorical Data.* Wiley, New York.

Agresti, A. (1990). *Categorical Data Analysis.* Wiley, New York.

Aguirre, N., and Nikulin, M.S. (1994). Chi-squared goodness-of-fit test for the family of logistic distributions. *Kybernetika,* **30** (3):214–222.

Akritas, M.G. (1988). Pearson-type goodness-of-fit tests: The invariate case. *J. Amer. Statist. Assoc.,* **83**:222–230.

Anderson, T.W., and Darling, D.A. (1952). Asymptotic theory of certain "goodness of fit" criteria based on stochastic processes. *Ann. Math. Statist.,* **23**:193–212.

Anderson, T.W., and Darling, D.A. (1954). A test of goodness of fit. *J. Amer. Statist. Assoc.,* **49**:765–769.

Arley, N., and Buch, K. (1950). *Introduction to the Theory of Probability and Statistics.* Wiley, New York.

Bagdonavičius, V., and Nikulin, M.S. (1994). *General semiparametric models in survival analysis,* Preprint 94-03, Mathématiques Stochastiques, Université Bordeaux 2, 29p.

Balakrishnan, N., ed. (1992). *Handbook of the Logistic Distribution.* Marcel Dekker, New York.

Balakrishnan, N., and Cohen, A.C. (1990). *Order Statistics and Inference: Estimation Methods.* Academic Press, Boston.

Bar-Lev, S.K., and Enis, P. (1986). Reproducibility and natural exponential families with power variance functions. *Ann. Statist.,* **14**:1507–1522.

Barndorff-Nielsen, O. (1978). *Information and Exponential Families.* Wiley, New York.

Barton, D.E. (1956). Neyman's ψ^2 test of goodness of fit when the null hypothesis is impossible. *Skand. Aktuarietidskr.,* **39**:216–245.

Bavarov, E.A., and Belyaev, P.F. (1961). On testing the randomness of sampling without replacement. *Theory Probabil. Appl.,* **4**:455–458.

Behnen, K. and Neuhaus, G. (1989). *Rank Tests with Estimated Scores and Their Applications.* B.G. Teubner, Stuttgart.

Bemis, K.G., and Bhapkar, V.P. (1983). On B.A.N. estimates for chi-squared test criteria. *Ann. Statist.,* **11**:183–186.

Beran, R. (1975). Tail probability of noncentral quadratic forms. *Ann. Statist.,* **3**:969–979.

Berkson, J. (1938). Some difficulties of interpretation encountered in the application of the chi-square table. *J. Amer. Statist. Assoc.*, **33**:526–536.

Berkson, J. (1978). In dispraise of the exact test. *J. Statist. Plan. Inf.*, **2**:27–42.

Bernstein, A.V. (1982). Asymptotically similar criteria. *J. Soviet Math.*, **17** (3):1825–1857.

Birch, M.W. (1964). A new proof of the Pearson-Fisher theorem. *Ann. Math. Statist.*, **35**:92–98.

Birnbaum, Z.W. (1953). Distribution free tests of fit for continuous distribution functions. *Ann. Math. Statist.*, **24**:1–8.

Blacher, R. (1988). A new form for the chi-squared test of independence. *Statistics*, **19** (4):519–536.

Blacher, R. (1990). *Quelques applications des fonctions orthogonales en probabilité et statistique*. Thèse de l'Univérsité Joseph Fourier, Grenoble 1.

Bofinger, E. (1973). Goodness-of-fit test using the sample quantile. *J. Roy. Statist. Soc.*, **B35**:277–284.

Bolshev, L.N. (1965). On a characteristic of the Poisson distribution and its statistical applications. *Theory Probabil. Appl.*, **10** (3):446–456.

Bolshev, L.N. (1966). On the question of testing "exponentiality." *Theory Probabil. Appl.*, **11** (3):542–544.

Bolshev, L.N. (1969a). Cluster analysis. *Bull. Intern. Statist. Inst.*, **43**:411–425.

Bolshev, L.N. (1969b). On exclusion criteria of outlying observations. *Tr. Inst. Prikl. Mat. Tbilissk. Gos. Univ.* **2**:161–172.

Bolshev, L.N. (1987). On the question of testing some composite statistical hypotheses. In *Theory of Probability and Mathematical Statistics. Selected Works.* Nauka, Moscow, 5–63.

Bolshev, L.N., and Mirvaliev, M. (1978). Chi square goodness-of-fit test for the Poisson, binomial, and negative binomial distributions. *Theory Probabil. Appl.*, **23**:461–474.

Bolshev, L.N., and Nikulin, M.S. (1975). One solution of the problem of homogeneity. *Serdika, Bulgarsko Mathematichesko Spicanie*, **1**:104–109.

Bolshev, L.N., and Smirnov, N.V. (1968). *Tables of Mathematical Statistics.* Nauka, Moscow (in Russian).

Borovkov, A.A. (1977). On the power of the χ^2 test with increasing number of class intervals. *Theory Probabil. Appl.*, **23**:366–370.

Borovkov, A.A. (1984). *Mathematical Statistics.* Nauka, Moscow (in Russian).

Bosq, D. (1978). Test Hilbertien et test du Chi-deux. *C.R. Acad. Sci., Paris*, **286**:946–948.

Bosq, D. (1983). Lois limites et efficacité asymptotique des tests Hilbertiens de dimension finie sous des hypothèses adjacentes. *Statistique et Analyse des Données*, **8** (1):1–40.

Chanda, K.C. (1981). Chi-square goodness-of-fit tests based on dependent observations. *Statistical Distributions in Scientific Work*, vol. 5. Riedel, Dordecht, 35–49.

Chase, G.R. (1972). Chi-square test when parameters are estimated independently of the sample. *J. Amer. Statist. Assoc.*, **67**:609–611.

Chen, J. (1975). Goodness of fit tests under random censorship. Ph.D. dissertation, Dept. Statistics, Oregon State University.

Chernoff, H., and Lehmann, E.L. (1954). The use of maximum likelihood estimates in χ^2 tests for goodness of fit. *Ann. Math. Statist.*, **25**:579–586.

Chibisov, D.M. (1961a). On asymptotic power and the effectiveness of the ω^2 criterion. *Dokl. Acad. Nauk SSSR*, **18** (2):322–325.

Chibisov, D.M. (1961b). On the test of fit based on sample spacings. *Theory Probabil. Appl.*, **6**:325–329.

Chibisov, D.M. (1962). Investigations of the power of some nonparametric criteria. *Theory Probabil. Appl.*, **8** (3):355–356.

Chibisov, D.M. (1965). An investigation of the asymptotic power of the tests of fit. *Theory Probabil. Appl.*, **10**:421–437.

Chibisov, D.M. (1971). Certain chi-square type tests for continuous distributions. *Theory Probabil. Appl.*, **18** (3):1–20.

Cochran, W.G. (1952). The χ^2 test of goodness of fit. *Ann. Math. Statist.*, **23**:315–342.

Cohen, A. and Sackrowitz, H.B. (1975). Unbiasedness of the chi-squared likelihood ratio, and other goodness-of-fit tests for the equal cell case. *Ann. Statist.*, **3**:959–969.

Conover, W.J. (1974). Some reasons for not using the Yates continuity correction on 2×2 contingency tables. *J. Amer. Statist. Assoc.*, **69**:374–382.

Consul, P.C., and Jain, G.C. (1973). A generalization of the Poisson distribution. *Technometrics*, **15**:791–799.

Cox, D.R. (1970). The continuity correction. *Biometrika*, **57**:217–219.

Cramer, H. (1946). *Mathematical Methods of Statistics*. Princeton University Press, Princeton, NJ.

Cressie, N. (1976). On the logarithms of high order spacings. *Biometrika*, **63**:343–355.

Cressie, N. (1978). Power results for tests based on high order gaps. *Biometrika*, **65**:214–218.

Cressie, N. (1979). An optimal statistic based on higher order gaps. *Biometrika*, **66**:619–627.

Cressie, N., and Read, T.R.C. (1989). Pearson's χ^2 and the loglikelihood ratio statistic G^2: a comparative review. *Int. Statist. Rev.*, **57**:19–43.

Csörgo, M. (1983). *Quantile Processes with Statistical Applications*, vol. 42. SIAM, Philadelphia.

D'Agostino, R.B., and Stephens, M.A., eds. (1986). *Goodness-of-fit Techniques.* Marcel Dekker, New York.

Dahiya, R.C., and Gurland, J. (1970a). Pearson chi-square test of fit with random intervals. I. Null case. MRC Technical Summary Report 1046, University of Wisconsin, Madison, WI.

Dahiya, R.C., and Gurland, J. (1970b). Pearson chi-square test of fit with random intervals. II. Non-null case. MRC Technical Summary Report 1051, University of Wisconsin, Madison, WI.

Dahiya, R.C., and Gurland, J. (1972a). Goodness-of-fit test for the gamma and exponential distribution. *Technometrics*, **14**:791–801.

Dahiya, R.C., and Gurland, J. (1972b). Pearson chi-square test-of-fit with random intervals. *Biometrika*, **59**:147–153.

Dahiya, R.C., and Gurland, J. (1973). How many classes in the Pearson chi-square test? *J. Amer. Statist. Assoc.*, **68**:707–712.

Darling, D.A. (1955). The Cramer-Smirnov test in the parametric case. *Ann. Math. Statist.*, **26**:1–20.

Darling, D.A. (1957). The Kolmogorov-Smirnov, Cramer-von Mises tests. *Ann. Math. Statist.*, **28**:823–838.

David, H. (1981). *Order Statistics*. Wiley, New York.

Del Pino, G.E. (1979). On the asymptotic distribution of k-spacings with applications to goodness-of-fit tests. *Ann. Statist.*, **7**:1058–1065.

Diaconis, P., and Efron, B. (1985). Testing for independence in a two-way table: new interpretation of the chi-squared statistic. *Ann. Statist.*, **13**:845–874.

Diamond, E.L., Mitra, S.K., and Roy, S.N. (1960). Asymptotic power and asymptotic independence in statistical analysis of categorical data. *Bull. inst. int. stat.*, **37**:3–23.

Doob, J.L. (1952). Heuristic approach to the Kolmogorov-Smirnov limit theorems. *Ann. Math. Statist.*, **20**:277–281.

Drost, F.C. (1988). *Asymptotics for Generalized Chi-squared Goodness-of-fit Tests*. Centre for Mathematics and Computer Science, CWI Tracts, vol. 48, Amsterdam.

Drost, F.C. (1989). Generalized chi-square goodness-of-fit tests for location-scale models when the number of classes tends to infinity. *Ann. Statist.*, **17**:1285–1300.

Drost, F.C., Kallenberg, W.C.M., and Oosterhoff, J. (1990). The power of EDF tests of fit under non-robust estimation of nuisance parameters. *Stat. Decisions*, **8** (2):167–182.

Dudley, R.M. (1976). Probabilities and metrics—convergence of laws on metric spaces with a view to statistical testing. Lecture Notes, vol 45. Aarhus Universitet, Denmark.

Dudley, R.M. (1979). On χ^2 tests of composite hypotheses. In *Probability Theory*, Banach Center Publications, **5**:75–87. Warsaw.

Durbin, J. (1973a). Weak convergence of the sample distribution function when parameters are estimated. *Ann. Statist.*, **1**:279–290.

Durbin, J. (1973b). *Distribution Theory for Tests Based on the Sample Distribution Function*, vol. 9. SIAM, Philadelphia.

Durbin, J. (1975). Kolmogorov-Smirnov tests when parameters are estimated with applications to tests of exponentiality and tests of spacings. *Biometrika*, **62**:5–22.

Dzhaparidze, K.O. (1983). On iterative procedures of asymptotic inference. *Statistics Neerlandica*, **37** (4):181–189.

Dzhaparidze, K.O., and Nikulin, M.S. (1974). On a modification of the standard statistics of Pearson. *Theory Probabil. Appl.*. **19**:851–852.

Dzhaparidze, K.O., and Nikulin, M.S. (1982a). Probability distributions of the Kolmogorov and omega-square statistics for continuous distributions with shift and scale parameters. *J. Soviet Math.*, **20**:2147–2163.

Dzhaparidze, K.O., and Nikulin, M.S. (1982b). Quasi-Newton methods of estimation and chi-squared type tests. *Proc. ICM Abstr.* **8**:Section 10, p. 36.

Dzhaparidze, K.O., and Nikulin, M.S. (1992). On the computation of chi-squared type statistics, Zapiski Nauchnykh Seminarov POMI im V.A. Steklova AN Russia. Studies Probabil. Distrib., **194** (12):59–90.

Feinberg, S.E. (1980). *The Analysis of Cross-Classified Categorical Data*. MIT Press, Cambridge, MA.

Feller, W. (1948a). On the Kolmogorov-Smirnov limit theorem for empirical distributions. *Ann. Math. Statist.*, **19**:177–180.

Feller, W. (1948b). On probability problem in the theory of counter, in *Courant Anniversary Volume*. Wiley-Interscience, New York, 105–115.

Feller, W. (1968). *An Introduction to Probability Theory and its Applications*, vols. 1 and 2. Wiley, New York.

Fisher, R.A. (1928). On a property connecting the χ^2 measure of discrepancy with the method of maximum likelihood. Atti de Congresso Internazionale dei Mathematici, Bologna, **6**:94–100.

Fix, E. (1944). Table of the noncentral χ^2. *Univ. Calif. Publ. Statistics*, **1** (2):15–19.

Fleiss, J. (1981). *Statistical Methods for Rates and Proportions*. Wiley, New York.

Fraser, D.A.S. (1976). *Probability and Statistics*. Duxbury Press, Boston, MA.

Freeman, G.H., and Malton, J.M. (1951). Note on the exact treatment of contingency, goodness of fit and other problems of significance. *Biometrika*, **38**:141–149.

Gail, M.H., and Gastwirth, J.L. (1978). A scale-free goodness-of-fit test for the exponential distribution based on the Gini statistic. *J. Roy. Statist. Soc.*, **B40**:350–357.

Geary, R.C. (1947). Testing for normality. *Biometrika*, **34**:209–242.

Gebert, J.R. and Kale, B.K. (1969). Goodness of fit tests based on discriminatory information. *Statistiche Hefte*, **10**:192–200.

Gihman, I.I. (1953). Some remarks on the consistency criterion of A.N. Kolmogorov. *Dokl. Akad. Nauk SSSR*, **91** (4):715–718.

Gihman, I.I. (1961). On the empirical distribution function in the case of grouping data. In *Selected Translations in Mathematical Statistics and Probability*, vol. 1. American Math. Soc., Providence, RI, 77–81.

Gleser, L.J., and Moore, D.S. (1983). The effect of dependence on chi-squared and empiric distribution tests of fit. *Ann. Statist.*, **11**:1100–1108.

Goodman, L.A., and Kruscal, W.H. (1979). *Measures of Association for Cross Classification*. Springer-Verlag, New York.

Greenwood, H. (1946). The statistical study of infectious disease. *J. Roy. Statist. Soc.*, **A109**:85–110.

Greenwood, P., and Nikulin, M.S. (1988). Application of tests of chi-square type. *J. Soviet Math.* **43** (6):2776–2791.

Greenwood, P.E., and Shiryayev, A.N. (1986). *Contiguity and the Statistical Invariance Principle*. Gordon and Breach, New York.

Grizzle, J.E. (1967). Continuity correction in the χ^2 test for 2×2 tables. *Amer. Statist.*, **21**:28–32.

Gupta, R.C. (1974). Modified power series distributions and some of its applications. *Sankhÿa*, **B35**:288–298.

Gupta, R.C. (1977). Minimum variance unbiased estimation in modified power series distribution and some of its applications. *Commun. Statist. Theory Meth.*, **A** (10):977–991.

Guttorp, P., and Lockhart, R.A. (1988). On the asymptotic distribution of quadratic forms in uniform order statistics. *Ann. Statist.*, **16**:433–449.

Gvanceladze, L.G., and Chibisov, D.M. (1979). On tests of fit based on grouped data. In *Contributions to Statistics, J. Hájek Memorial Volume* (editor J. Jurečková). Academia, Prague, 79–89.

Haber, M. (1980). A comparison of some continuity corrections for the chi-square test on 2 × 2 tables. *J. Amer. Statist. Assoc.*, **75** (373):510–515.

Haberman, S.J. (1974). *The Analysis of Frequency Data*. University of Chicago Press, Chicago.

Haberman, S.J. (1977). Log-linear models and frequency tables with small expected cell counts. *Ann. Statist.*, **5**:1148–1169.

Habib, M.G., and Thomas, D.R. (1986). Chi-square goodness-of-fit tests for randomly censored data. *Ann. Statist.*, **14**:759–765.

Haight, F.A., and Breuer, M.A. (1960). The Borel-Tanner distribution. *Biometrika*, **47**:143–150.

Hajek, J., and Sidak, Z. (1967). *Theory of Rank Tests*. Academic Press, New York.

Haldane, J.B.S. (1937). The exact value of the moments of the distribution of χ^2, used as a test of goodness of fit. *Biometrika*, **29**:133–143.

Hall, P. (1986). On powerful distributional tests based on sample spacings. *J. Multivariate Analysis*, **19**:201–204.

Hamdan, M.A. (1963). The number and width of classes in the chi-square test. *J. Amer. Statist. Assoc.*, **58**:678-689.

Harrison, R.H. (1985). Choosing the optimum number of classes in the chi-square test for arbitrary power levels. *Sankyha*, **B47**:319–324.

Hartley, H.O., and Pfaffenberger, R.C. (1972). Quadratic forms in orders statistics used as goodness-of-fit criteria. *Biometrika*, **59** (3):605–611.

Hegazy, Y.A. and Green, J.R. (1975). Some new goodness-of-fit tests using order statistics. *Appl. Statist.*, **24**:299–308.

Hjort, N.L. (1984). Weak convergence of cumulative intensity processes when parameters are estimated, with applications to goodness-of-fit tests in models with censoring. Research Report 763, Norwegian Computing Centre, Oslo.

Hjort, N.L. (1990). Goodness of fit tests in models for life history data based on cumulative hazard rates. *Ann. Statist.*, **18**:1221–1258.

Hoaglin, D. (1977). Exploring a table of percentage points of χ^2. *J. Amer. Statist. Assoc.*, **72**:508–515.

Hollander, M., and Peña, E.A. (1992). Randomly censored data. *J. Amer. Statist. Assoc.*, **87** (418):458–463.

Holst, L. (1979). Asymptotic normalities of sum-functions of spacings. *Ann. Prob.*, **7**:1066–1072.

Holst, L. (1981). Some conditional limit theorems in exponential families, Ann. Prob., **9**:818–830.

Holtzman, G.I., and Good, I.J. (1986). The Poisson and chi-squared approximations as compared with the true upper-tail probability of Pearson's χ^2 equiprobable multinomials. *J. Statist. Planning Infer.*, **13**:283–295.

Horn, S.D. (1977). Goodness of fit tests for discrete data: a review and an application to a health impairment scale. *Biometrics*, **33**:237–248.

Hougaard, P. (1986). Survival models for heterogeneous populations derived from stable distributions. *Biometrika*, **73**:387–396.

Huber, C. and Nikulin, M.S. (1991). Quelques problèmes statistiques liés aux promenades aléatoires. In *Statistique des processus en milieu médical*:Séminaire 90–91 de Statistique Médical de l'Université Paris V.

Hutchinson, T.P. (1979). The validity of the chi-squared test when expected frequencies are small: a list of recent research references. *Commun. Statist. Theory Methods*, **8** (3):327–335.

Ibragimov, I.A., and Hasminskii, R.Z. (1981). *Statistical Estimation Asymptotic Theory*. Springer-Verlag, Heidelberg.

Jackson, J.E. (1991). *A User's Guide to Principal Components*. Wiley, New York.

Jacque, C.M., and Bera, A.K. (1987). A test for normality of observations and regression residuals. *Int. Statist. Rev.*, **55**:163–177.

Jain, G.C. (1974). A Borel-Tanner distribution and its approximation to the negative binomial distribution. *Math. Operationforsch. Statist.*, **H3** (B5):207–212.

Jain, G.C., and Consul, P.C. (1971). A generalized negative binomial distribution. *SIAM J. Appl. Math.*, **21**:501–513.

Jain, G.C., and Gupta, R.C. (1973). A logarithmic type distribution. *Trab. Estadist.*, **24**:99–105.

Jammalamadaka, S.R., and Tiwaru, R.C. (1985). Asymptotic comparison of three tests for goodness of fit. *J. Statist. Plan. Inf.*, **12**:295–304.

Jammalamadaka, S.R., and Tiwaru, R.C. (1987). Efficiencies of some disjoint spacings tests relative to χ^2 test. In *New Perspectives in Theoretical and Applied Statistics*. Wiley, New York, 311–317.

Jani, P.N. (1977). Minimum variance unbiased estimation for certain left-truncated modified power distributions. *Sankhy̅a*, **B39**:258–278.

Johnson, N.L., and Kotz, S. (1969). *Discrete Distributions*. Houghton-Mifflin, Boston.

Johnson, N.L., and Kotz, S. (1970). *Distribution in statistics: Continuous Univariate Distributions*, vol. 2. Houghton-Mifflin, Boston.

Johnson, N.L., and Kotz, S. (1982). Developments in discrete distributions, 1969–1980. *Inter. Statist. Rev.*, **50**:71–101.

Joshi, S.W., and Park, C.J. (1974). Minimum variance unbiased estimation for truncated power series distributions. *Sankhy̅a, Indian J. Statist.*, **A36**:305–314.

Kac, M., Kiefer, J., and Wolfowitz, J. (1955). On tests of normality and other tests of goodness of fit based on distance methods. *Ann. Math. Statist.* **26**:189–211.

Kallenberg, W.C.M., Oosterhoff, J., and Schriever, B.F. (1985). The number of classes in chi-squared goodness-of-fit tests. *J. Amer. Statist. Assoc.*, **80**:959–968.

Karakostas, R.X. (1985). On minimum variance unbiased estimator. *Amer. Statist.*, **39**:303–305.

Katti, S.K. (1973). Exact distribution for the chi-square test in the one-way tables. *Commun. Statist. Theory Methods*, **2**:435–447.

Kaufman, H., Mathai, A.M., and Rathie, P.N. (1972). A mathematical foundation for Pearson's χ^2 goodness-of-fit statistic. *Sankhyā*, **A34**:441–442.

Kempthorne, O. (1966). The classical problem of inference-goodness of fit. In *Proceedings of the Fifth Berkeley Symposium on Mathematical Statistics and Probability*, **1**:235–249.

Kempthorne, O. (1979). In dispraise of the exact test: reactions. *J. Statist. Planning Infer.*, **3**:199–213.

Kendall, D., and Stuart, A. (1963). *The Advanced Theory of Statistics, Inference and Relationship*, vols. 2 and 3. Hafner, New York.

Kendall, D., and Stuart, A. (1965). *The Advanced Theory of Statistics*, vol. 1. *Distributions Theory*. Hafner, New York.

Kim, J.H. (1993). Chi-square goodness-of-fit tests for randomly censored data. *Ann. Statist.*, **21**:1621–1639.

Kimball, B.F. (1947). Some basic theorems for developing tests of fit for the case of the non-parametric probability distribution function. *Ann. Math. Statist.*, **18**:540–548.

Kimball, B. (1950). On the asymptotic distribution of the sum of powers of unit frequency differences. *Ann. Math. Statist.*, **21**:263–271.

Koehler, K.J. (1977). *Goodness of Fit Statistics for Large Sparse Multinomials*. Unpublished Ph.D. Dissertation, School of Statistics, University of Minnesota.

Koehler, K.J., and Larntz, K. (1980). An empirical investigation of goodness-of-fit statistics for sparse multinomials. *J. Amer. Statist. Assoc.*, **75**:336–344.

Kolchin, V.F., Sevastianov, B.A., and Chistyakov, V.P. (1978). *Random Allocations*. Nauka, Moscow.

Kolmogorov, A.N. (1933). Sulla determinazione empirica di une legge di distribuzione. *Giorna. Ist. Attuari*, **4**:83–91.

Koziol, J.A. (1987). An alternative formulation of Neyman's smooth goodness of fit tests under composite alternatives. *Metrika*, **34**:17–24.

Lancaster, H.O. (1969). *The Chi-Squared Distribution*. Wiley, New York.

Larntz, K. (1973). Small sample comparisons of chi-square statistics. *Amer. Statist. Assoc. Proc., Social Statist. Sect.*:336–339.

Larntz, K. (1978). Small sample comparisons of exact levels for chi-square goodness-of-fit statistics. *Amer. Statist. Assoc. Proc., Social Statist. Sect.*, **73**:253–263.

Latscha, R. (1953). Significance tests in a 2×2 contingency table. *Biometrika*, **40**:74–86.

Lawal, H.B. (1980). Tables of percentage points of Pearson's goodness-of-fit statistic for use with small expectations. *Appl. Statist.*, **29** (3):292–298.

Lawal, H.B., and Upton, G.J.G. (1980). An approximation to the distributionof the χ^2 goodness-of-fit statistic for use with small expectations. *Biometrika*, **67** (2):442–453.

Lawless, J.F. (1982). *Statistical Models and Methods of Lifetime Data.* Wiley, New York.

LeCam, L. (1956). On the asymptotic theory of estimation and testing hypotheses. In *Proceedings of the Third Berkeley Symposium on Mathematical Statistics and Probability*, **1**:129–156.

LeCam, L., Mahan, C., and Singh, A. (1983). An extension of a theorem of H. Chernoff and E.L. Lehmann. In *Recent Advances in Statistics.* Academic Press, Orlando, FL, 303–332.

Lehmann, E.H. (1973). On two modifications of the Cramer-von Mises statistic. *J. Roy. Statist. Soc.*, **B35**, No. 3, 523.

Lehmann, E.H. (1991). *Testing Statistical Hypotheses.* Wiley, New York.

Li, G., and Doss, H. (1993). Generalized Pearson-Fisher chi-square goodness-of-fit tests, with applications to models with life history data. *Ann. Statist.*, **21**:772–797.

Linhart, H., and Zucchini, W. (1986). *Model Selection.* Wiley, New York.

Mäkelänen, T. (1966). On quadratic forms in normal variables. *Commentationes Phisico-Mathematicae, Societas Scientarium Fennica*, **31**:1–6.

Mann, H.B., and Wald, A. (1942). On the choice of the number of intervals in the application of the chi-square test. *Ann. Math. Statist.*, **13**:306–317.

Mardia, K.V., and Zemroch, P.J. (1978). *Tables of F- and Related Distributions with Algorithms.* Academic Press, London.

Marsden, E., and Barrat, T. (1911). The probability distribution of the time intervals of particles with applications to the number of particles emitted by uranium. *Proc. Phys. Soc. Lond.*, **23**:50–61.

Martynov, G.V. (1977). *Omega-Square Criteria.* Nauka, Moscow (in Russian).

McCulloch, C.E. (1985). Relationships among some chi-square goodness of fit statistics. *Commun. Statist. Theory Meth.*, **14**:593–603.

Michel, R. (1979). On the asymptotic efficiency of conditional tests for exponential families. *Ann. Statist.*, **7** (6):1256–1263.

Mihalko, D.P. and Moore, D.S. (1980). Chi-square tests of fit for type II censored data. *Ann. Statist.*, **8**:625–644.

Miller, F.L., and Quesenberry, C.P. (1979). Power studies of some tests for uniformity. II. *Commun. Statist.-Simul. Comp.*, **8**:271–290.

Mirvaliev, M. (1978). The rejection of outlying observations in regression analysis. *Theory Probabil. Appl.*, **25**:619–629.

Mirvaliev, M. (1979). Chi-square goodness-of-fit tests with one degree of freedom for verifying the "Poisson," "binomial" and "negative binomial" property. *Theory Probabil. Appl.*, **24**(3):666-667.

Mirvaliev, M. (1987). Chi-square tests for two-dimensional Poisson, binomial and negative triomial distributions. In *Probability Methods and Mathematical Statistics.* FAN, Tashkent, 106–117.

Mirvaliev, M. (1988). Chi-square goodness-of-fit test in contingency tables for Poisson distribution. *Izv. Akad. Nauk Uzb. SSR*, 32–37.

Mirvaliev, M. (1989). Chi-square goodness-of-fit tests for a family of multidimensional discrete distributions. *Theory Probabil. Appl.*, **39**(4):728–732.

Miyamoto, Y. (1976). Optimum spacing for goodness of fit tests based on sample quantiles. In *Essays in Probability and Statistics – A Volume in Honor of Professor Junioro Ogawa* (S. Ikeda, T. Hayakama, H. Hudimoto, M. Okamoto, M. Siotani, and S. Vemamoto, eds.). Shinko Tsushio, Tokyo, 475–483.

Moore, D.S. (1971). A chi-square statistic with random cell boundaries. *Ann. Math Statist.*, **42**:147–156.

Moore, D.S. (1977). Generalized inverses, Wald's method and the construction of chi-squared tests of fit. *J. Amer. Statist. Assoc.*, **72**:131–137.

Moore, D.S. (1982). The effect of dependence on chi-squared test of fit. *Ann. Statist.*, **10**:1163–1171.

Moore, D.S., and Spruill, M.C. (1975). Unified large-sample theory of general chi-squared statistics for tests of fit. *Ann. Statist.*, **3**:599–616.

Moran, P.A. (1947). The random division of an interval. *J. Roy. Statist. Soc.*, Supp. 9, 92–98.

Moran, P.A.P. (1951). The random division of an interval: Part II. *J. Roy. Statist. Soc.*, **B13**:147–150.

Morris, C. (1975). Central limit theorem for multinomial sums. *Ann. Statist.*, **3** (1):165–188.

Morris, C. (1982). Natural exponential families with quadratic variance functions. *Ann. Statist.*, **10**:65–80.

Morris, C.N. (1983). Natural exponential families with quadratic variance functions: statistical theory. *Ann. Statist.*, **11** (2):515–529.

Mosteller, F. (1946). On some useful "inefficient statistics." *Ann. Math Statist.*, **17**:377–407.

Murthy, V.K., and Gafarian, A.V. (1970). Limiting distributions of some variations of the chi-square statistics. *Ann. Math. Statist.*, **41**:188–194.

Neyman, J. (1937). "Smooth" test for goodness of fit. *Skand. Aktuarietidskr.*, **20**:150–199.

Neyman, J. (1949). Contribution to the theory of the χ^2 test. *Proc. Berkeley Symp. on Mathematical Statistics and Probability*, **1**:239–273.

Nikulin, M.S. (1973a). Chi-square test for normality. International *Vilnius Conference on Probability Theory and Mathematical Statistics*, **2**:119–122.

Nikulin, M.S. (1973b). Chi-square test for continuous distributions with shift and scale parameters. *Theory Probabil. Appl.*, **18**:559–568.

Nikulin, M.S. (1973c). On a chi-square test for continuous distributions. *Theory Probabil. Appl.*, **18**:638–639.

Nikulin, M.S. (1974). On a quantile test. *Theory Probabil. Appl.*, **19**:410–413.

Nikulin, M.S. (1979). Hypothesis testing for a parameter difference in binomial distributions. *Theory Probabil. Appl.*, **24**:392–396.

Nikulin, M.S. (1984a). Applications of the model of two-factor analysis of variance without interaction. *J. Soviet Math.*, **25** (3):1196–1207.

Nikulin, M.S. (1984b). A test for detecting the outlying observations in a multivariate normal family. *J. Soviet Math.*, **24** (5):604–607.

Nikulin, M.S. (1991). Some recent results on chi-square tests. Queen's Papers in Pure and Applied Mathematics, #86, Queen's University, Kingston, Ontario, Canada.

Nikulin, M.S. (1992a). Gihman statistic and goodness-of-fit tests for grouped data. *Mathematical Reports of the Academy of Science of the Royal Society of Canada*, **14** (4):151–156.

Nikulin, M.S. (1992b). A variant of the generalized omega-square statistic. *J. Soviet Math*, **61** (4):1896:1900, (translation from *Zapiski nauchnikh seminarov LOMI*, (1989), **177**:108–113.

Nikulin, M.S., and Huber, C. (1993). Classical random walks and some statistical problems. In *Rings and Modules. Limit Theorems of Probability Theory*, vol. 3. St. Petersburg University Press, 204–216.

Nikulin, M.S., and Osidze, A.G. (1987). Some probabilistic properties of the generalized omega-square statistic. *J. Soviet Math.*, **36** (4):520–524.

Nikulin, M.S., and Seddik-Ameur, N. (1991). Analyse statistique des données binomiales. In *Statistique processus en milieu médical*, Séminaire 90–91 de Statistique Médicale de l'Université Paris V.

Nikulin, M.S., and Voinov, V.G. (1983). A remark on the paper of Bolshev and Mirvaliev about the exponential model of emission of α-particles, Leningrad, LOMI (Leningrad Branch of the Steklov Mathematical Institute), Preprint E-5-83.

Nikulin, M.S., and Voinov, V.G. (1987). A chi-square goodness-of-fit test for exponential distributions of the first order. Leningrad, LOMI (Leningrad Branch of the Steklov Mathematical Institute), Preprint E-8-87.

Nikulin, M.S., and Voinov, V.G. (1989). A chi-square goodness-of-fit test for exponential distributions of the first order. *Springer-Verlag Lecture Notes in Mathematics*, **1412**:239–258.

Nikulin, M.S., and Voinov, V.G. (1993). Unbiased estimators of multivariate discrete distributions and chi-square goodness-of-fit test. *Qüestiio*, **17** (3):301–326.

Nikulin, M.S., and Yusas, J.S. (1976). Comment on G. Chase "The χ^2 test in case when the parameters are estimated independently of the sample," *Litov. Mat. Sb.*, **16**:221–222.

Nikulin, M.S., and Yusas, J.S. (1983). Distribution of a χ^2 statistic with additional observations used to estimate the unknown parameter. *J. Soviet Math.*, **23**:2290–2295.

Noack, A. (1950). A class of random variables with discrete distributions. *Ann. Math. Statist.*, **24**:127–132.

Ogasawara, T., and Takahashi, M. (1951). Independence of quadratic forms in normal systems. *J. Sci. Hiroshima Univ.*, **15**:1–9.

Ogawa, J. (1976). A note on the optimal spacing of the systematic statistic – normal distribution. In *Essays in Probability and Statistics – A Volume in Honor of Professor Junioro Ogawa*. (S. Ikeda, T. Hayakama, H. Hudimoto, M. Okamoto, M. Siotani, and S. Vemamoto, eds.) Shinko Tsushio, Tokyo.

Oluyede, B.O. (1994). A modified chi-square test of independence against a class of ordered alternatives in an $r \times c$ contingency table. *Canadian J. Statist.*, **22** (1):75–87.

O'Reilly, F.J., and Quesenberry, C.P. (1973). The conditional probability integral transformation and applications to obtain composite chi-square goodness-of-fit tests. *Ann. Statist.*, **1**:74–85.

Ortega, J.M., and Rheinboldt, W.C. (1979). *Iterative Solutions of Non-linear Equations in Several Variables.* Academic Press, New York.

Park, C.J. (1973a). The distribution of frequency counts of the geometric distribution. *Sankhỹa*, **A35**:106–111.

Park, C.J. (1973b). The power series distributions with unknown truncation parameters. *Ann. Statist.*, **1**:395–399.

Patel, S.R. (1978). Minimum variance unbiased estimation of multivariate modified power series distributions. *Metrika*, **25** (3):155–161.

Patil, G.P. (1963). Minimum variance unbiased estimation and certain problems of additive number theory. *Ann. Math. Statist.*, **34**:1050–1056.

Patil, G.P. (1965). On multivariate generalized power series distribution and its application to the multinomial and negative multinomial. *Sankhya*, Ser. A:225–239.

Patnaik, P.B. (1949). The non-central χ^2- and F-distributions and their applications. *Biometrika*, **36**:202–232.

Pearson, E.S. (1959). Note on an approximation to the distribution of noncentral χ^2. *Biometrika*, **46**:364–379.

Pearson, E.S., and Hartley, H.O. (1956). *Biometric Tables for Statisticians*, vol. 1. Cambridge University Press, London.

Pearson, K. (1900). On the criterion that a given system of deviation is such that it can be reasonably supposed to have arisen from random sampling. *Phil. Mag.*, **50**:157.

Pearson, K. (1932). Experimental discussion of the χ^2 test for goodness-of-fit. *Biometrika*, **24**:351–381.

Plackett, R.L. (1964). The continuity correction in 2×2 tables. *Biometrika*, **5**:327–337.

Plackett, R.L. (1983). Karl Pearson and the chi squared test. *Int. Statist. Rev.*, **51**:59–72.

Pollard, D. (1979). General chi-square goodness-of-fit tests with data-dependent cells. *Z. Wahr.*, **50**:317–331.

Pollard H. (1946). The representations of $\exp\{-x^\lambda\}$ as a Laplace transform. *Bull. Amer. Math. Soc.*, **52** (10):908–910.

Portnoy, S.L. (1977). Asymptotic efficiency of minimum variance unbiased estimators. *Ann. Statist.*, **5**:522–529.

Prokhorov, Yu. V. (1965). Characterization of a class of distributions by the distribution of some statistics. *Theory Probabil. Appl.*, **10** (3):479–487.

Puri, M.L., and Russel, C.T. (1984). Convergence of generalized inverses with applications to asymptotic hypothesis testing. *Sankhỹa*, **A46**, Pt. 2, 278–286.

Pyke, R. (1965). "Spacings". *J. Roy. Statist. Soc.*, **B27** (3):395–449.

Quesenberry. C.P. (1986). Some transformation methods in goodness-of-fit. In *Goodness-of-fit techniques* (D'Agostino, R.B., and Stephens, M.A. eds.). Marcel Dekker, New York. 235–277.

Quesenberry, C.P., and Miller, F.L., Jr. (1977). Power studies of some tests for uniformity. *J. Statist. Comput. Simulation*, **5**:161–169.

Quine, M.P., and Robinson, J. (1985). Efficiencies of chi-square and likelihood ratio goodness-of-fit tests. *Ann. Statist.*, **13**:727–742.

Radlow, R., and Alf, E.F. (1975). An alternative multinomial assessment of the accuracy of the χ^2 test of goodness of fit. *J. Amer. Statist. Assoc.*, **70**:811–813.

Rajarshi, M.B. (1986). Chi-squared type goodness-of-fit tests for stochastical models through conditional least squares estimation. *J. Indian Statist. Assoc.*, **24**:65–76.

Rao, C.R. (1973). *Linear Statistical Inference and its Application*, Wiley, New York.

Rao, J.S., and Kuo, N. (1984). Asymptotic results on the Greenwood statistic and some of its generalization. *J. Roy. Statist. Soc.*, **B46**:228–237.

Rao, J.S., and Sethuraman, J. (1975). Weak convergence of empirical distribution functions of random variable subject to perturbation and scale factors. *Ann. Statist.*, **3**:299–313.

Rao, K.C., and Robson, D.S. (1974). A chi-squared statistic for goodness-of-fit tests within the exponential family. *Commun. Statist.*, **3**:1139–1153.

Rayner, J.C.W. (1982). The choice of class probabilities and number of classes for the simple χ^2 goodness of fit test. *Sankhȳa*, **44B**:28–37.

Rayner, J.C.W., and Best, D.J. (1986). Neyman-type smooth tests for location-scale families. *Biometrika*, **77**:437–446.

Rayner, J.C.W., and Best, D.J. (1989). *Smooth Tests of Goodness of Fit.* Oxford University Press, Oxford.

Read, T.R.C., and Cressie, N.A.C. (1988). *Goodness-of-fit Statistics for Discrete Multivariate Data.* Springer-Verlag, New York.

Roscoe, J.T., and Byars, J.A. (1971). An investigation of the restraints with respect to sample size commonly imposed on the use of the chi-square statistic. *J. Amer. Statist. Assoc.*, **66**:755–759.

Rosenblatt, N. (1952). Remark on a multivariate transformation. *Ann. Math. Statist.*, **23**:470–472.

Roy, A.R. (1956). On χ^2-statistics with variable intervals, Technical Report N1, Stanford University, Statistics Department, Palo Alto, CA.

Roy, J., and Mitra, S. (1957). Unbiased minimum variance estimation in a case of discrete distributions. *Sankhȳa*, **18**:371–378.

Rutherford, E., Chadwick, J., and Ellis, C.D. (1930). *Radiations from Radioactive Substances.* Cambridge University Press, London.

Ruymgaart, F.H. (1974). Asymptotic normality of nonparametric tests for independence. *Ann. Statist.*, **2**:892–910.

Ruymgaart, F.H. (1975). A note on chi-square statistics with random cell boundaries. *Ann. Statist.*, **3**:965–968.

Scheffe, H. (1947). A useful convergence theorem for probability distributions. *Ann. Math. Statist.*, **18**:434–438.

Schorr, B. (1974). On the choice of the class intervals in the application of the chi-square test. *Math. Operationsforsch. Statist.*, **5**:357–377.

Selby, B. (1965). The index of dispersion as a test statistic. *Biometrika*, **52**:627–629.

Serfling, R.J. (1980). *Approximation Theorems of Mathematical Statistics.* Wiley, New York.

Sethuraman, J. (1961). Some limit theorems for joint distributions. *Sankhÿa*, **A23**:379–386.

Shanmugan, R. (1988). A goodness-of-fit test for the Lagrangian logarithmic series distribution. *J. Statist. Res.*, **22**:31–36.

Shapiro, S.S., and Wilk, M.B. (1965). An analysis of variance test for normality. *Biometrika*, **52**:591–611.

Shapiro, S.S., Wilk, M.B., and Chen, H.Y. (1968). A comparative study of various tests for normality. *J. Amer. Statist. Assoc.*, **63**:1343–1372.

Sharkadi, K. (1966). On testing for normality. In *Proceedings of the Fifth Berkeley Symposium on Mathematical Statistics and Probability*, **1**:373–387.

Sherman, B. (1950). A random variable related to the spacing of sample values. *Ann. Math. Statist.*, **21**:783–786.

Shorack, G., and Wellner, J. (1986). *Empirical Processes with Applications to Statistics.* Wiley, New York.

Singh, J. (1975). A note on the Stirling distribution of the second kind. *Commun. Statist.*, **4** (8):753–759.

Smirnov, N.V. (1936). Sur la distribution de ω^2. *C.R. Acad. Sci. Paris*, **202**:449-452.

Smirnov, N.V. (1937). On the distribution of the ω^2 criterion of von Mises. *Rec. Math.*, (NS), **2**:973-993.

Smirnov, N.V. (1939). Sur les écarts de la courbe de distribution empirique. *Rec. Math.*, **6**:3–26.

Smirnov, N.V. (1970). *Probability Theory and Mathematical Statistics. Selected Works.* Nauka, Moscow (in Russian).

Smith, P.I., Rae, D.S., Manderscheid, R.W. and Silbergold, S. (1979). Exact and approximate distribution of the chi-square statistic for equiprobability. *Comm. Statist. – Simula. Computa*, **B8** (2):131–149.

Steck, G.P. (1957). Limit theorem for conditional distributions. *Univ. Calif. Publ. Statist.*, **2**:237–284.

Stephens, M.A. (1966). Statistics connected with the uniform distribution: percentage points and applications to testing for randomness of directions. *Biometrika*, **53**:235–239.

Stephens, M.A. (1986a). Test for the uniform distribution. In *Goodness-of-fit techniques* (D'Agostino, R.B., and Stephens, M.A., eds.). Marcel Dekker, New York, 331–366.

Stephens, M.A. (1986b). Tests based on EDF statistics. In *Goodness-of-fit techniques* (D'Agostino, R.B., and Stephens, M.A., eds.). Marcel Dekker, New York, 97–193.

Stephens, M.A. (1986c). Tests for the exponential distribution. In *Goodness-of-fit techniques* (D'Agostino, R.B., and Stephens, M.A., eds.). Marcel Dekker, New York, 421–459.

Stephens, M.A. (1986d). Tests for the normal distribution. In *Goodness-of-fit techniques* (D'Agostino, R.B., and Stephens, M.A., eds.). Marcel Dekker, New York, 367–419.

Stigler, S.M. (1992). Studies in the history of probability and statistics XLIII; Karl Pearson and quasi-independence. *Biometrika*, **79**:563–575.

Stroud, T.W.F. (1971). On obtaining large-sample tests from asymptotically normal estimates. *Ann. Math. Statist.*, **42**:1412–1429.

Sturges, H.A. (1926). The choice of a class interval. *J. Amer. Statist. Assoc.*, **21**:65–66.

Sutrick, K.H. (1986). Asymptotic power comparison of the chi-square and likelihood ratio test. *Ann. Inst. Statist. Math.*, **38** (Part A):503–511.

Svensson, A. (1985). On χ^2 test of goodness-of-fit for a class of discrete multivariate models. *Statist. Prob. Letters*, **3**:331–336.

Sverdrup, E. (1967). *Law and Chance Variations*, Vol. **2**. North Holland, Amsterdam.

Sweeting, T.J. (1989). On conditional weak convergence. *J. Theor. Prob.*, **2**:461–474.

Tate, M.W., and Hyer, L.A. (1973). Inaccuracy of the χ^2 test of goodness of fit when expected frequencies are small. *J. Amer. Statist. Assoc.*, **68**:836–841.

Thomas, D.R., and Pierce, D.A. (1979). Neyman's smooth goodness-of-fit test when the hypothesis is composite. *J. Amer. Statist. Assoc.*, **74**:441–445.

Tseranbat, O. (1990). On a modification of a chi-squared type statistic for testing the hypothesis of multivariate normality. *Izv. Akad. Nauk Uzb. SSR*, **4**:40–47.

Tumanyan, S.Kh. (1956). Asymptotic distribution of χ^2 criterion when the numbers of observations and classes increase simultaneously. *Theory Probabil. Appl.*, **1**:131–145.

Turnbull, B.W., and Weiss, L. (1978). A likelihood ratio statistic for testing goodness-of-fit with randomly censored data. *Biometrika*, **34**:367–375.

Vessereau, A. (1958). Sur les conditions d'application du criterium χ^2 de Pearson. *Bull. Inst. Int. Statist.*, **36**:81–101.

Voinov, V.G., and Nikulin, M.S. (1984). A remark on the paper by Bolshev and Mirvaliev on an exponential model of radiation of α-particles. *Theory Probabil. Appl.*, **29**:174–175.

Voinov, V.G., Neumann, A.R., and Nikulin, M.S. (1986). The Lagrange method of evaluation of quantiles and noncentrality parameters of probability distributions, *Theory Probabil. Appl.*, **31**:185–187.

Voinov, V.G., and Nikulin, M.S. (1989). *Unbiased Estimators and their Applications.* Nauka, Moscow.

Voinov, V.G., and Nikulin, M.S. (1993). *Unbiased Estimators and their Applications. Vol. 1: Univariate case.* Kluwer, Dordrecht.

Volodin, I.N. (1965). On the distinction between the Polya and the Poisson distributions when a large number of small samples is available. *Theory Probabil. Appl.*, **10** (2):335–338.

von Mises, R. (1936). *Wahrscheinlichkeit, Statistic und Wahrheit.* Springer-Verlag, Heidelberg.

von Mises, R. (1964). *Mathematical Theory of Probability and Statistics* (Hilda Geiringer, ed.), Academic Press, New York.

Wald, A. (1943). Tests of statistical hypotheses concerning several parameters when the number of observations is large. *Amer. Math. Soc. Trans.*, **54**:426–482.

Walsh, J.E. (1960). Probabilities for the Cramer-von Mises-Smirnov test using grouped data. *Ann. Inst. Statist. Math.*, **12**:143–145.

Washio, V., Morimoto, H., and Ikeda, N. (1956). Unbiased estimation based on sufficient statistics. *Bull. Math. Statist.*, **6**:69–94.

Watanabe, Y. (1952). On the w^2-distribution. *Gakugei* [Tokushima University], **2**:21–30.

Watson, G.S. (1957). On the chi-square goodness-of-fit test for normal distributions. *Biometrika*, **44**:336–348.

Watson, G.S. (1958). On the chi-square goodness-of-fit test for continuous distributions. *J. Roy. Statist. Soc.*, **B20**:44–61.

Watson, G.S. (1959). Some recent results in chi-square goodness-of-fit tests. *Biometrics*, **15**:440–468.

Weiss, L. (1956). A certain class of tests of fit. *Ann. Math. Statist.*, **27**:1165–1170.

Weiss, L. (1957). The asymptotic power of certain tests of fit based on sample spacings. *Ann. Math. Statist.*, **28**:783–786.

Weiss, L. (1974). The asymptotic sufficiency of a relatively small number of order statistics in goodness of fit. *Ann. Statist.*, **2**:795–802.

Wilks, S.S. (1962). *Mathematical Statistics*, Wiley, New York.

Williams, C.A. (1950). On the choice of the number and width of classes for the chi-square test of goodness of fit. *J. Amer. Statist. Assoc.*, **45**:77–86.

Witting, H. (1959). Über einen χ^2-Test, dessen Klassen durch geordnete Stichprobenfunktionen festgelegt werden. *Ark. Mat.*, **10**:468–479.

Witting, H., and Nölle, G. (1970). *Angewandte Mathematische Statistik*. Teubner, Stuttgart.

Yarnold, J.K. (1970). The minimum expectation in χ^2 goodness of fit tests and the accuracy of approximations for the null distribution. *J. Amer. Statist. Assoc.*, **65**:864–886.

Yates, F. (1934). Contingency tables involving small numbers and the χ^2 test. *J. Roy. Statist. Soc.*, Supp. **1** (2):217–235.

Zacks, S. (1971). *The Theory of Statistical Inference*. Wiley, New York.

INDEX

Author entries are in SMALL CAPITALS.

WILEY SERIES IN PROBABILITY AND STATISTICS

Probability and Statistics

ANDERSON · An Introduction to Multivariate Statistical Analysis, *Second Edition*
*ANDERSON · The Statistical Analysis of Time Series
ARNOLD, BALAKRISHNAN, and NAGARAJA · A First Course in Order Statistics
BACCELLI, COHEN, OLSDER, and QUADRAT · Synchronization and Linearity: An Algebra for Discrete Event Systems
BERNARDO and SMITH · Bayesian Theory
BHATTACHARYYA and JOHNSON · Statistical Concepts and Methods
BILLINGSLEY · Convergence of Probability Measures
BILLINGSLEY · Probability and Measure, *Third Edition*
BOROVKOV · Asymptotic Methods in Queuing Theory
BRANDT, FRANKEN, and LISEK · Stationary Stochastic Models
CAINES · Linear Stochastic Systems
CAIROLI and DALANG · Sequential Stochastic Optimization
CHEN · Recursive Estimation and Control for Stochastic Systems
CONSTANTINE · Combinatorial Theory and Statistical Design
COOK and WEISBERG · An Introduction to Regression Graphics
COVER and THOMAS · Elements of Information Theory
*DOOB · Stochastic Processes
DUDEWICZ and MISHRA · Modern Mathematical Statistics
ETHIER and KURTZ · Markov Processes: Characterization and Convergence
FELLER · An Introduction to Probability Theory and Its Applications, Volume 1, *Third Edition,* Revised; Volume II, *Second Edition*
FREEMAN and SMITH · Aspects of Uncertainty: A Tribute to D. V. Lindley
FULLER · Introduction to Statistical Time Series, *Second Edition*
FULLER · Measurement Error Models
GIFI · Nonlinear Multivariate Analysis
GUTTORP · Statistical Inference for Branching Processes
HALD · A History of Probability and Statistics and Their Applications before 1750
HALL · Introduction to the Theory of Coverage Processes
HANNAN and DEISTLER · The Statistical Theory of Linear Systems
HEDAYAT and SINHA · Design and Inference in Finite Population Sampling
HOEL · Introduction to Mathematical Statistics, *Fifth Edition*
HUBER · Robust Statistics
IMAN and CONOVER · A Modern Approach to Statistics
JUREK and MASON · Operator-Limit Distributions in Probability Theory
KAUFMAN and ROUSSEEUW · Finding Groups in Data: An Introduction to Cluster Analysis
LARSON · Introduction to Probability Theory and Statistical Inference, *Third Edition*
LESSLER and KALSBEEK · Nonsampling Error in Surveys
LINDVALL · Lectures on the Coupling Method
MANTON, WOODBURY, and TOLLEY · Statistical Applications Using Fuzzy Sets
MARDIA · The Art of Statistical Science: A Tribute to G. S. Watson
MORGENTHALER and TUKEY · Configural Polysampling: A Route to Practical Robustness

*Now available in a lower priced paperback edition in the Wiley Classics Library.

Probability and Statistics (Continued)

MUIRHEAD · Aspects of Multivariate Statistical Theory
OLIVER and SMITH · Influence Diagrams, Belief Nets and Decision Analysis
*PARZEN · Modern Probability Theory and Its Applications
PRESS · Bayesian Statistics: Principles, Models, and Applications
PUKELSHEIM · Optimal Experimental Design
PURI and SEN · Nonparametric Methods in General Linear Models
PURI, VILAPLANA, and WERTZ · New Perspectives in Theoretical and Applied Statistics
RAO · Asymptotic Theory of Statistical Inference
RAO · Linear Statistical Inference and Its Applications, *Second Edition*
*RAO and SHANBHAG · Choquet-Deng Type Functional Equations and Applications to Stochastic Models
RENCHER · Methods of Multivariate Analysis
ROBERTSON, WRIGHT, and DYKSTRA · Order Restricted Statistical Inference
ROGERS and WILLIAMS · Diffusions, Markov Processes, and Martingales, Volume I: Foundations, *Second Edition;* Volume II: Îto Calculus
ROHATGI · An Introduction to Probability Theory and Mathematical Statistics
ROSS · Stochastic Processes
RUBINSTEIN · Simulation and the Monte Carlo Method
RUBINSTEIN and SHAPIRO · Discrete Event Systems: Sensitivity Analysis and Stochastic Optimization by the Score Function Method
RUZSA and SZEKELY · Algebraic Probability Theory
SCHEFFE · The Analysis of Variance
SEBER · Linear Regression Analysis
SEBER · Multivariate Observations
SEBER and WILD · Nonlinear Regression
SERFLING · Approximation Theorems of Mathematical Statistics
SHORACK and WELLNER · Empirical Processes with Applications to Statistics
SMALL and McLEISH · Hilbert Space Methods in Probability and Statistical Inference
STAPLETON · Linear Statistical Models
STAUDTE and SHEATHER · Robust Estimation and Testing
STOYANOV · Counterexamples in Probability
STYAN · The Collected Papers of T. W. Anderson: 1943–1985
TANAKA · Time Series Analysis: Nonstationary and Noninvertible Distribution Theory
THOMPSON and SEBER · Adaptive Sampling
WHITTAKER · Graphical Models in Applied Multivariate Statistics
YANG · The Construction Theory of Denumerable Markov Processes

Applied Probability and Statistics

ABRAHAM and LEDOLTER · Statistical Methods for Forecasting
AGRESTI · Analysis of Ordinal Categorical Data
AGRESTI · Categorical Data Analysis
AGRESTI · An Introduction to Categorical Data Analysis
ANDERSON and LOYNES · The Teaching of Practical Statistics
ANDERSON, AUQUIER, HAUCK, OAKES, VANDAELE, and WEISBERG · Statistical Methods for Comparative Studies
*ARTHANARI and DODGE · Mathematical Programming in Statistics
ASMUSSEN · Applied Probability and Queues
*BAILEY · The Elements of Stochastic Processes with Applications to the Natural Sciences
BARNETT and LEWIS · Outliers in Statistical Data, *Third Edition*
BARTHOLOMEW, FORBES, and McLEAN · Statistical Techniques for Manpower Planning, *Second Edition*
BATES and WATTS · Nonlinear Regression Analysis and Its Applications

*Now available in a lower priced paperback edition in the Wiley Classics Library.

Applied Probability and Statistics (Continued)

BECHHOFER, SANTNER, and GOLDSMAN · Design and Analysis of Experiments for Statistical Selection, Screening, and Multiple Comparisons
BELSLEY · Conditioning Diagnostics: Collinearity and Weak Data in Regression
BELSLEY, KUH, and WELSCH · Regression Diagnostics: Identifying Influential Data and Sources of Collinearity
BERRY · Bayesian Analysis in Statistics and Econometrics: Essays in Honor of Arnold Zellner
BERRY, CHALONER, and GEWEKE · Bayesian Analysis in Statistics and Econometrics: Essays in Honor of Arnold Zellner
BHAT · Elements of Applied Stochastic Processes, *Second Edition*
BHATTACHARYA and WAYMIRE · Stochastic Processes with Applications
BIEMER, GROVES, LYBERG, MATHIOWETZ, and SUDMAN · Measurement Errors in Surveys
BIRKES and DODGE · Alternative Methods of Regression
BLOOMFIELD · Fourier Analysis of Time Series: An Introduction
BOLLEN · Structural Equations with Latent Variables
BOULEAU · Numerical Methods for Stochastic Processes
BOX · R. A. Fisher, the Life of a Scientist
BOX and DRAPER · Empirical Model-Building and Response Surfaces
BOX and DRAPER · Evolutionary Operation: A Statistical Method for Process Improvement
BOX, HUNTER, and HUNTER · Statistics for Experimenters: An Introduction to Design, Data Analysis, and Model Building
BROWN and HOLLANDER · Statistics: A Biomedical Introduction
BUCKLEW · Large Deviation Techniques in Decision, Simulation, and Estimation
BUNKE and BUNKE · Nonlinear Regression, Functional Relations and Robust Methods: Statistical Methods of Model Building
CHATTERJEE and HADI · Sensitivity Analysis in Linear Regression
CHATTERJEE and PRICE · Regression Analysis by Example, *Second Edition*
CLARKE and DISNEY · Probability and Random Processes: A First Course with Applications, *Second Edition*
COCHRAN · Sampling Techniques, *Third Edition*
*COCHRAN and COX · Experimental Designs, *Second Edition*
CONOVER · Practical Nonparametric Statistics, *Second Edition*
CONOVER and IMAN · Introduction to Modern Business Statistics
CORNELL · Experiments with Mixtures, Designs, Models, and the Analysis of Mixture Data, *Second Edition*
COX · A Handbook of Introductory Statistical Methods
*COX · Planning of Experiments
COX, BINDER, CHINNAPPA, CHRISTIANSON, COLLEDGE, and KOTT · Business Survey Methods
CRESSIE · Statistics for Spatial Data, *Revised Edition*
DANIEL · Applications of Statistics to Industrial Experimentation
DANIEL · Biostatistics: A Foundation for Analysis in the Health Sciences, *Sixth Edition*
DAVID · Order Statistics, *Second Edition*
*DEGROOT, FIENBERG, and KADANE · Statistics and the Law
*DEMING · Sample Design in Business Research
DILLON and GOLDSTEIN · Multivariate Analysis: Methods and Applications
DODGE and ROMIG · Sampling Inspection Tables, *Second Edition*
DOWDY and WEARDEN · Statistics for Research, *Second Edition*
DRAPER and SMITH · Applied Regression Analysis, *Second Edition*
DUNN · Basic Statistics: A Primer for the Biomedical Sciences, *Second Edition*
DUNN and CLARK · Applied Statistics: Analysis of Variance and Regression, *Second Edition*

*Now available in a lower priced paperback edition in the Wiley Classics Library.

Applied Probability and Statistics (Continued)

ELANDT-JOHNSON and JOHNSON · Survival Models and Data Analysis
EVANS, PEACOCK, and HASTINGS · Statistical Distributions, *Second Edition*
FISHER and VAN BELLE · Biostatistics: A Methodology for the Health Sciences
FLEISS · The Design and Analysis of Clinical Experiments
FLEISS · Statistical Methods for Rates and Proportions, *Second Edition*
FLEMING and HARRINGTON · Counting Processes and Survival Analysis
FLURY · Common Principal Components and Related Multivariate Models
GALLANT · Nonlinear Statistical Models
GLASSERMAN and YAO · Monotone Structure in Discrete-Event Systems
GREENWOOD and NIKULIN · A Guide to Chi-Squared Testing
GROSS and HARRIS · Fundamentals of Queueing Theory, *Second Edition*
GROVES · Survey Errors and Survey Costs
GROVES, BIEMER, LYBERG, MASSEY, NICHOLLS, and WAKSBERG · Telephone Survey Methodology
HAHN and MEEKER · Statistical Intervals: A Guide for Practitioners
HAND · Discrimination and Classification
*HANSEN, HURWITZ, and MADOW · Sample Survey Methods and Theory, Volume 1: Methods and Applications
*HANSEN, HURWITZ, and MADOW · Sample Survey Methods and Theory, Volume II: Theory
HEIBERGER · Computation for the Analysis of Designed Experiments
HELLER · MACSYMA for Statisticians
HINKELMAN and KEMPTHORNE: · Design and Analysis of Experiments, Volume 1: Introduction to Experimental Design
HOAGLIN, MOSTELLER, and TUKEY · Exploratory Approach to Analysis of Variance
HOAGLIN, MOSTELLER, and TUKEY · Exploring Data Tables, Trends and Shapes
HOAGLIN, MOSTELLER, and TUKEY · Understanding Robust and Exploratory Data Analysis
HOCHBERG and TAMHANE · Multiple Comparison Procedures
HOCKING · Methods and Applications of Linear Models: Regression and the Analysis of Variance
HOEL · Elementary Statistics, *Fifth Edition*
HOGG and KLUGMAN · Loss Distributions
HOLLANDER and WOLFE · Nonparametric Statistical Methods
HOSMER and LEMESHOW · Applied Logistic Regression
HØYLAND and RAUSAND · System Reliability Theory: Models and Statistical Methods
HUBERTY · Applied Discriminant Analysis
IMAN and CONOVER · Modern Business Statistics
JACKSON · A User's Guide to Principle Components
JOHN · Statistical Methods in Engineering and Quality Assurance
JOHNSON · Multivariate Statistical Simulation
JOHNSON and KOTZ · Distributions in Statistics
Continuous Univariate Distributions—2
Continuous Multivariate Distributions
JOHNSON, KOTZ, and BALAKRISHNAN · Continuous Univariate Distributions, Volume 1, *Second Edition;* Volume 2, *Second Edition*
JOHNSON, KOTZ, and KEMP · Univariate Discrete Distributions, *Second Edition*
JUDGE, GRIFFITHS, HILL, LÜTKEPOHL, and LEE · The Theory and Practice of Econometrics, *Second Edition*
JUDGE, HILL, GRIFFITHS, LÜTKEPOHL, and LEE · Introduction to the Theory and Practice of Econometrics, *Second Edition*
JUREČKOVÁ and SEN · Robust Statistical Procedures: Aymptotics and Interrelations
KADANE · Bayesian Methods and Ethics in a Clinical Trial Design
KADANE and SCHUM · A Probabilistic Analysis of the Sacco and Vanzetti Evidence

*Now available in a lower priced paperback edition in the Wiley Classics Library.

Applied Probability and Statistics (Continued)

KALBFLEISCH and PRENTICE · The Statistical Analysis of Failure Time Data
KASPRZYK, DUNCAN, KALTON, and SINGH · Panel Surveys
KISH · Statistical Design for Research
*KISH · Survey Sampling
LAD · Operational Subjective Statistical Methods
LANGE, RYAN, BILLARD, BRILLINGER, CONQUEST, and GREENHOUSE · Case Studies in Biometry
LAWLESS · Statistical Models and Methods for Lifetime Data
LEBART, MORINEAU., and WARWICK · Multivariate Descriptive Statistical Analysis: Correspondence Analysis and Related Techniques for Large Matrices
LEE · Statistical Methods for Survival Data Analysis, *Second Edition*
LEPAGE and BILLARD · Exploring the Limits of Bootstrap
LEVY and LEMESHOW · Sampling of Populations: Methods and Applications
LINHART and ZUCCHINI · Model Selection
LITTLE and RUBIN · Statistical Analysis with Missing Data
MAGNUS and NEUDECKER · Matrix Differential Calculus with Applications in Statistics and Econometrics
MAINDONALD · Statistical Computation
MALLOWS · Design, Data, and Analysis by Some Friends of Cuthbert Daniel
MANN, SCHAFER, and SINGPURWALLA · Methods for Statistical Analysis of Reliability and Life Data
MASON, GUNST, and HESS · Statistical Design and Analysis of Experiments with Applications to Engineering and Science
McLACHLAN · Discriminant Analysis and Statistical Pattern Recognition
McLACHLAN and KRISHNAN · The EM Algorithm and Extensions
MILLER · Survival Analysis
MONTGOMERY and MYERS · Response Surface Methodology: Process and Product in Optimization Using Designed Experiments
MONTGOMERY and PECK · Introduction to Linear Regression Analysis, *Second Edition*
NELSON · Accelerated Testing, Statistical Models, Test Plans, and Data Analyses
NELSON · Applied Life Data Analysis
OCHI · Applied Probability and Stochastic Processes in Engineering and Physical Sciences
OKABE, BOOTS, and SUGIHARA · Spatial Tesselations: Concepts and Applications of Voronoi Diagrams
OSBORNE · Finite Algorithms in Optimization and Data Analysis
PANKRATZ · Forecasting with Dynamic Regression Models
PANKRATZ · Forecasting with Univariate Box-Jenkins Models: Concepts and Cases
PORT · Theoretical Probability for Applications
PUTERMAN · Markov Decision Processes: Discrete Stochastic Dynamic Programming
RACHEV · Probability Metrics and the Stability of Stochastic Models
RÉNYI · A Diary on Information Theory
RIPLEY · Spatial Statistics
RIPLEY · Stochastic Simulation
ROSS · Introduction to Probability and Statistics for Engineers and Scientists
ROUSSEEUW and LEROY · Robust Regression and Outlier Detection
RUBIN · Multiple Imputation for Nonresponse in Surveys
RYAN · Modern Regression Methods
RYAN · Statistical Methods for Quality Improvement
SCHUSS - Theory and Applications of Stochastic Differential Equations
SCOTT · Multivariate Density Estimation: Theory, Practice, and Visualization
SEARLE · Linear Models
SEARLE · Linear Models for Unbalanced Data
SEARLE · Matrix Algebra Useful for Statistics
SEARLE, CASELLA, and McCULLOCH · Variance Components

Applied Probability and Statistics (Continued)

SKINNER, HOLT, and SMITH · Analysis of Complex Surveys
STOYAN, KENDALL, and MECKE · Stochastic Geometry and Its Applications, *Second Edition*
STOYAN and STOYAN · Fractals, Random Shapes and Point Fields: Methods of Geometrical Statistics
THOMPSON · Empirical Model Building
THOMPSON · Sampling
TIERNEY · LISP-STAT: An Object-Oriented Environment for Statistical Computing and Dynamic Graphics
TIJMS · Stochastic Modeling and Analysis: A Computational Approach
TITTERINGTON, SMITH, and MAKOV · Statistical Analysis of Finite Mixture Distributions
UPTON and FINGLETON · Spatial Data Analysis by Example, Volume 1: Point Pattern and Quantitative Data
UPTON and FINGLETON · Spatial Data Analysis by Example, Volume II: Categorical and Directional Data
VAN RIJCKEVORSEL and DE LEEUW · Component and Correspondence Analysis
WEISBERG · Applied Linear Regression, *Second Edition*
WESTFALL and YOUNG · Resampling-Based Multiple Testing: Examples and Methods for *p*-Value Adjustment
WHITTLE · Optimization Over Time: Dynamic Programming and Stochastic Control, Volume I and Volume II
WHITTLE · Systems in Stochastic Equilibrium
WONNACOTT and WONNACOTT · Econometrics, *Second Edition*
WONNACOTT and WONNACOTT · Introductory Statistics, *Fifth Edition*
WONNACOTT and WONNACOTT · Introductory Statistics for Business and Economics, *Fourth Edition*
WOODING · Planning Pharmaceutical Clinical Trials: Basic Statistical Principles
WOOLSON · Statistical Methods for the Analysis of Biomedical Data

Tracts on Probability and Statistics

BILLINGSLEY · Convergence of Probability Measures
KELLY · Reversibility and Stochastic Networks
TOUTENBURG · Prior Information in Linear Models